THE INSECTS AND ARACHNIDS OF CANADA

PART 8

The Plant Bugs of the Prairie Provinces of Canada

Heteroptera: Miridae

Leonard A. Kelton

Biosystematics Research Institute
Ottawa, Ontario

Research Branch
Agriculture Canada

Publication 1703 1980

Available in Canada through

Authorized Bookstore Agents
and other bookstores

or by mail from

Canadian Government Publishing Centre
Supply and Services Canada
Hull, Quebec, Canada K1A 0S9

Catalogue No. A 42-42/1980-8	Canada: $9.95
ISBN 0-660-10613-2	Other countries: $11.95

Price subject to change without notice

Part 1. Collecting, Preparing, and Preserving Insects, Mites, and Spiders, compiled by J. E. H. Martin, Biosystematics Research Institute, Ottawa, 1977.

Part 2. The Bark Beetles of Canada and Alaska (Coleoptera: Scolytidae), by D. E. Bright, Jr., Biosystematics Research Institute, Ottawa, 1976.

Part 3. The Aradidae of Canada (Hemiptera: Aradidae), by R. Matsuda, Biosystematics Research Institute, Ottawa, 1977.

Part 4. The Anthocoridae of Canada and Alaska (Heteroptera: Anthocoridae), by L. A. Kelton, Biosystematics Research Institute, Ottawa, 1978.

Part 5. The Crab Spiders of Canada and Alaska (Araneae: Philodromidae and Thomisidae), by C. D. Dondale and J. H. Redner, Biosystematics Research Institute, Ottawa, 1978.

Part 6. The Mosquitoes of Canada (Diptera: Culicidae), by D. M. Wood, P. T. Dang, and R. A. Ellis, Biosystematics Research Institute, Ottawa, 1979.

Partie 7. Genera des Trichoptères du Canada et des États adjacents, par F. Schmid, Institut de recherches biosystématiques, Ottawa, 1980.

Contents

Acknowledgments

Most of the specimens for this faunal work were collected with the assistance of the late Arthur R. Brooks, Agriculture Canada Research Station, Saskatoon, Sask., and are in the Canadian National Collection of Insects, Ottawa, Ont. and Agriculture Canada's Research Station, Saskatoon, Sask. The other specimens and records were obtained from the Canadian National Collection of Insects, and from the Agriculture Canada's Research Stations at Lethbridge, Alta. and Winnipeg, Man.

I am grateful to the following people and their institutions for making available the collections in their care: Dr. R. C. Froeschner, U.S. National Museum, Washington, D.C.; Dr. J. L. Herring, SEL, USDA, U.S. National Museum, Washington, D.C.; and Dr. P. H. Arnaud, Jr., California Academy of Sciences, San Francisco, Calif.

I gratefully acknowledge the cooperation of the superintendents and wardens of Waterton, Banff, Jasper, and Riding Mountain national parks while I was working in the parks in 1970 and 1972, and of D. Allen, Director, and G. Schumann, technical supervisor, Whiteshell Provincial Park, Falcon Beach, Man. while I was working there in 1974, 1975, and 1978.

Special thanks are due to D. Brown, my technician, for preparing the distribution maps, and to S. Rigby of this Institute, for preparing several illustrations of adult Miridae.

Introduction

The Miridae, or plant bugs, are of great economic importance to man because they feed on plants, or prey on other arthropods. The bugs may also act as vectors of plant virus diseases. Damage to crops by these bugs has often resulted in losses of millions of dollars.

This group of bugs is not well known in Canada, particularly in the northern areas of the Prairie Provinces, the Yukon, and the Northwest Territories. Despite their large numbers, the plant bugs are poorly represented in collections, and confusion exists in naming them. In recent years special efforts have been made to collect the Miridae from coast to coast in Canada, and especially in the farming and ranching areas of the Prairie Provinces. As a result, knowledge of the range and distribution of the species has been expanded and new information on their habits and ecology has been obtained. The aim of this report is to help the economic entomologists and other field workers in the Prairie Provinces and elsewhere to identify the species of this important group of plant bugs so often encountered in field surveys and control programs.

This faunal work treats the species of Miridae known to occur in the Prairie Provinces. The work was initiated by the late Arthur R. Brooks and the author in 1949 at Agriculture Canada's Research Station, Saskatoon, Sask. At the outset, many difficulties were encountered in identifying species because the only publications available were Knight (1923*b*, 1941*c*) and Blatchley (1926). With the publication of *A catalogue of the Miridae of the world* by Carvalho (1957-1959), and subsequent publications by Kelton (see References), Kelton and Knight (1962), Knight (1974), and Kelton and Herring (1978), 45 species have been reported from the Prairie Provinces. This work shows that 314 species of Miridae occur in this area. Most of the species are restricted to the Nearctic region, but 27 species are Holarctic in distribution. Most of the Holarctic species are native to North America, but several have been introduced by man.

In economic terms the effect of harmful species seems to far outweigh the beneficial effect of the predaceous species. The bugs suck juices from plants and when large populations are allowed to develop, they may destroy or greatly reduce the yield of grain, seed, and vegetable crops. Their feeding may also reduce plant vigor, deform foliage of ornamental flowers and shrubs, and cause malformed fruit, resulting in reduced yields and lower commercial grades of apples, plums, currants, and strawberries. Certain species may also destroy plants by transmitting virus diseases.

There are several species of Miridae that are predators. However, the use of the Miridae for biological control of arthropod pests and weeds in the Prairie Provinces has not been fully explored.

This faunal work includes brief descriptions of adults and selected illustrations of adults and of the male genital claspers. The known habitats and the distribution of the species are given. Keys to subfamilies, tribes, genera, and species are also included.

Collecting and preserving specimens

There are three ways to collect Miridae, and the collecting methods depend on the habitat, type of vegetation, and type of species to be collected. The three ways are sweeping the plants, using a regular sweep net; beating the branches, using a beating stick and sheet; and attracting the species, using a light trap.

The sweeping method is used to collect mirids living on grasses, sedges, herbaceous plants, flexible shrubs, and delicate foliage of deciduous trees. This is the most productive method for general collecting and, when plants are isolated, may give accurate host associations. Because the mirids are fragile and delicate insects, sweeping must be done carefully so as not to damage the bugs in the net. Flower heads, leaves, and other debris often picked up in sweeping can damage the bugs in the net if sweeping is prolonged; therefore the bugs should be picked out of the net frequently with an aspirator. Sweeping should be done under dry conditions as moisture in the net will mat and ruin the specimens.

The beating stick and sheet method is used for collecting mirids on branches of coniferous trees, tough or spiny deciduous trees, and fruit trees and is the best method for obtaining accurate host records. The sheet is held under a branch and the branch is sharply struck with the stick. The bugs are jarred loose and fall on the sheet and should be picked off the sheet quickly with an aspirator.

The use of a black light on a warm, calm, and humid night will often attract mirids not collected by sweeping or beating. They are usually night-flying mirids and for many of these the host plant is unknown. A black light may be either permanently mounted on a container designed to funnel insects to poison below or temporarily set up against a white sheet and the mirids picked up with an aspirator as soon as they alight on the sheet. Since moths, beetles, and other insects are also attracted to the lights, mirids collected by this method are less likely to be damaged by larger insects or covered with moth scales than those collected in permanently mounted traps.

The collected specimens are killed promptly in cyanide and mounted. If they cannot be mounted immediately, they may be stored for several weeks in pill boxes between layers of cellulose cotton. Each pill box is labeled with pertinent information about the specimens such as place collected, date, collector, and host plant if known. Before mounting the stored specimens, the pill boxes are placed in a relaxing container and the bugs relaxed.

Mirids should be mounted on narrow triangular bristol board points. The tip of the point is bent to fit the angle of the thorax so that the specimen will be level when mounted. Only the tip of the point should be covered with glue and the point attached to the right side of the thorax above the middle coxa. Miridae should not be pinned through the body, and they should never be placed in alcohol.

For additional details on collecting and preserving techniques, see Martin (1977).

Biology

Most mirids pass the winter in the egg stage. The eggs are normally inserted in a tender part of the host plant and hatch early in the spring when the host plant is sprouting new shoots. The nymphs, which feed on the new growth by sucking out the sap, pass through five stages of development, and at the fifth molt become adults. The adults mate, the males die soon after, and the females oviposit and die. Very few species have more than one generation a summer.

Relatively few species hibernate as adults. These adults seek shelter in the fall close to the host plant. The hibernating adults emerge in the spring, commence feeding on the tender new shoots of plants, mate, lay eggs, and gradually die off. The eggs hatch during the summer, the nymphs feed throughout their development, become adults, and continue feeding until hiberation.

Most mirids feed on ferns, herbaceous plants, shrubs, and trees. Many species are host specific or are limited to a group of related plants. Thus, the distribution of the host-specific species may be limited to the areas where the host plant grows. Other species are omnivorous, readily disperse from plant to plant, and generally have extensive distributions.

Several species of Miridae are now known to be partly or chiefly predaceous. They prey on soft bodied arthropods and may substantially reduce large colonies of aphids, psyllids, leafhoppers, and mites feeding on the host plant. Most species of *Phytocoris, Deraeocoris, Pilophorus, Ceratocapsus*, and *Hyaliodes* are beneficial predators.

Morphology

The Miridae are distinguished by the four-segmented antennae, the four-segmented rostrum, and the lack of ocelli, except the subfamily Isometopinae. The hemelytron is typically separated into clavus, corium, embolium, cuneus, and wing membrane. However, in a number of species

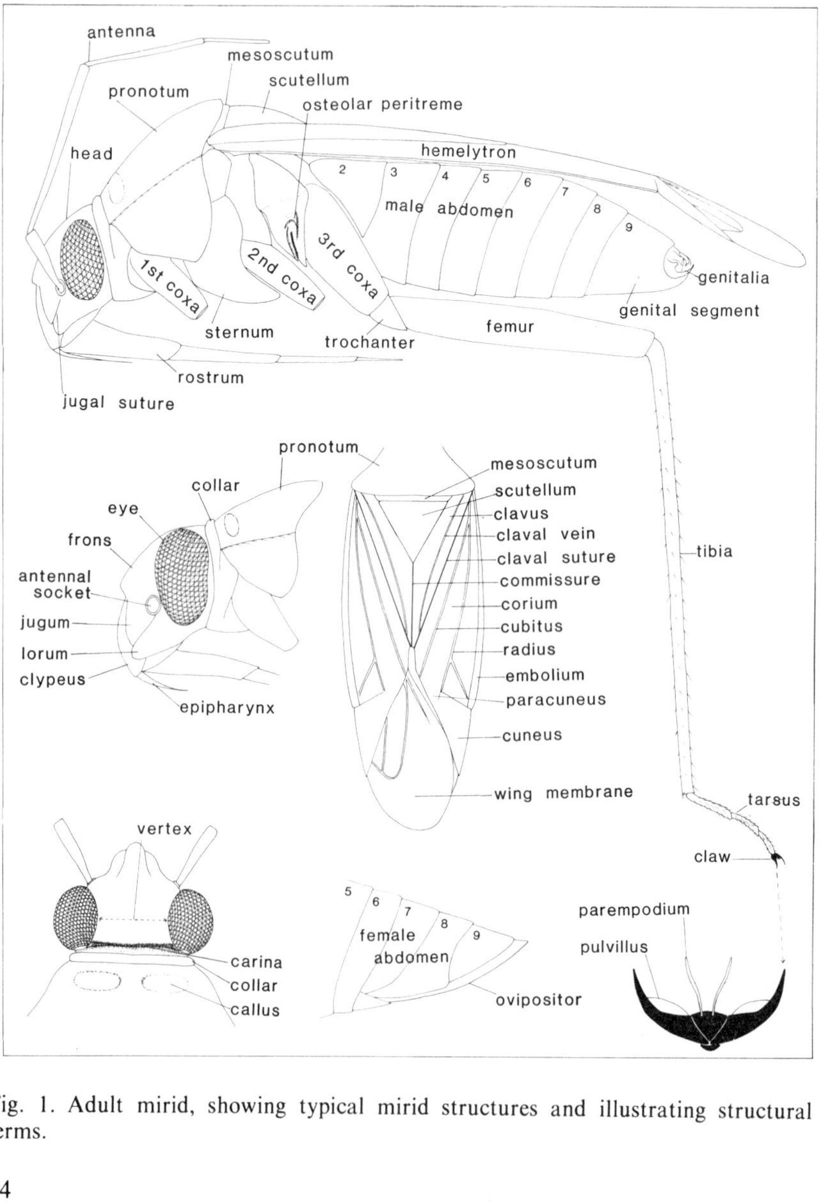

Fig. 1. Adult mirid, showing typical mirid structures and illustrating structural terms.

the hemelytra may be absent or reduced to mere pads, and these apterous or brachypterous adults may be mistaken for nymphs. Nymphs, however, do not have male or female genital structures, as shown in Fig. 1. The abdomen consists of nine segments, but only eight are visible. Each leg consists of a coxa, trochanter, femur, tibia, and usually a three-segmented tarsus. The claws and the structures between them, the parempodia and the pulvilli, provide reliable characters for separating the subfamilies. Fig. 1 shows the typical mirid structures and illustrates the structural terms.

Classification

The classification of the Miridae is based primarily on the tarsal claws proposed by Reuter (1910). The Reuter classification was further elaborated by Knight (1918*a*) and modified and condensed by Carvalho (1955*a*). The six subfamilies in the Carvalho classification system are Mirinae, Orthotylinae, Phylinae, Deraeocorinae, Bryocorinae, and Cylapinae. Kelton (1959) followed this classification in large part, but suggested that Dicyphinae be considered as a distinct subfamily. Carayon (1958) showed that the Isometopinae belong in the Miridae.

Schuh (1976) proposed major changes in the classification of the Miridae. He suggested that the Orthotylinae and the Deraeocorinae be recognized as tribes within Phylinae and Mirinae respectively, and that Dicyphinae be recognized as a tribe within Bryocorinae. For practical reasons Schuh's proposals are not adopted in this faunal work, but rather the more generally accepted classification of Carvalho (1955*a*), in part, and Kelton (1959). The Cylapinae and Isometopinae are not represented in the Prairie Provinces.

Key to subfamilies

1. Parempodia large and membraneous (Figs. 2, 3) 2
   Parempodia slender and hairlike (Figs. 4–8) 3
2. Parempodia divergent toward apices (Fig. 2); pronotal collar distinct (Fig. 9) .... **Mirinae** Hahn (p. 16)
   Parempodia parallel or convergent at apices (Fig. 3); pronotal collar depressed, inconspicuous (Fig. 10) .... **Orthotylinae** Van Duzee (p. 188)
3. Pronotal collar absent .... **Phylinae** Douglas & Scott (p. 279)
   Pronotal collar present (Fig. 9) 4
4. Pulvilli absent (Figs. 6, 7) .... **Deraeocorinae** Douglas & Scott (p. 349)
   Pulvilli present (Figs. 4, 5, 8) 5
5. Wing membrane with two cells (Fig. 11); slender species (Fig. 271) .... **Dicyphinae** Reuter (p. 372)
   Wing membrane with one cell (Fig. 12); oval species (Fig. 278) .... **Bryocorinae** Baerensprung (p. 379)

Subfamily Mirinae Hahn

In North America this is the largest subfamily. There are approximately 80 genera and probably over 650 species. Most of the species are phytophagous, but many are predaceous, especially in the genus *Phytocoris*.

The following are the subfamily characteristics: 1) large, free parempodia, diverging toward apices; 2) usually prominent pronotal collar; and 3) male genitalia basically with membranous lobes, and flexible ductus seminis.

In the Prairie Provinces the subfamily is represented by 4 tribes, 34 genera, and approximately 138 species.

Key to tribes of Mirinae

1. First segment of hind tarsus as long as second and third together (Fig. 13) 2
   First segment of hind tarsus shorter than second and third together (Fig. 14) .. 3
2. Antlike species; cuneus and wing membrane usually absent (Fig. 17) .... **Pithanini** (p. 16)
   Not antlike species; cuneus and wing membrane usually present .... **Stenodemini** (p. 20)
3. Pronotum and hemelytra velvety .... **Resthenini** (p. 48)
   Pronotum and hemelytra shiny, not velvety .... **Mirini** (p. 55)

Tribe Pithanini

In the Prairie Provinces the tribe is represented by two genera and two species.

Key to genera of Pithanini

1. First antennal segment shorter than width of vertex .... ***Pithanus* Fieber** (p. 16)
   First antennal segment longer than width of vertex ...... ***Mimoceps* Uhler** (p. 18)

Genus *Pithanus* Fieber

Elongate, subcylindrical, black species. Head vertical, eyes spherical, carina between them absent. First antennal segment short. Pronotum cylindrical in brachypterous forms, trapeziform in macropterous forms, calli strongly swollen. Hemelytra glabrous.

Kelton (1966*b*) found this Holarctic species widely distributed in North America.

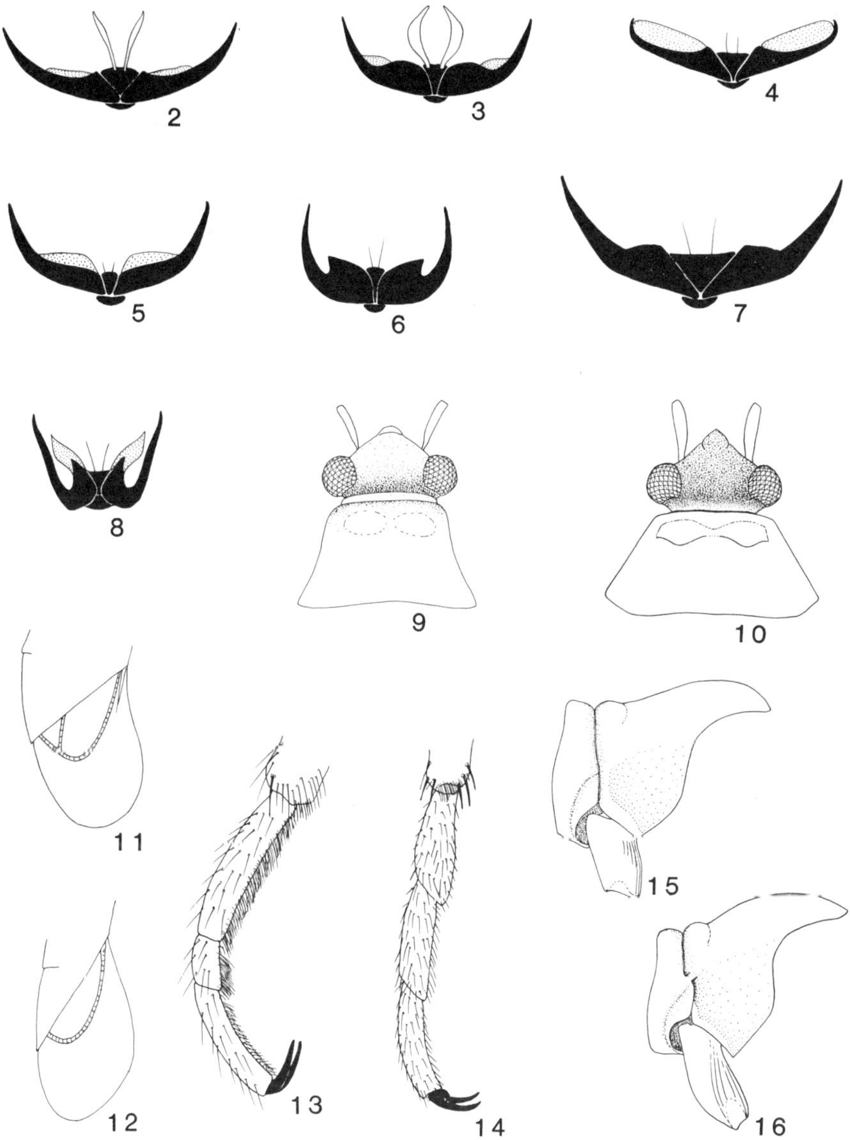

Figs. 2–16. Miridae structures. 2–8, Claws of Miridae; 2, Mirinae; 3, Orthotylinae; 4,5, Phylinae; 6,7, Deraeocorinae; 8, Dicyphinae; 9, Pronotum of Dicyphinae; 10, Pronotum of Phylinae; 11, Typical wing membrane; 12, Wing membrane of Bryocorinae; 13, Tarsus of Pithanini and Stenodemini; 14, Tarsus of Resthenini and Mirini; 15, Pronotum of *Opistheurista*; 16, Pronotum of *Prepops*.

Pithanus maerkeli (Herrich-Schaeffer)

Fig. 17; Map 1

Capsus maerkeli Herrich-Schaeffer, 1838:78.
Pithanus maerkeli: Reuter, 1875*d*:103.

Length 3.50–5.32 mm; width 0.98–1.40 mm. Head, pronotum, and scutellum black. Hemelytra black, costal margin pale. Ventral surface black, legs yellow. Both sexes brachypterous, females rarely macropterous.

Remarks. The species is distinguished by its shape and form (Fig. 17).

Habitat. Collected on grasses and sedges in damp areas.

Distribution. Transcontinental in northern States and southern Canada; reported from the Prairie Provinces (Map 1).

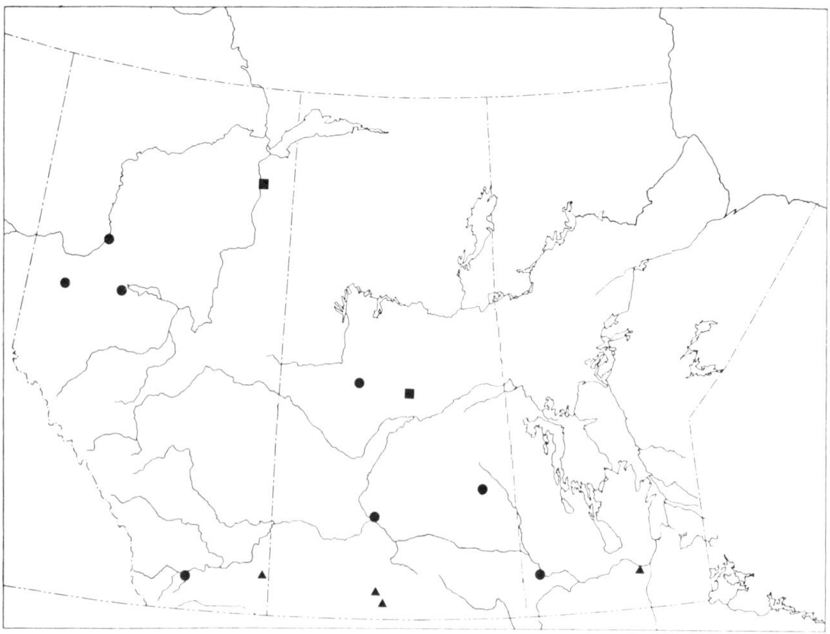

Map 1. Collection localities for *Pithanus maerkeli* (▲), *Mimoceps insignis* (●), and *Actitocoris signatus* (■).

Genus *Mimoceps* Uhler

Elongate, subcylindrical, glabrous species. Head oblique, carina between eyes absent. Pronotum cylindrical, calli strongly swollen. Hemelytra green, banded with black.

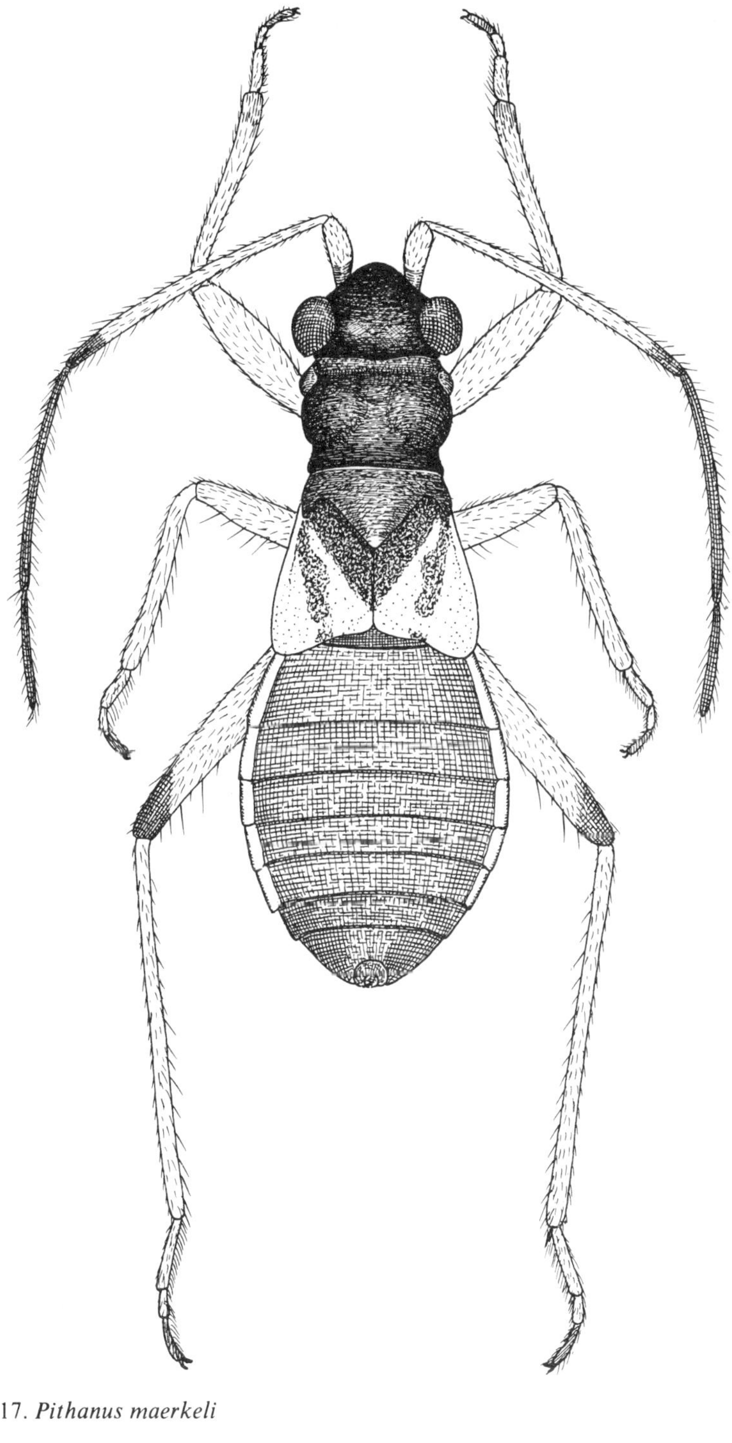

Fig. 17. *Pithanus maerkeli*

One species occurs in North America and is found in the Prairie Provinces.

Mimoceps insignis Uhler

Fig. 18; Map 1

Mimoceps insignis Uhler, 1890:84.
Mimoceps gracilis Uhler, 1890:85.

Length 3.50–5.04 mm; width 0.98–1.54 mm. Head black, two spots behind eyes pale. Pronotum black. Hemelytra brachypterous or macropterous. Legs long, slender, femora reddish.

Remarks. This species is distinguished by the banded hemelytra (Fig. 18).

Habitat. Collected on sedges.

Distribution. New York to Ohio, west to New Mexico, North Central States, Alaska, northern Canada, British Columbia, Ontario, Quebec; now known to occur in the Prairie Provinces (Map 1).

Tribe Stenodemini

In the Prairie Provinces the tribe is represented by 7 genera and 20 species.

Key to genera of Stenodemini

1. Second antennal segment clavate (Fig. 19) .... ***Actitocoris* Reuter** (p. 22)
   Second antennal segment linear (Fig. 20) 2
2. Lateral margins of pronotum rounded; head strongly elongate, eyes removed from anterior margin of pronotum (Fig. 20) ...... ***Collaria* Provancher** (p. 22)
   Lateral margins of pronotum angulate or carinate; head not elongate 3
3. Pronotum with deep, prominent punctures 4
   Pronotum impunctate 5
4. First antennal segment thickened with long, dense pubescence (Fig. 21) .... ***Stenodema* Laporte** (p. 24)
   First antennal segment slender, almost glabrous (Fig. 22) .... ***Litomiris* Slater** (p. 29)
5. First antennal segment with dense, long pubescence (Fig. 23) .... ***Leptopterna* Fieber** (p. 31)
   First antennal segment with short, sparse pubescence 6
6. Head long, pointed; first antennal segment shorter than head width (Fig. 24) .... ***Trigonotylus* Fieber** (p. 34)
   Head short, rounded in front; first antennal segment longer than head width (Fig. 25) .... ***Teratocoris* Fieber** (p. 44)

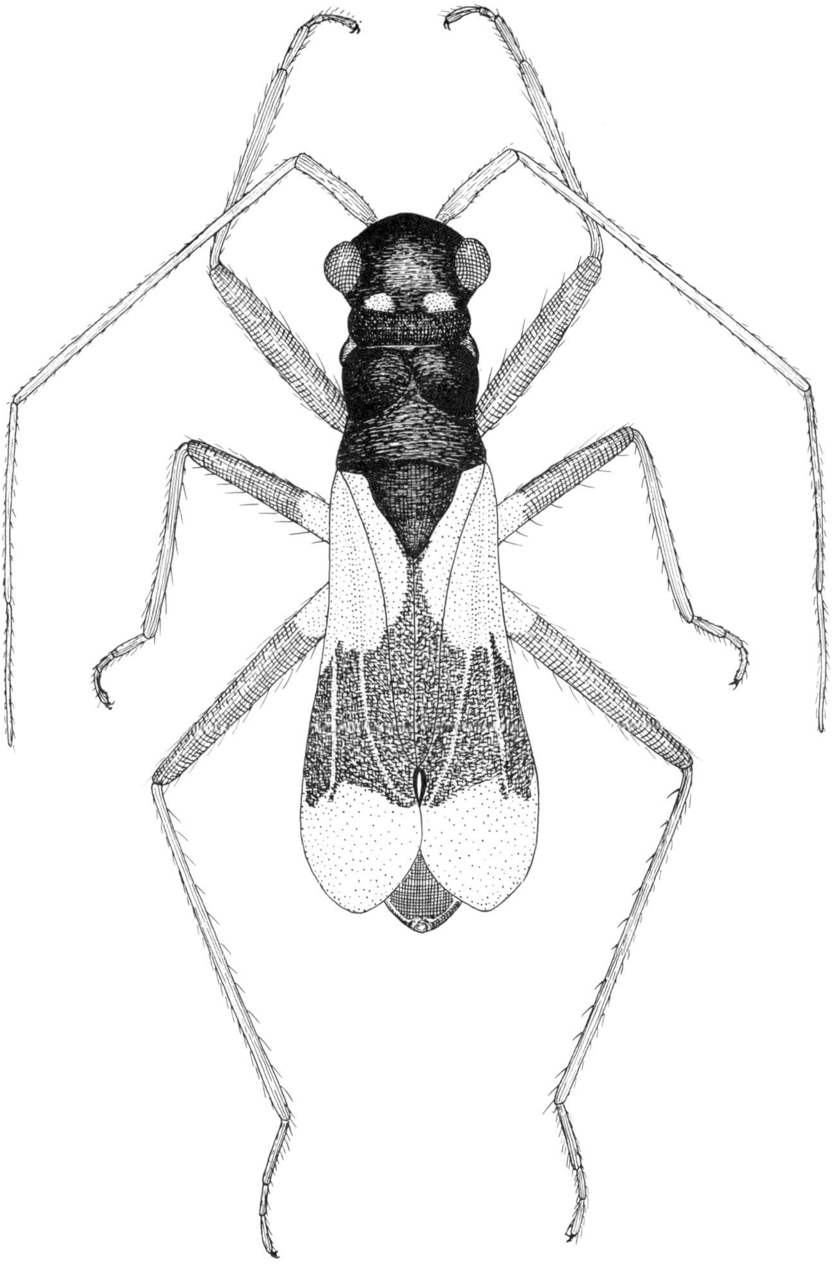

Fig. 18. *Mimoceps insignis*

Genus *Actitocoris* Reuter

Elongate, cylindrical, pubescent species. Head vertical, eyes spherical, carina between them absent. Pronotum trapeziform in macropterous forms, subcylindrical in brachypterous forms, lateral margins carinate. Hemelytra impunctate, pubescent. Tibiae strongly pilose.

Kelton (1966*b*) reported this Holarctic species from North America.

Actitocoris signatus Reuter

Fig. 19; Map 1

Actitocoris signatus Reuter, 1878:194.

Length 4.06–6.02 mm; width 1.40–1.82 mm. Head and pronotum yellowish green marked with reddish brown. Second antennal segment clavate. Hemelytra yellowish green marked with longitudinal fuscous lines.

Remarks. This species is distinguished by the clavate second antennal segment and by the pilose tibiae (Fig. 19).

Habitat. Collected on grasses and sedges.

Distribution. Northwest Territories; Alberta, Saskatchewan (Map 1).

Genus *Collaria* Provancher

Elongate, slender, subcylindrical species with long, slender legs. Head oblique, carina between eyes absent, base of head narrowed into a neck. Pronotum subcampanulate, calli strongly swollen, basal angles with large, velvety black spots. Hemelytra pale, marked with black. Legs pale, femora spotted with black.

Three species occur in North America, one in the Prairie Provinces.

Collaria meilleurii Provancher

Fig. 20; Map 2

Collaria meilleurii Provancher, 1872:79.
Nabidea coracina Uhler, 1878*a*:398.

Length 6.02–7.00 mm; width 1.40–1.82 mm. Head black with distinct neck, crescent mark behind eyes pale. Pronotum black. Legs long, slender.

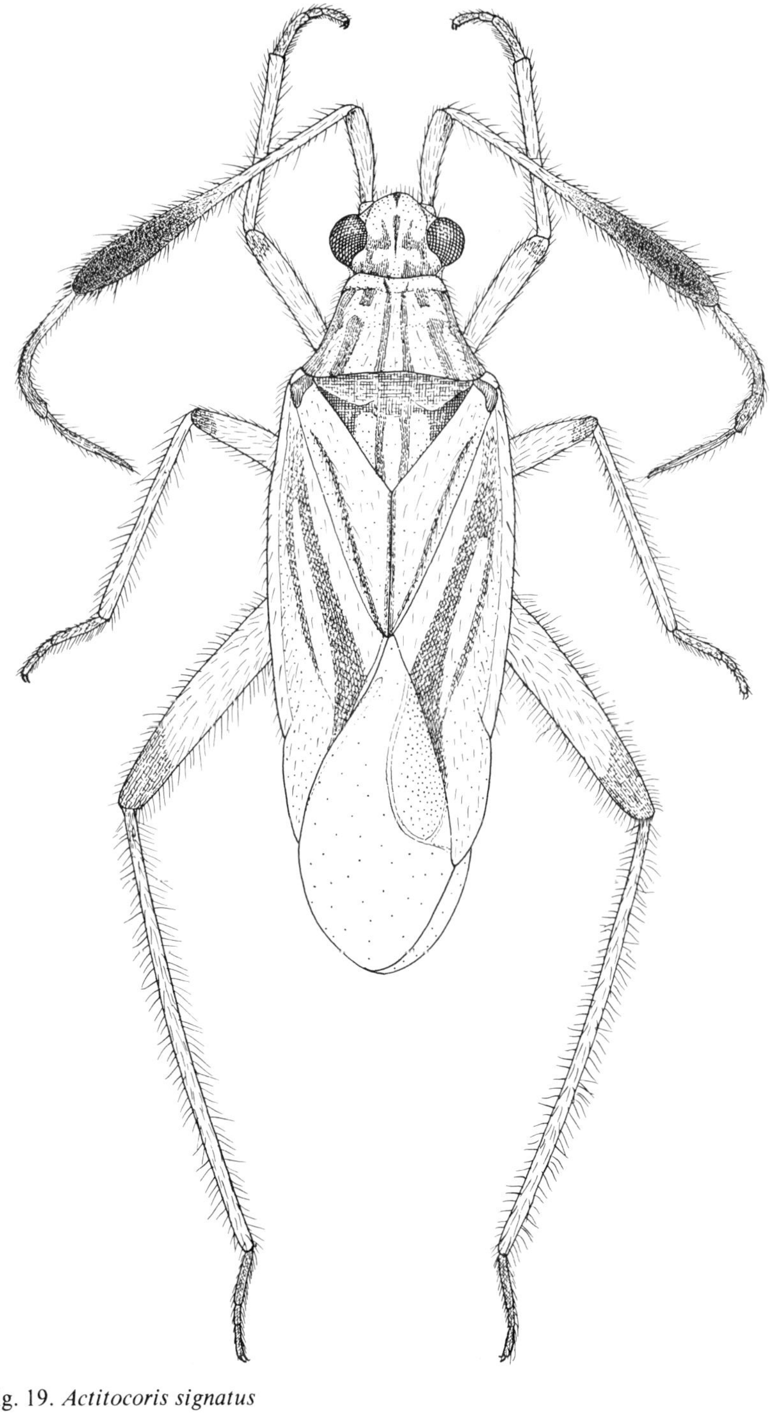

Fig. 19. *Actitocoris signatus*

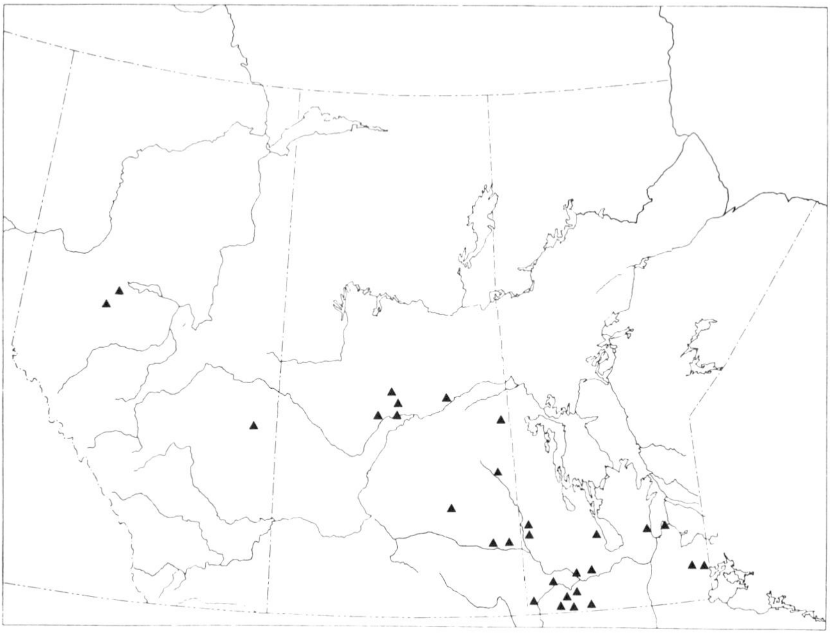

Map 2. Collection localities for *Collaria meilleurii*.

Remarks. This species is distinguished by the elongate head and by the velvety black spots on the pronotum (Fig. 20).

Habitat. Collected on grasses.

Distribution. Northeastern and North Central States, West Virginia to Illinois, Eastern Canada; Alberta, now known to occur in Saskatchewan and Manitoba (Map 2).

Genus *Stenodema* Laporte

Elongate, flattened, yellowish green species. Head straight, frons striate and flattened, vertex with longitudinal groove; eyes small, carina between them absent. Pronotum trapeziform, pale yellow, punctate; lateral margins carinate and with longitudinal median carina. Hemelytra pubescent.

The Nearctic species were reviewed by Kelton (1961*a*). Six species have been described from North America, three occur in the Prairie Provinces.

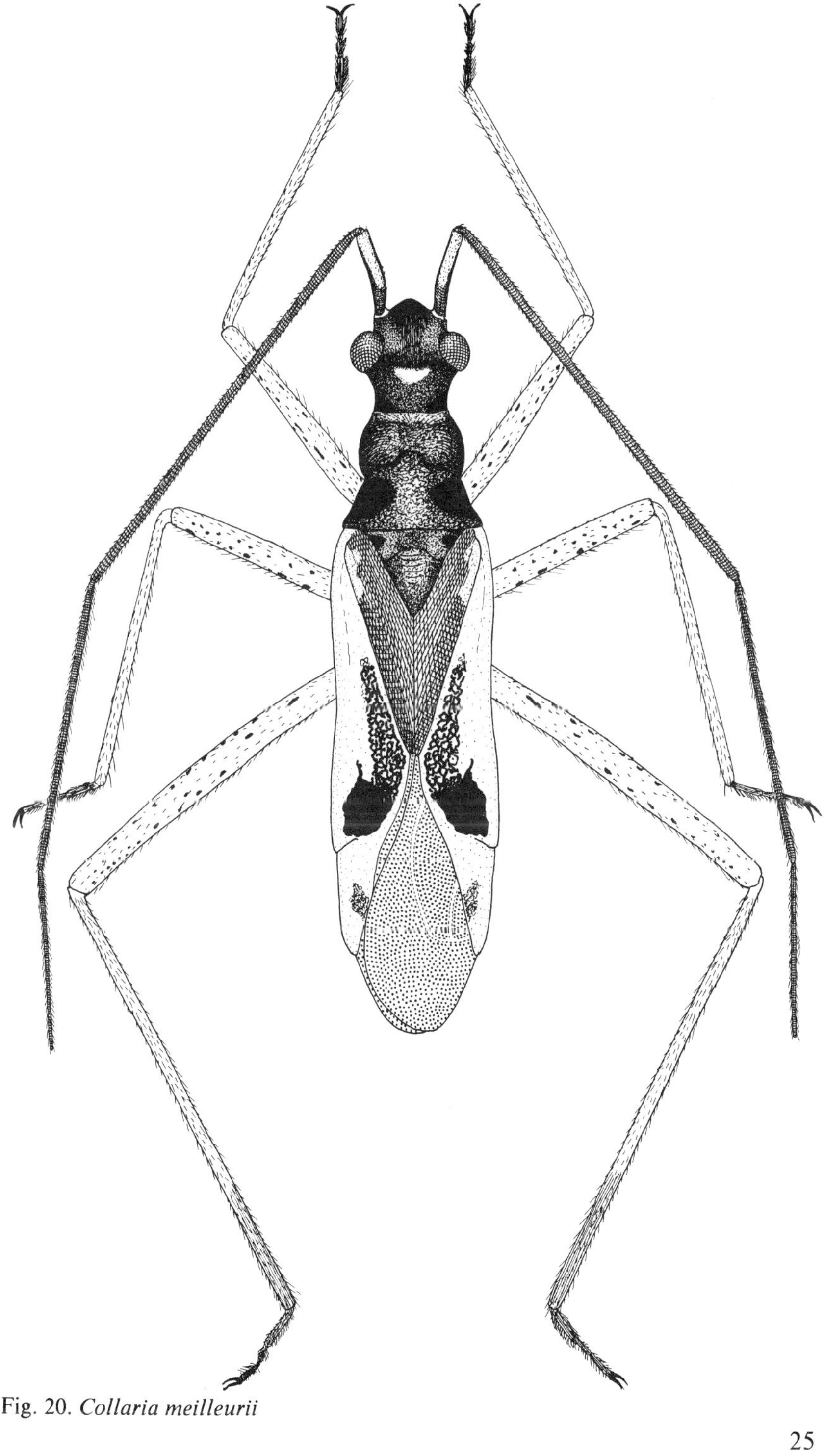

Fig. 20. *Collaria meilleurii*

Key to species of *Stenodema*

1. Hind femur with three short spines near apex .... ***trispinosa* Reuter** (p. 26)
   Hind femur without spines 2
2. Species greenish; antennae greenish brown; hind tibia strongly pilose .... ***pilosipes* Kelton** (p. 27)
   Species reddish brown; antennae reddish; hind tibia with short, slanting bristles .... ***vicina* (Provancher)** (p. 27)

Stenodema trispinosa Reuter

Map 3

Stenodema trispinosum Reuter, 1904:4.

Length 6.51–8.47 mm; width 1.54–2.03 mm. Straw yellow or green. Hind femur with three short spines near apex.

Remarks. The three spines on the femur readily distinguish the species.

Habitat. Collected on grasses in marshy areas, and along slough margins.

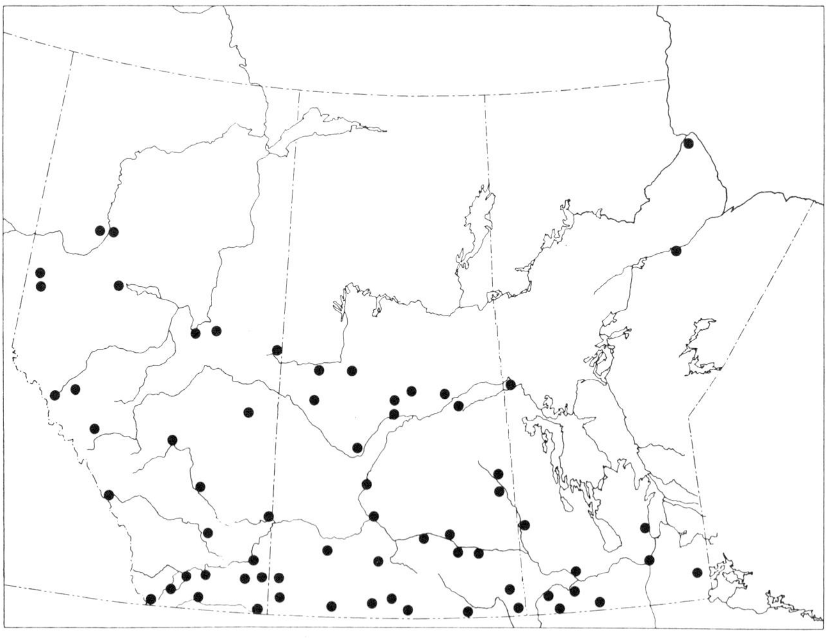

Map 3. Collection localities for *Stenodema trispinosa*.

Distribution. Holarctic, widespread in North America; reported from the Prairie Provinces (Map 3).

Stenodema pilosipes Kelton

Map 4

Stenodema pilosipes Kelton, 1961*a*:453.

Length 6.93–8.68 mm; width 1.68–2.17 mm. Olive green. Second antennal segment brown. Hind tibia strongly pilose.

Remarks. The pilose hind tibia is distinctive.

Habitat. Collected on grasses in meadows.

Distribution. Western North America; reported from the Prairie Provinces (Map 4).

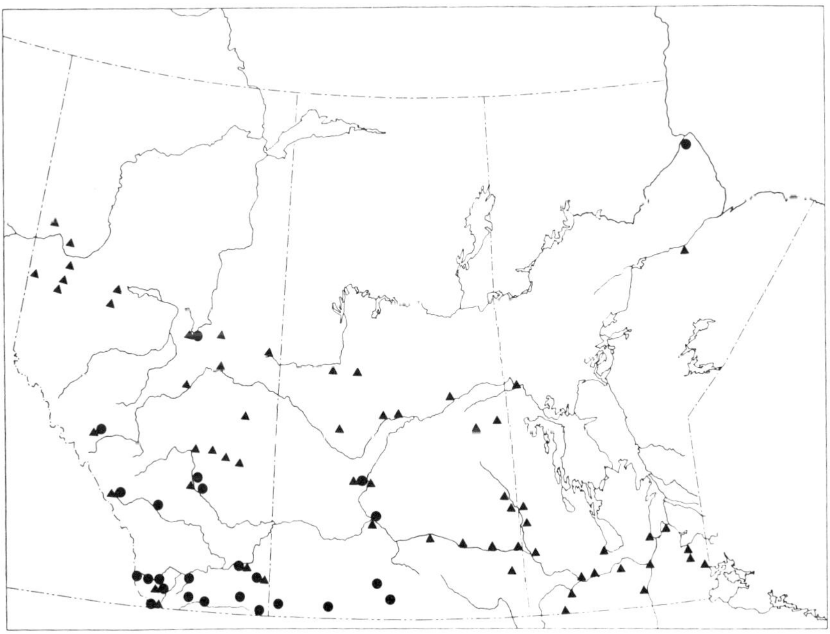

Map 4. Collection localities for *Stenodema pilosipes* (●) and *S. vicina* (▲).

Stenodema vicina (Provancher)

Fig. 21; Map 4

Miris vicinus Provancher, 1872:77.
Miris instabilis Uhler, 1875:836.

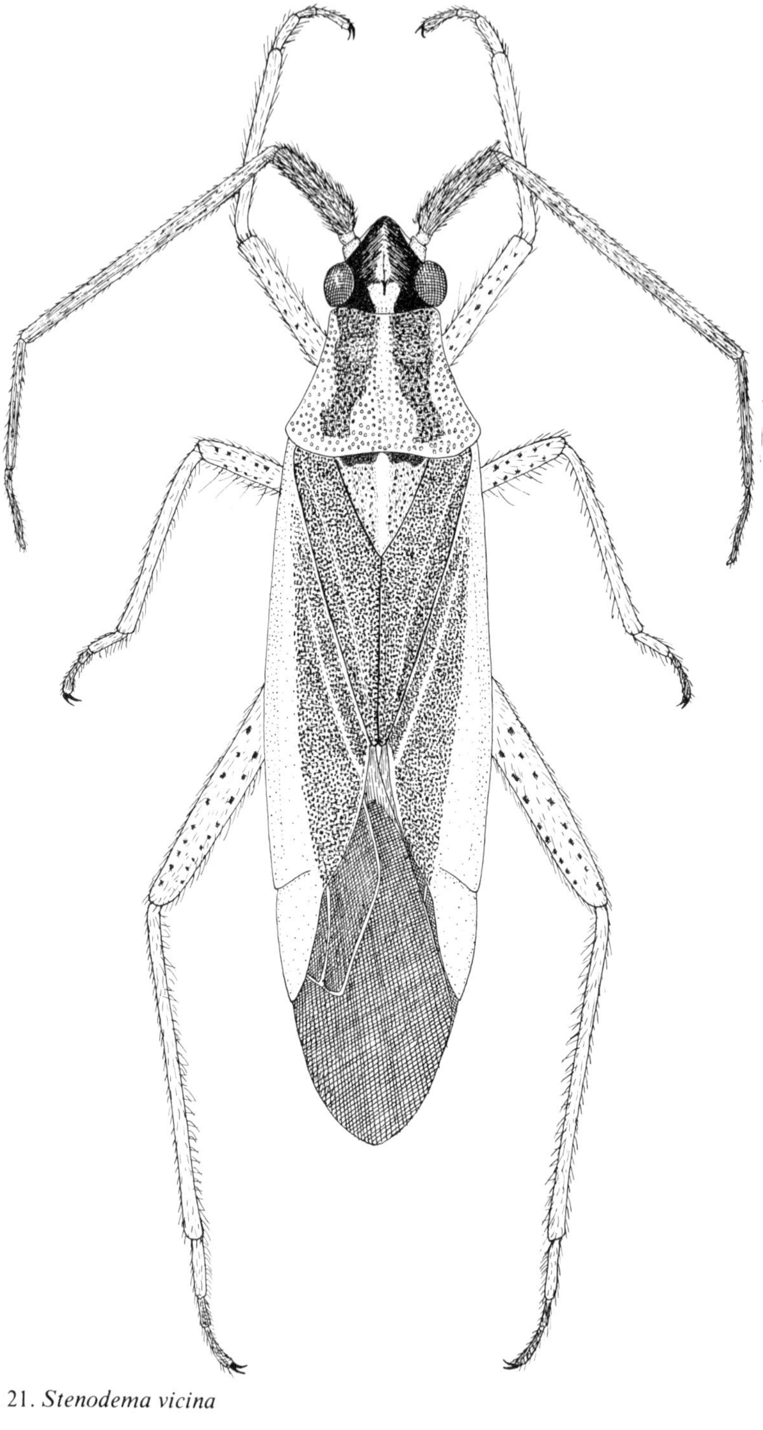

Fig. 21. *Stenodema vicina*

Miris affinis Reuter, 1875*d*:59.
Stenodema vicinum: Van Duzee, 1917:304.

Length 6.65–8.54 mm; width 1.57–2.03 mm. Hemelytra tinged with reddish brown. Second antennal segment reddish. Hind tibia with short, slanting hairs.

Remarks. This species is separated from *pilosipes* by the reddish second antennal segment, by the reddish brown hemelytra, and by the short, slanting hairs on hind tibia (Fig. 21).

Habitat. Collected on grasses in open fields and moist meadows, and often on cereal crops such as oats, wheat, and rye.

Distribution. Widespread in North America; reported from the Prairie Provinces (Map 4).

Genus *Litomiris* Slater

Elongate, pale green species. Head horizontal, frons smooth, eyes large, carina between them absent, vertex with longitudinal groove. First antennal segment long, practically glabrous. Pronotum trapeziform, punctate on basal half with longitudinal median carina. Legs long, slender.

Five species are known from North America, one occurs in the Prairie Provinces.

Litomiris debilis (Uhler)

Fig. 22, Map 5

Megaloceroea debilis Uhler, 1872:408.
Litomiris debilis: Slater, 1956:120.

Length 7.35–12.11 mm; 1.82–2.24 mm. Pale yellowish. First antennal segment long, slender. Pronotum with two sublateral black lines. Hemelytra fuscous, wide costal margin pale green.

Remarks. This species is distinguished by the long and slender first antennal segment (Fig. 22).

Habitat. Collected on native and cultivated grasses.

Distribution. District of Columbia, Iowa, Colorado to Montana, British Columbia; Alberta, now known to occur in Saskatchewan and Manitoba (Map 5).

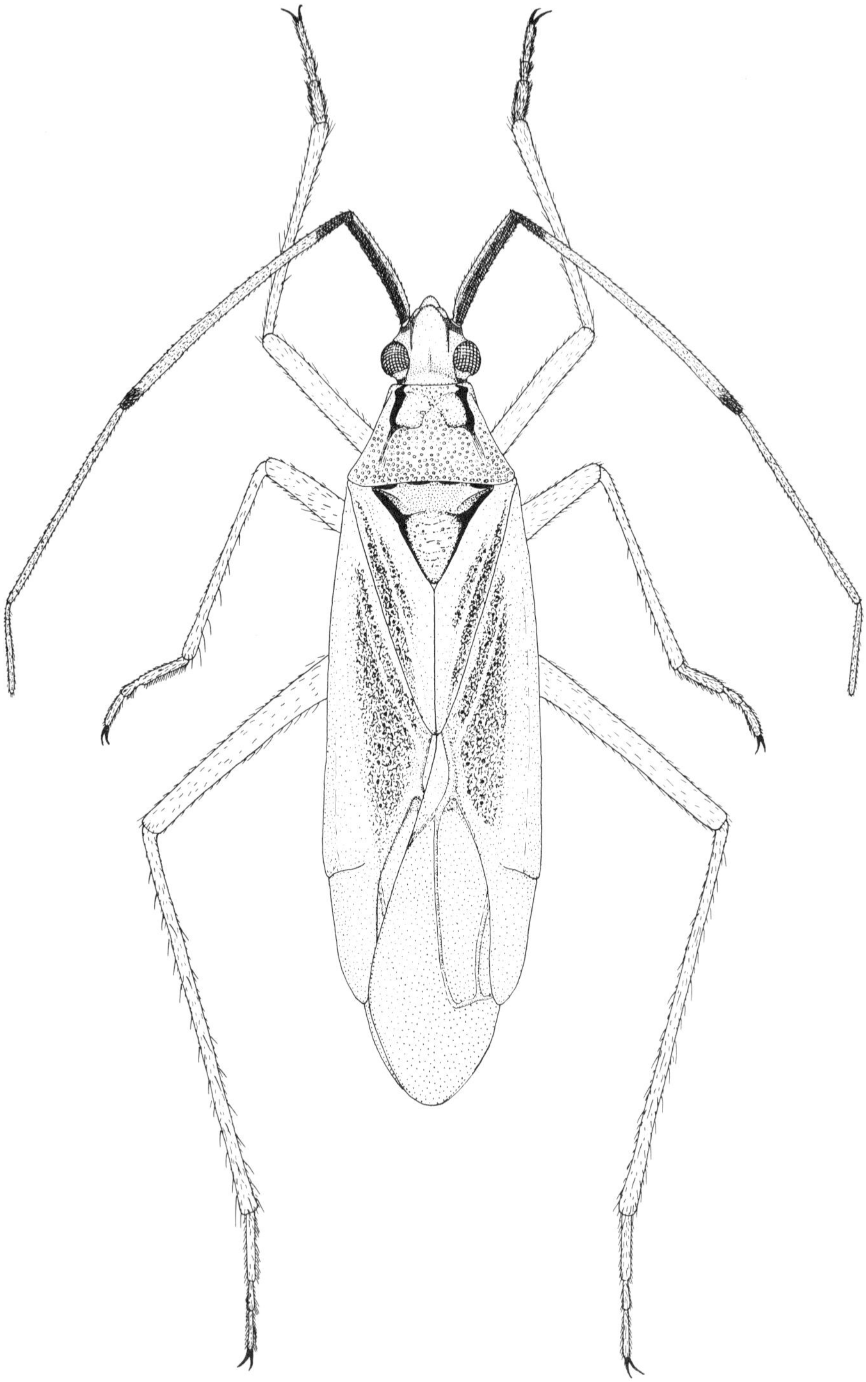

Fig. 22. *Litomiris debilis*

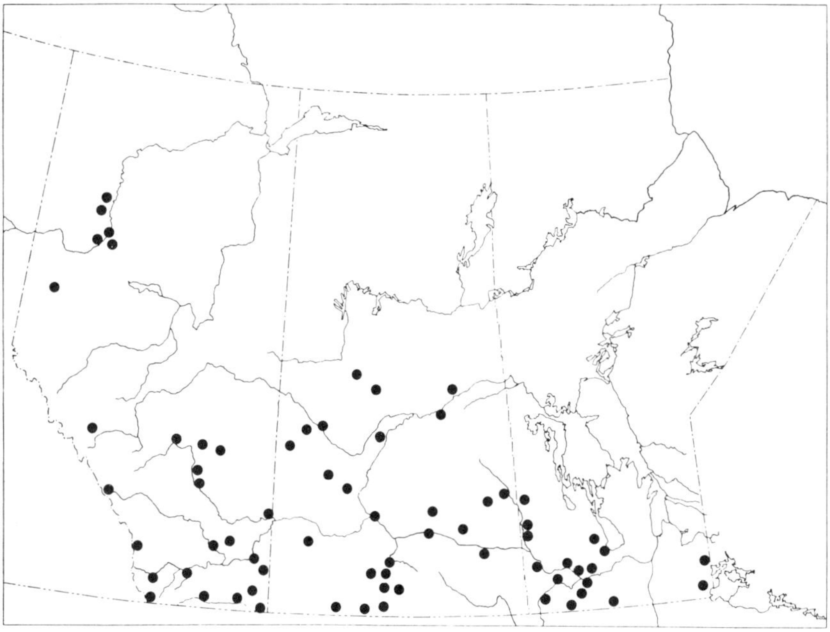

Map 5. Collection localities for *Litomiris debilis*.

Genus *Leptopterna* Fieber

Elongate, yellowish or reddish and black, pubescent species. Head oblique, frons tumid, eyes small, carina between them absent. First and second antennal segments with long, dense pubescence. Pronotum trapeziform, lateral margins carinate. Hemelytra reduced in female. Pubescence pale, long, erect. Legs long, slender, pilose.

Two species, both Holarctic in distribution, occur in North America.

Key to species of *Leptopterna*

1. First antennal segment about as long as width of head; second antennal segment much thinner than front tibia .... ***dolabrata* (Linnaeus)** (p. 31)
   First antennal segment much longer than width of head; second antennal segment almost as thick as front tibia (Fig. 23) ...... ***ferrugata* (Fallén)** (p. 32)

Leptopterna dolabrata (Linnaeus)

Map 6

Cimex dolabratus Linnaeus, 1758:449.
Leptopterna dolabrata: Fieber, 1861:245.
Miris belangeri Provancher, 1872:78.

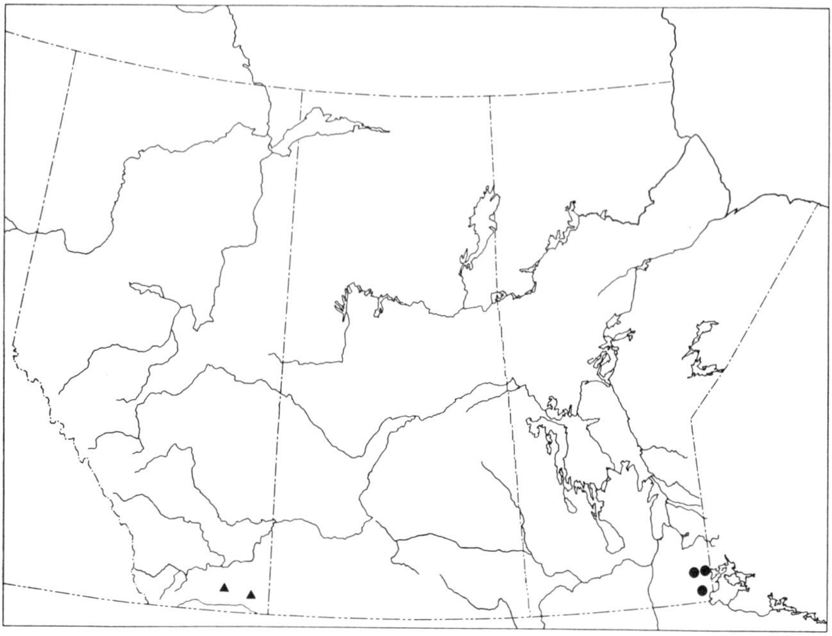

Map 6. Collection localities for *Leptopterna dolabrata* (●) and *L. ferrugata* (▲).

Length 7.30–9.30 mm; width 1.90–2.40 mm. First antennal segment about as long as width of head; second segment much thinner than front tibia. Hemelytra reddish brown in male, fuscous in female, costal margins yellowish green.

Remarks. The males are reddish and black, and the female hemelytra are long, extending to tip of abdomen. The nymphs of this species emerge very early in the spring.

Habitat. Collected on grasses in meadows.

Distribution. Northeastern States, Oregon, Ontario, Quebec, British Columbia; now known to occur in Manitoba (Map 6).

Leptopterna ferrugata (Fallén)

Fig. 23; Map 6

Miris ferrugatus Fallén, 1807:107.
Leptopterna amoena Uhler, 1872:409.
Leptopterna ferrugata: Reuter, 1875*a*:14.

Length 7.70–9.80 mm; width 1.80–2.50 mm. Males yellowish. First antennal segment much longer than width of head; second segment almost

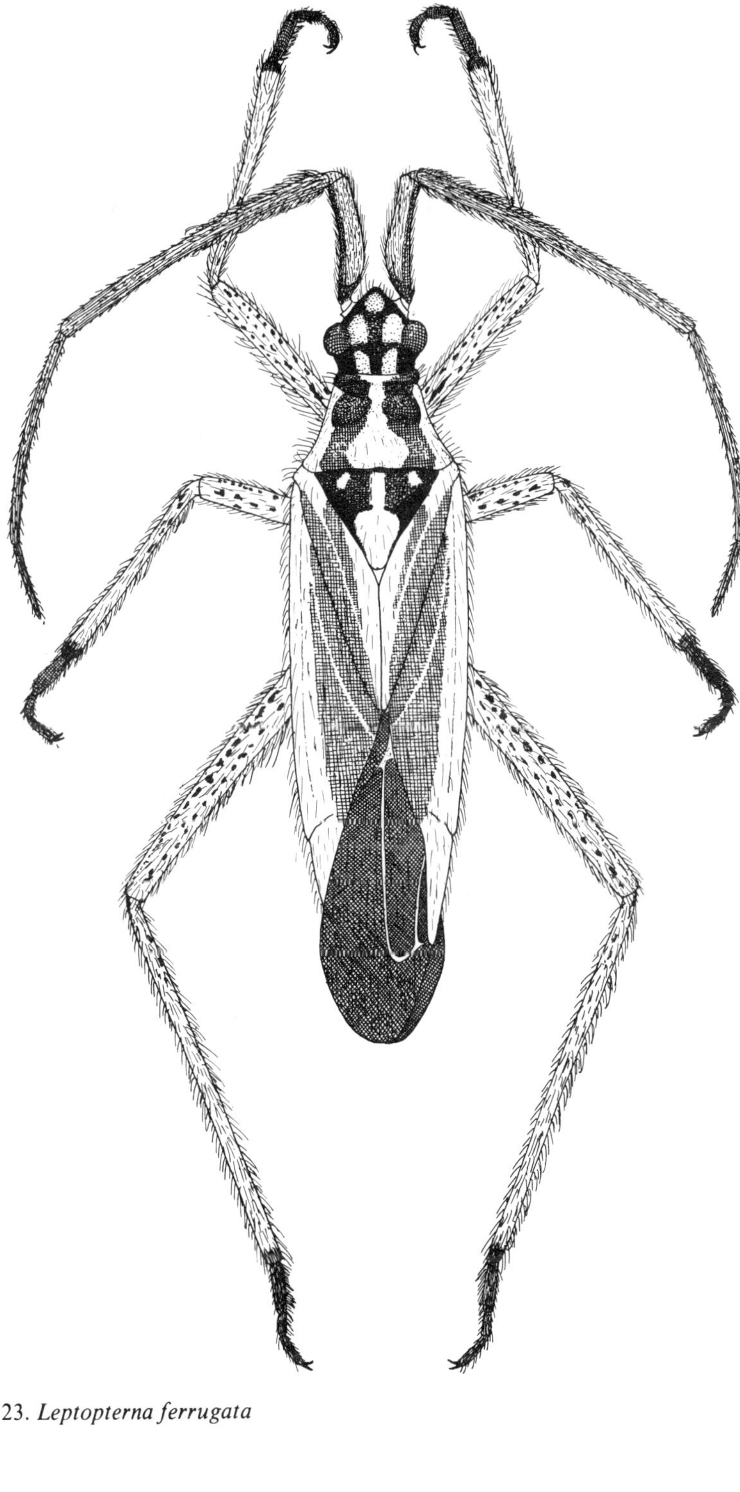

Fig. 23. *Leptopterna ferrugata*

as thick as front tibia. Hemelytra fuscous in both sexes, costal margin pale green (Fig. 23); in female hemelytra short.

Remarks. The males of this species are yellowish, those of *dolabrata* are reddish and black; the female hemelytra of *ferrugata* are short, extending to middle of abdomen.

Habitat. Collected on grasses.

Distribution. Alaska, northwestern States, Colorado to Kentucky, Quebec; now known to occur in Alberta (Map 6).

Genus *Trigonotylus* Fieber

Elongate, slender, sparsely pubescent species. Head horizontal, frons smooth, vertex with longitudinal median groove, clypeus prominent, extending forward. Pronotum finely punctate. Legs long, slender.

The North American species were revised by Kelton (1971*a*). Eighteen species are known from North America, eight occur in the Prairie Provinces, one of which is a Holarctic species.

Key to species of *Trigonotylus*

1. Frons not projecting above base of clypeus (Fig. 26) .... ***tarsalis* (Reuter)** (p. 37)
   Frons projecting above base of clypeus (Fig. 27) 2
2. First antennal segment thickened, spindle-shaped .... ***antennatus* Kelton** (p. 39)
   First antennal segment slender 3
3. First antennal segment black .... ***brooksi* Kelton** (p. 39)
   First antennal segment green, brown, or marked with red 4
4. Second antennal segment less than 1.9 mm in length; yellow, orange, or green; hind tibia green 5
   Second antennal segment more than 2.0 mm in length; reddish; hind tibia reddish 6
5. Second antennal segment yellowish orange, first segment green; rostrum 1.27 mm long .... ***flavicornis* Kelton** (p. 39)
   Second antennal segment greenish brown, first segment brown; rostrum 1.15 mm long .... ***canadensis* Kelton** (p. 40)
6. Rostrum more than 1.7 mm in length; first antennal segment with reddish stripes .... ***coelestialium* (Kirkaldy)** (p. 42)
   Rostrum less than 1.7 mm in length; first antennal segment diffuse reddish 7
7. First antennal segment with stiff, black hairs; spiculum thick, curved .... ***ruficornis* (Geoffroy)** (p. 42)
   First antennal segment with fine, black hairs; spiculum thin, straight .... ***americanus* Carvalho** (p. 43)

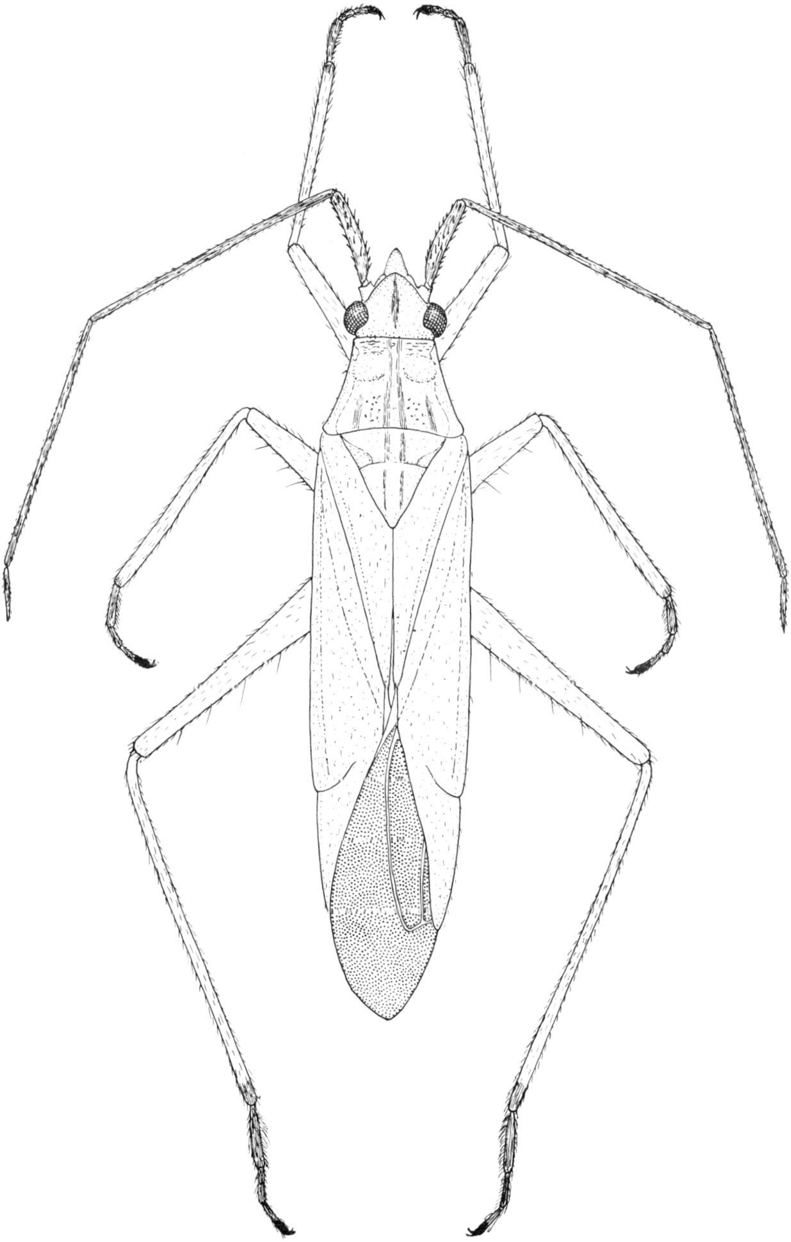

Fig. 24. *Trigonotylus ruficornis*

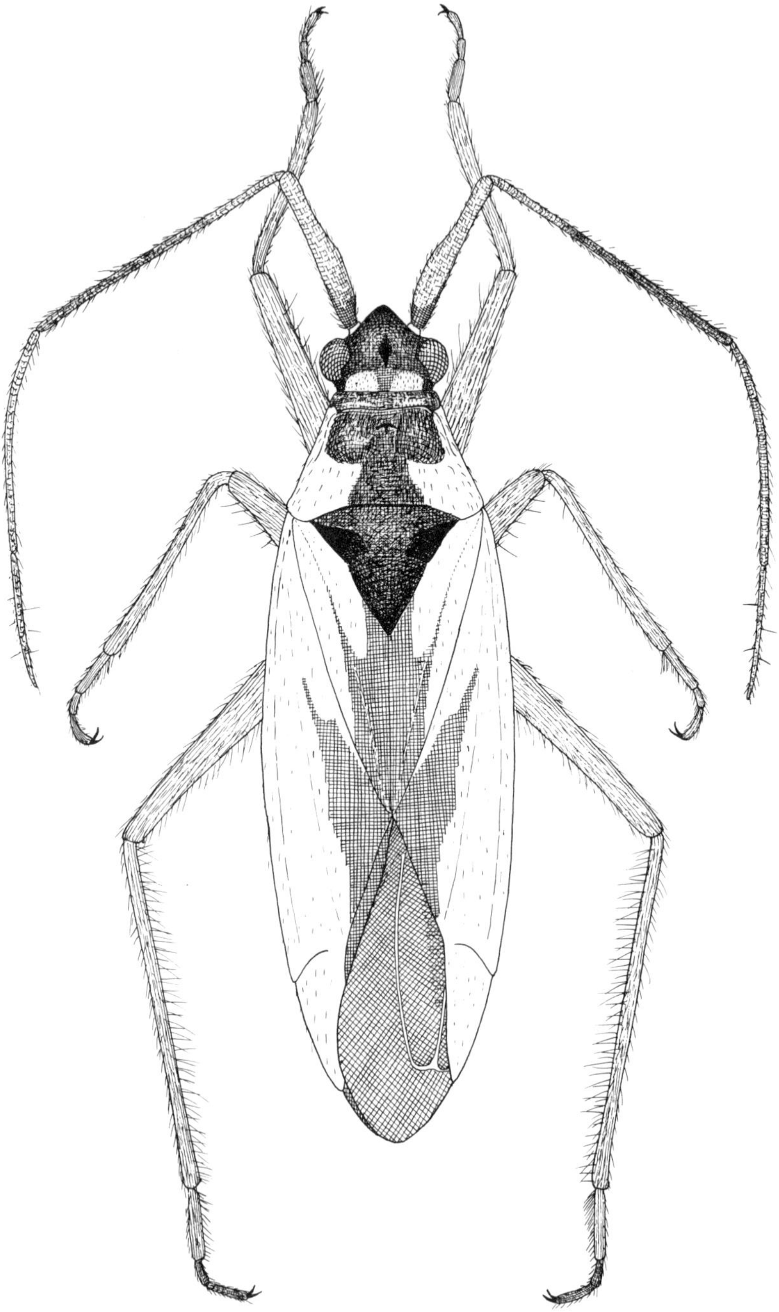

Fig. 25. *Teratocoris discolor*

Trigonotylus tarsalis (Reuter)

Figs. 26, 28; Map 7

Callimiris tarsalis Reuter, 1875*a*:60.
Trigonotylus tarsalis: Reuter, 1909:6.

Length 5.18–5.74 mm; width 1.26–1.40 mm. Head pale green, frons gently sloping to base of clypeus (Fig. 26), clypeus bulbous in front. First antennal segment long, thickened with stout, black bristles. Hemelytra light green. Hind tibia and tarsus often black, tibia pilose.

Remarks. This species is distinguished by the long and thickened first antennal segment, by the sloping frons, and by the black hind tibia. The genitalia (Fig. 28) are distinctive.

Habitat. Collected on grasses along slough margins.

Distribution. Northeastern and North Central States, Utah to Texas, Saskatchewan to Nova Scotia; now known to occur in Alberta (Map 7).

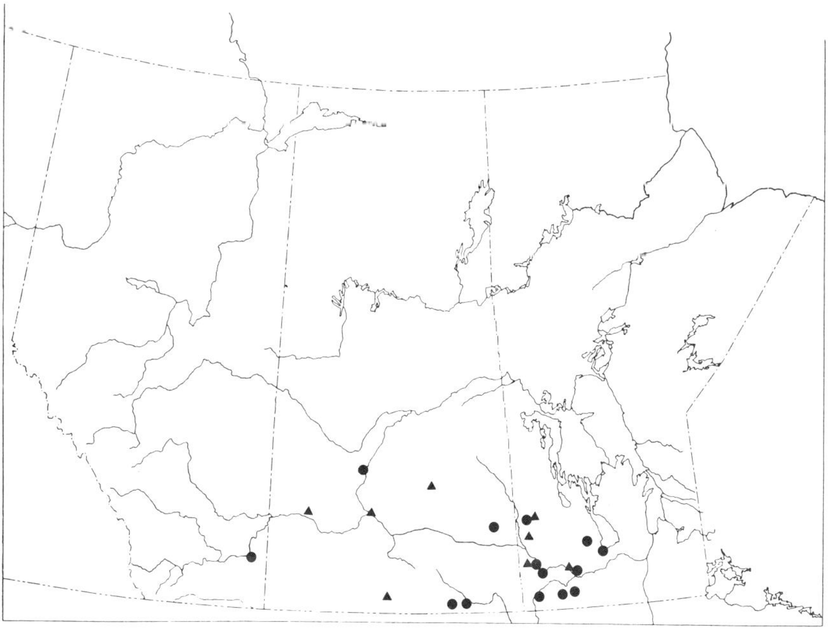

Map 7. Collection localities for *Trigonotylus tarsalis* (●) and *T. antennatus* (▲).

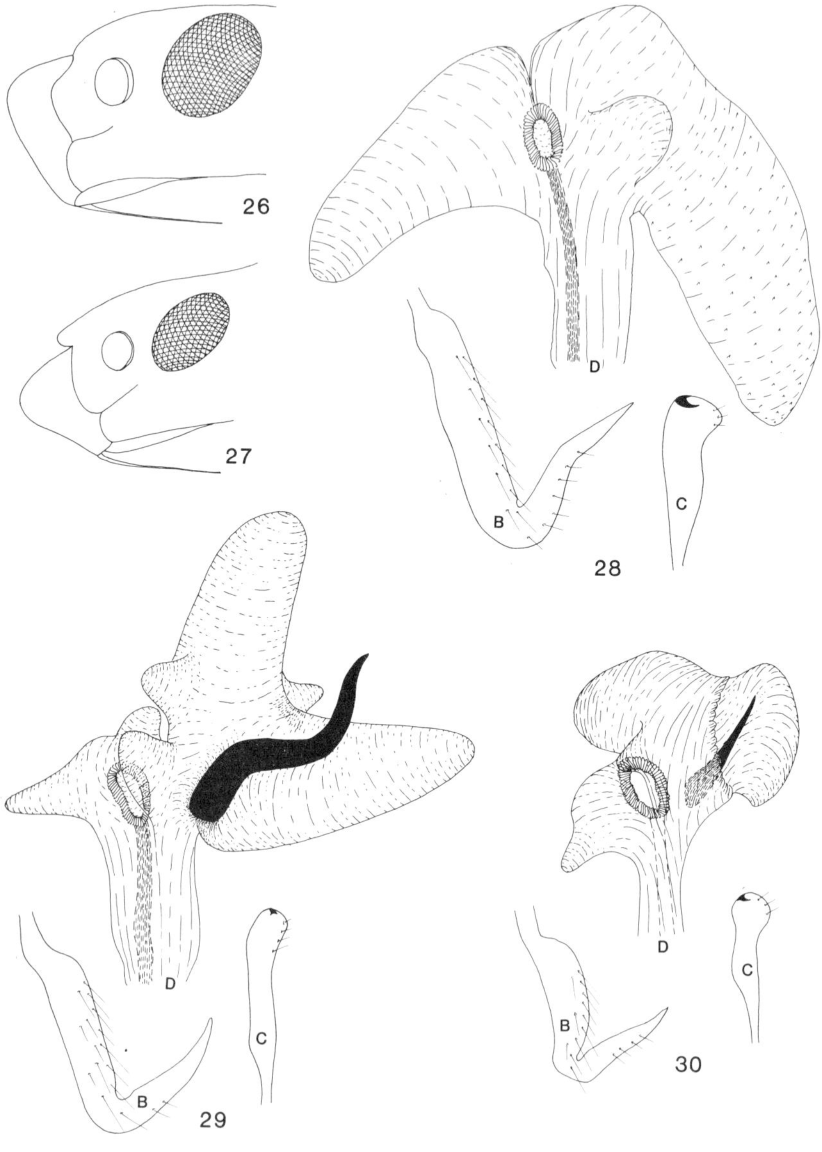

Figs. 26–30. Heads and genitalia of *Trigonotylus* spp. 26, *tarsalis*; 27, *flavicornis*; 28, *tarsalis*; 29, *antennatus*; 30, *brooksi*.

Trigonotylus antennatus Kelton

Fig. 29; Map 7

Trigonotylus antennatus Kelton, 1970:337.

Length 5.18–5.95 mm; width 1.12–1.40 mm. Pale green. Frons projecting over base of clypeus, clypeus acutely rounded in front. Antennae reddish orange, first segment thick, spindle-shaped. Apex of hind tibia reddish.

Remarks. This species is distinguished by the thick and spindle-shaped first antennal segment, and by the reddish orange antennae. The genitalia (Fig. 29) are distinctive.

Habitat. Collected on grasses.

Distribution. Western States, British Columbia; Saskatchewan, Manitoba (Map 7).

Trigonotylus brooksi Kelton

Fig. 30; Map 8

Trigonotylus brooksi Kelton, 1970:334.

Length 4.20–5.11 mm; width 0.98–1.12 mm. Head and pronotum pale green, longitudinal lines black; clypeus rounded in front. First antennal segment black or brown, other segments orange.

Remarks. This species is distinguished by the black or brown first antennal segment. The genitalia (Fig. 30) are distinctive.

Habitat. Collected on rangeland grasses.

Distribution. British Columbia; Saskatchewan (Map 8).

Trigonotylus flavicornis Kelton

Fig. 31; Map 8

Trigonotylus flavicornis Kelton, 1970:335.

Length 4.06–4.25 mm; width 0.94–1.05 mm. Head and pronotum yellowish green, longitudinal lines black. First antennal segment pale green, slender, with prominent stiff bristles, other segments orange.

Remarks. This species resembles *brooksi*, but the first antennal segment is green, with prominent stiff bristles. The genitalia (Fig. 31) are distinctive.

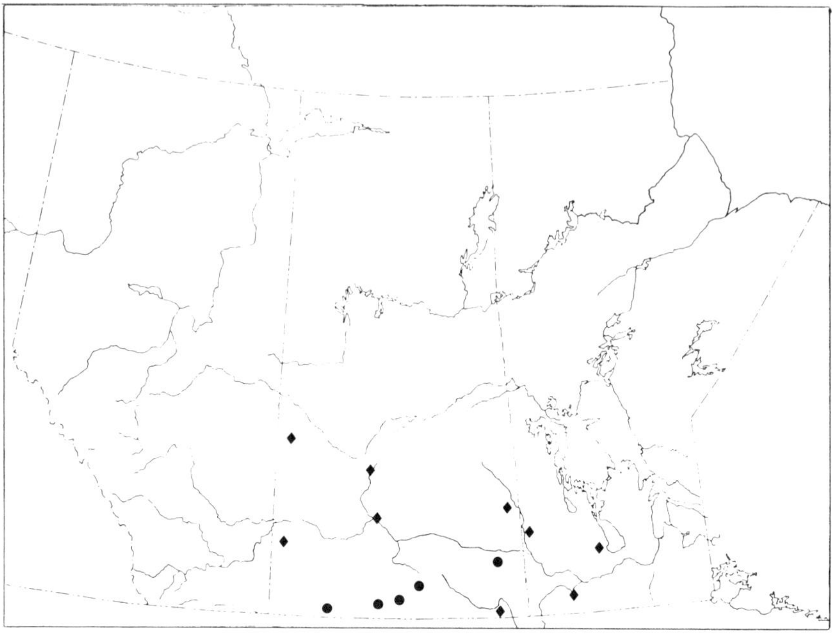

Map 8. Collection localities for *Trigonotylus brooksi* (●) and *T. flavicornis* (◆).

Habitat. Collected on prairie grasses.

Distribution. Known only from Saskatchewan and Manitoba (Map 8).

Trigonotylus canadensis Kelton

Fig. 32; Map 9

Trigonotylus canadensis Kelton, 1970:336.

Length 4.20–4.90 mm; width 0.91–1.12 mm. Head and pronotum yellowish green, longitudinal lines black. Antennal segments brown.

Remarks. This species is distinguished by the brown antennae. The male genitalia (Fig. 32) are distinctive.

Habitat. Collected on prairie grasses.

Distribution. Known only from the Prairie Provinces (Map 9).

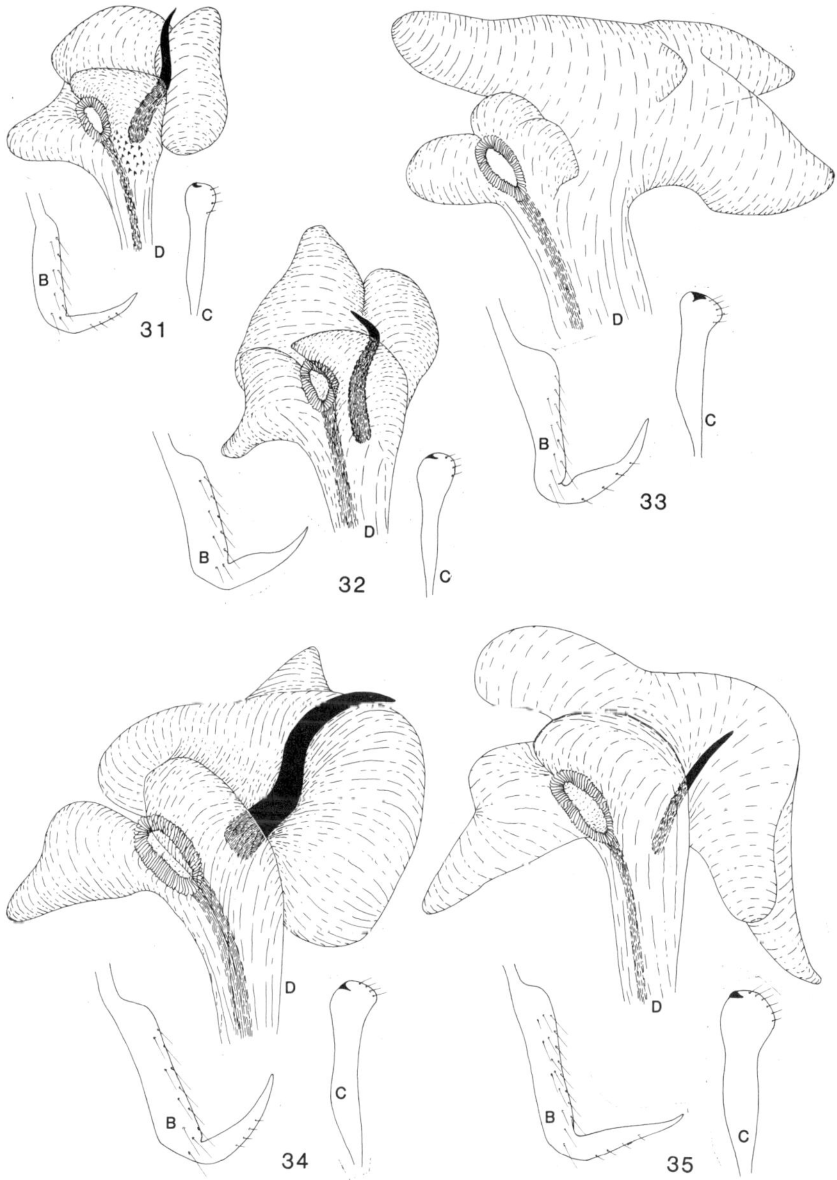

Figs. 31–35. Genitalia of *Trigonotylus* spp. 31, *flavicornis*; 32, *canadensis*; 33, *coelestialium*; 34, *ruficornis*; 35, *americanus*.

Trigonotylus coelestialium (Kirkaldy)

Fig. 33; Map 9

Megaloceroea coelestialium Kirkaldy, 1902:266.
Trigonotylus coelestialium: Reuter, 1903:1.

Length 5.22–6.16 mm; width 1.26–1.47 mm. Yellowish green. First antennal segment greenish with three longitudinal reddish stripes and short, fine, black hairs; other segments reddish. Hind tibia reddish near apex.

Remarks. This Holarctic species is distinguished by the reddish stripes on the first antennal segment, and by the reddish tip of the hind tibia. The male genitalia (Fig. 33) are distinctive.

Habitat. Collected on prairie grasses.

Distribution. Eastern States, Eastern Canada; Prairie Provinces (Map 9).

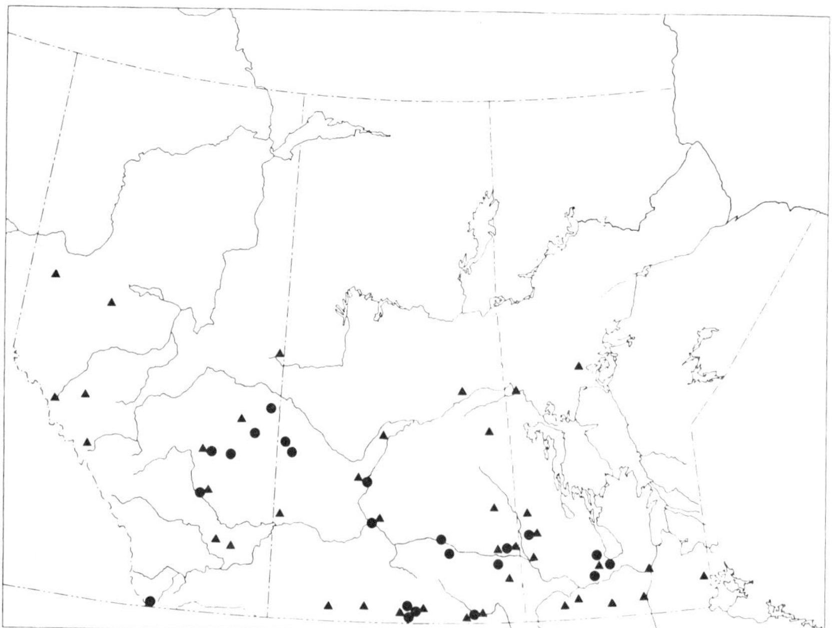

Map 9. Collection localities for *Trigonotylus canadensis* (●) and *T. coelestialium* (▲).

Trigonotylus ruficornis (Geoffroy)

Figs. 24, 34; Map 10

Cimex ruficornis Geoffroy, 1785:209.
Trigonotylus ruficornis: Fieber, 1861:243.

Miris viridis Provancher, 1872:78.
Trigonotylus montanus Carvalho, 1957:137.

Length 5.60–6.65 mm; width 1.33–1.61 mm. Yellowish green. First antennal segment green to brownish, often with reddish tinge, long, slender, with short, stiff, black bristles. Hind tibia reddish near apex.

Remarks. This Holarctic species is distinguished by the short, stiff, black bristles on the long and slender first antennal segment (Fig. 24). The male genitalia (Fig. 34) are distinctive.

Habitat. Collected on rangeland grasses.

Distribution. Widespread in North America; Prairie Provinces (Map 10).

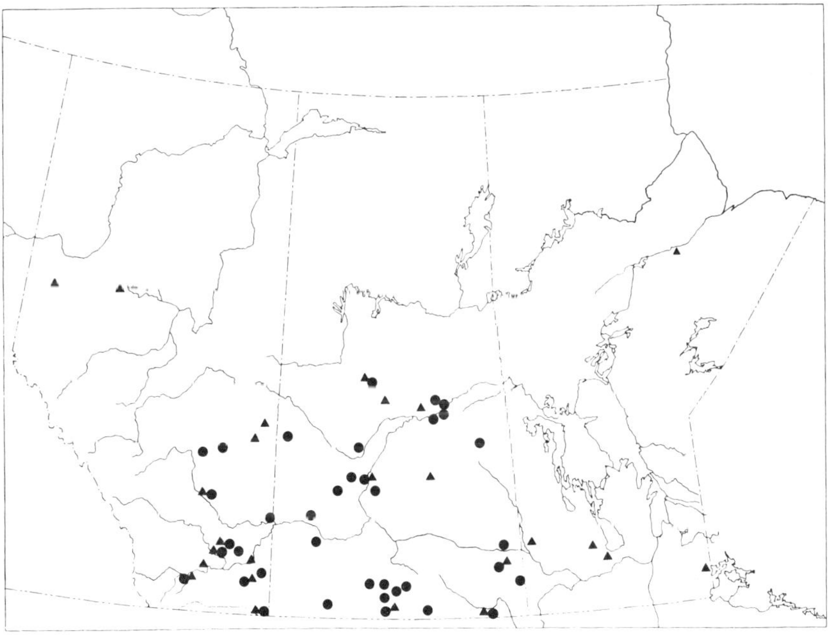

Map 10. Collection localities for *Trigonotylus ruficornis* (▲) and *T. americanus* (●).

Trigonotylus americanus Carvalho

Fig. 35; Map 10

Trigonotylus americanus Carvalho, 1957:125.

Length 5.32–6.16 mm; width 1.19–1.40 mm. Yellowish green. Antennal segments mostly reddish, first segment slender with short, fine, black hairs. Hind tibia reddish near apex.

Remarks. This species is similar in appearance to *coelestialium*, but lacks the reddish stripes. The male genitalia (Fig. 35) are distinctive.

Habitat. Collected on rangeland grasses.

Distribution. Western States, British Columbia; Alberta, Saskatchewan (Map 10).

Genus *Teratocoris* Fieber

Elongate, slender, and flattened, green or green and black species. Head horizontal, short, vertex with longitudinal median groove. First antennal segment longer than width of head. Pronotum smooth, lateral margins carinate. Legs long, slender.

Six species are known from North America, four occur in the Prairie Provinces, two of which are Holarctic species. Kelton (1966*a*) reviewed the species in the Nearctic region.

Key to species of *Teratocoris*

1. Hemelytra black along inner margins of clavus and corium (Fig. 25); genitalia (Fig. 36) .... ***discolor* Uhler** (p. 44)

   Hemelytra greenish throughout 2
2. First and second antennal segments and hind tibia reddish; genitalia (Fig. 37) .... ***paludum* Sahlberg** (p. 45)

   First and second antennal segments and hind tibia not reddish 3
3. Left margin of genital segment with slender process (Fig. 38A); fifth abdominal segment of female with small median process (Fig. 38E) .... ***saundersi* Douglas & Scott** (p. 46)

   Left margin of genital segment with broad process (Fig. 39A); fifth abdominal segment of female entire .... ***caricis* Kirkaldy** (p. 46)

Teratocoris discolor Uhler

Fig. 36; Map 11

Teratocoris discolor Uhler, 1887:68.

Length 4.20–5.95 mm; width 1.19–1.75 mm. Head and pronotum mostly black, hemelytra green, black along inner margins. Ventral surface black, legs reddish, hind tibia strongly pilose.

Remarks. This species is distinguished by the black head and pronotum, by the green hemelytra with black inner margins, and by the pilose hind tibia. The genital structures (Fig. 36) are distinctive.

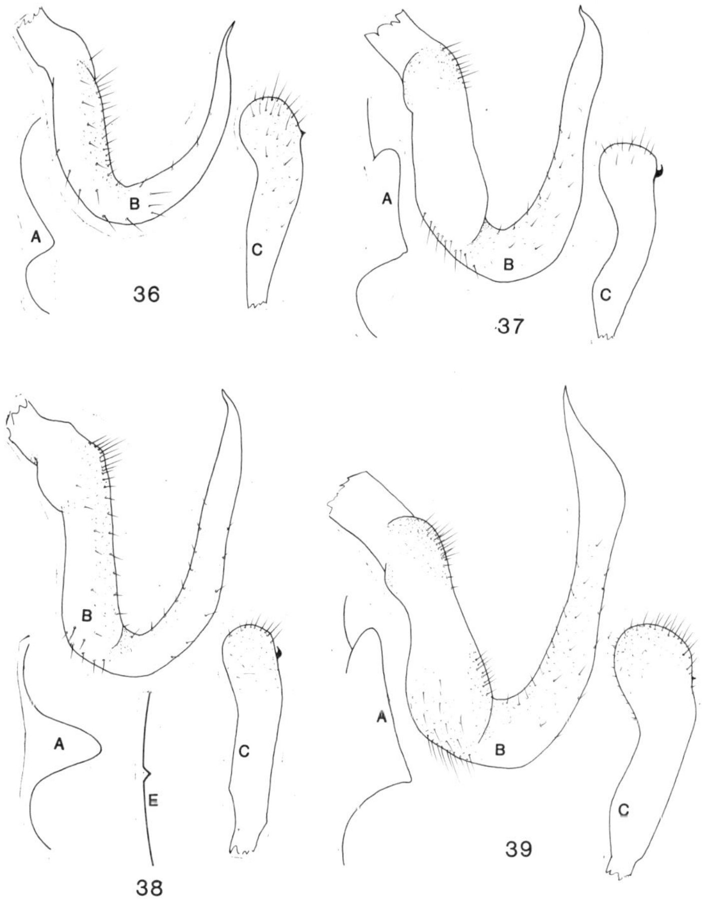

Figs. 36–39. Genitalia of *Teratocoris* spp. 36, *discolor*; 37, *paludum*; 38, *saundersi*; 39, *caricis*.

Habitat. Collected on *Carex* spp.

Distribution. Northeastern and Central States, British Columbia, Ontario, Quebec; Prairie Provinces (Map 11).

Teratocoris paludum Sahlberg

Fig. 37; Map 11

Teratocoris paludum Sahlberg, 1870:291.

Length 4.97–6.30 mm; width 1.22–1.47 mm. Green. First and second antennal segments reddish. Hind tibia reddish and strongly pilose.

Remarks. The reddish first and second antennal segments, the reddish hind tibia, and the genital structure (Fig. 37) readily distinguish the species.

Habitat. Collected on *Carex* spp.

Distribution. Alaska, New York, Central States, California, Yukon, British Columbia, Ontario, Quebec, Newfoundland; Prairie Provinces (Map 11).

Teratocoris saundersi Douglas & Scott

Fig. 38; Map 12

Teratocoris saundersi Douglas & Scott, 1869:260.
Teratocoris herbaticus Uhler, 1887*b*:67.
Teratocoris longicornis Uhler, 1895:29

Length 4.55–6.30 mm; width 1.22–1.68 mm. Pale green; median longitudinal line on head and pronotum, collar, and scutellum often black. Hind tibia pilose. Left margin of genital segment with slender process (Fig. 38A). Fifth abdominal segment of female with small, median process (Fig. 38E).

Remarks. This species is distinguished by the characters given in the key.

Habitat. Collected on *Scirpus* spp., *Carex* spp., and *Calamagrostis* spp.

Distribution. Alaska, Colorado, northern Canada, British Columbia, Newfoundland; Prairie Provinces (Map 12).

Teratocoris caricis Kirkaldy

Fig. 39; Map 12

Teratocoris caricis Kirkaldy, 1909:390.

Length 4.69–6.79 mm; width 1.33–1.89 mm. Pale green, median longitudinal line on head and pronotum, collar and scutellum often black. Hind tibia pilose.

Remarks. This species is similar to *saundersi* in color and appearance, but the process on the genital segment is broad (Fig. 39A).

Habitat. Collected on *Carex* spp.

Distribution. Alaska, Wyoming, Colorado, California, British Columbia, Newfoundland; Prairie Provinces (Map 12).

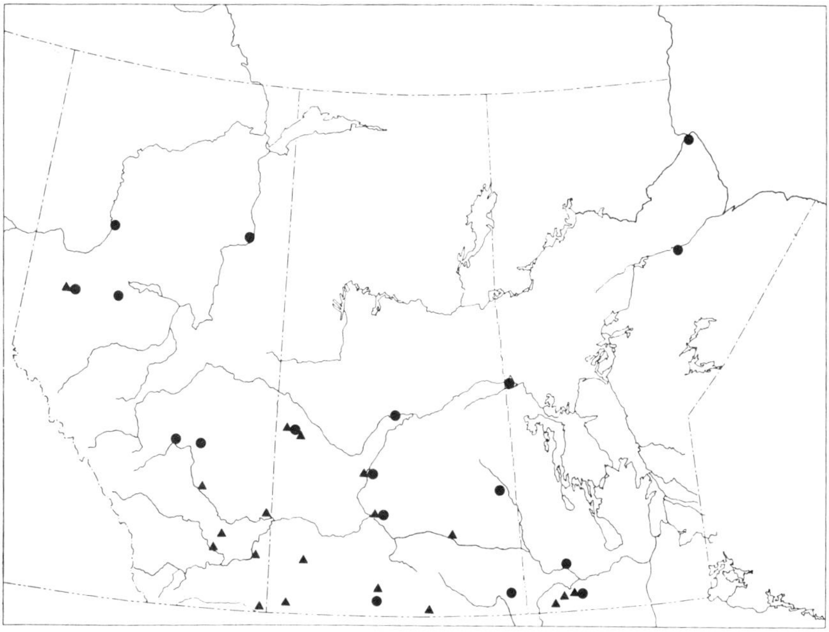

Map 11. Collection localities for *Teratocoris discolor* (▲) and *T. paludum* (●).

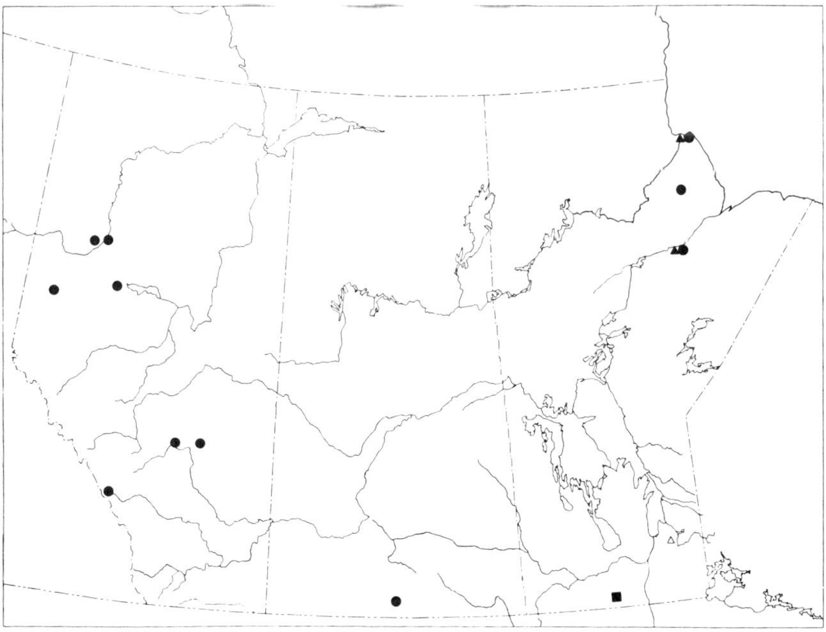

Map 12. Collection localities for *Teratocoris saundersi* (▲), *T. caricis* (●), and *Opistheurista clandestina* (■).

Tribe Resthenini

In the Prairie Provinces the tribe is represented as two genera and eight species.

Key to genera of Resthenini

1. Stricture of pronotal collar not interrupted at side (Fig. 15) .... ***Opistheurista* Carvalho** (p. 48)
   Stricture of pronotal collar interrupted at side (Fig. 16) .... ***Prepops* Reuter** (p. 48)

Genus *Opistheurista* Carvalho

Elongate, velvety, black and orange and red species. Head vertical, short. Pronotum trapeziform, lateral margins rounded, collar prominent, stricture of collar not interrupted at side. Osteolar peritreme small, indistinct. Head, pronotum, and hemelytra velvety.

This Nearctic genus is monobasic.

Opistheurista clandestina (Van Duzee)

Fig. 16; Map 12

Opistheuria clandestina Van Duzee, 1915:110.
Opistheurista clandestina: Carvalho, 1959:347.

Length 6.30–7.70 mm; width 2.52–3.22 mm. Head black, jugum orange, antennae black. Pronotum black, collar and side margins orange or red. Scutellum black. Hemelytra black, costal margin and cuneus orange or red; pubescence short. Ventral surface mostly black with some orange, legs black, pilose.

Remarks. Carvalho (1959) transferred this species to the present genus. It is similar to those of *Prepops* in texture and appearance, but the collar stricture is not interrupted at the side (Fig. 16)

Habitat. Collected on *Vicia* spp.

Distribution. New York, Florida, Central States, Ontario; now known to occur in Manitoba (Map 12).

Genus *Prepops* Reuter

Elongate, velvety, black, or black and red species. Head vertical, short. Pronotum trapeziform, lateral margins rounded, collar prominent, stricture of collar interrupted at side. Osteolar peritreme small, indistinct. Head, pronotum, and hemelytra velvety.

Approximately 35 species are known from North America, seven occur in the Prairie Provinces.

Key to species of *Prepops*

1. Large species 8.54–9.45 mm .... ***fraternus* (Knight)** (p. 49)
   Smaller species less than 8.0 mm 2
2. Costal margins of hemelytra red (Fig. 40) .... ***zonatus* (Knight)** (p. 50)
   Hemelytra black 3
3. Second antennal segment short, only 1.3 times longer than head width .... ***eremicola* (Knight)** (p. 51)
   Second antennal segment over 1.5 times longer than head width 4
4. Rostrum over 2.10 mm long; second antennal segment 1.6 times head width or longer 5
   Rostrum 2.10 mm or shorter; second antennal segment 1.5 times head width or shorter 6
5. Pubescence on red areas of pronotum black .... ***nigripilus* (Knight)** (p. 52)
   Pubescence on red areas of pronotum pale .... ***rubellicollis* (Knight)** (p. 52)
6. Pronotum and scutellum black .... ***borealis* (Knight)** (p. 53)
   Median line on pronotum and scutellum red .... ***bivittis* (Stål)** (p. 54)

Prepops fraternus (Knight)

Map 13

Platytylellus fraternus Knight, 1923*b*:557.
Prepops fraternus: Carvalho, 1959:335.

Length 8.54–9.45 mm; width 3.08–3.71 mm. Head red, clypeus and frons black. Antennae black. Pronotum black along wide median line, lateral margins red. Scutellum black. Hemelytra black along median line, costal margins red. Ventral surface mostly red, legs black, pilose. Male genital segment with two prominent tubercles.

Remarks. This species varies considerably in color, and many color combinations have been given varietal names. The color form in the Prairie Provinces is *rubromarginatus* Knight.

Habitat. Collected on *Quercus macrocarpa.*

Distribution. Eastern States; now known to occur in Manitoba (Map 13).

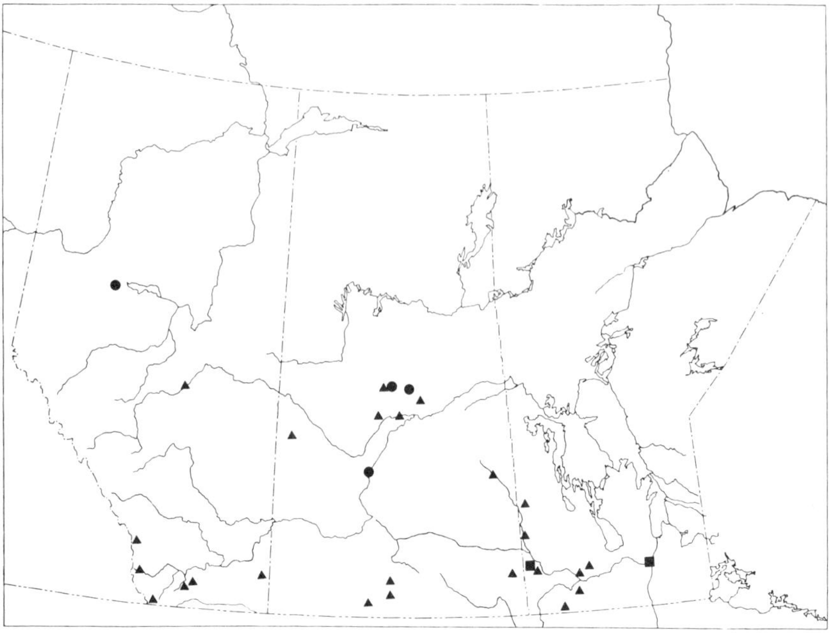

Map 13. Collection localities for *Prepops fraternus* (■), *P. zonatus* (●), and *P. eremicola* (▲).

Prepops zonatus (Knight)

Fig. 40; Map 13

Platytylellus zonatus Knight, 1926*c*:254.
Prepops zonatus: Carvalho, 1959:343.

Length 5.60–6.30 mm; width 2.38–2.66 mm. Head orange red, clypeus and frons black. Antennae black. Pronotum red, calli and wide rays behind them black. Scutellum red, basal angles black. Hemelytra black, narrow median line and costal margins red. Ventral surface black to orange, legs mostly orange.

Remarks. This species is distinguished by the red costal margin on the hemelytra (Fig. 40), and by the orange legs.

Habitat. Collected on herbaceous plants.

Distribution. North Central States; now known to occur in Alberta and Saskatchewan (Map 13).

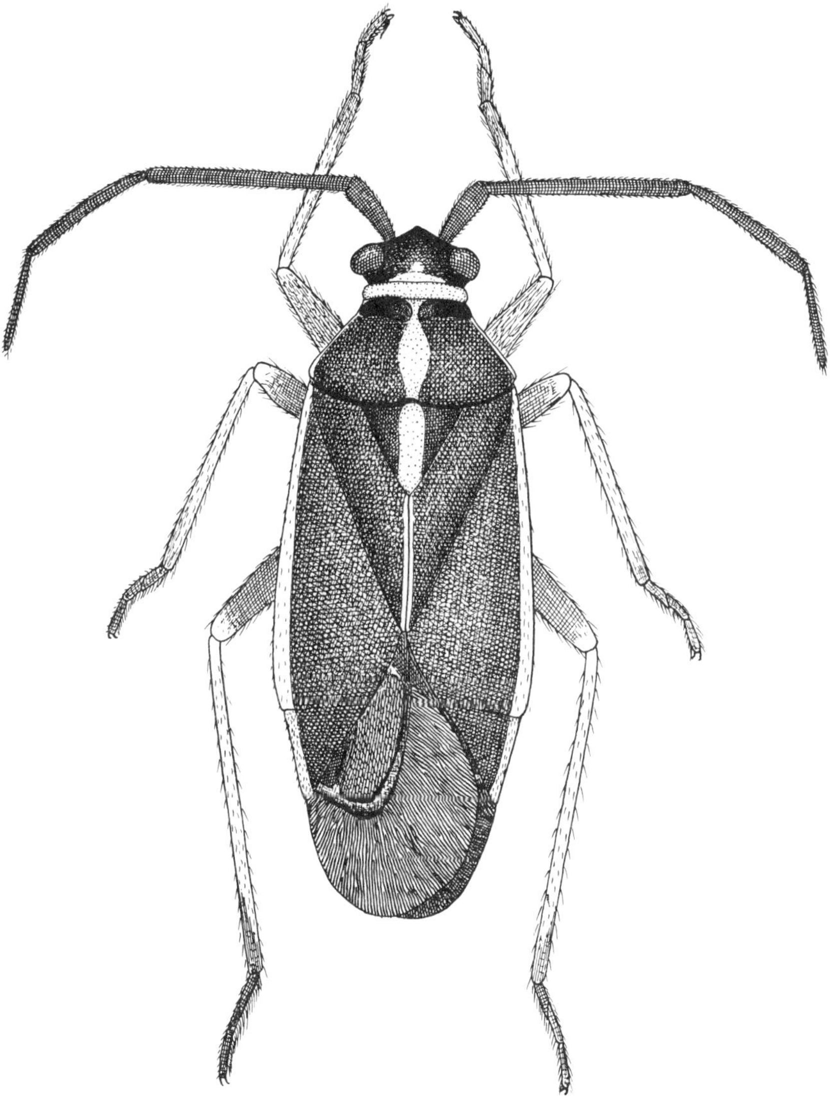

Fig. 40. *Prepops zonatus*

Prepops eremicola (Knight)

Map 13

Platytylellus eremicola Knight, 1929*b*:189.
Prepops eremicola: Carvalho, 1959:335.

Length 6.16–7.00 mm; width 2.24–2.80 mm. Head black, base red. Pronotum black, sometimes collar, side margins, and median line red.

Scutellum black, median line red. Hemelytra black. Ventral surface black, side margins red; legs black.

Remarks. This species is distinguished by the relatively short second antennal segment, otherwise it is similar in color and appearance to *bivittis* and *rubellicollis*.

Habitat. Collected on herbaceous plants.

Distribution. Northwestern States, New Mexico; Alberta, now known to occur in Saskatchewan and Manitoba (Map 13).

Prepops nigripilus (Knight)

Map 13

Platytylellus nigripilus Knight, 1929*b*:189.
Prepops nigripilus: Carvalho, 1959:339.

Length 6.23–7.42 mm; width 2.10–2.80 mm. Head black, base reddish. Pronotum black, collar with black hairs, lateral margins and longitudinal median line red. Scutellum red, basal angles black. Hemelytra black. Ventral surface black and red; legs black.

Remarks. This species is distinguished by the black hairs on the collar and by the red median line on the pronotum.

Habitat. Unknown.

Distribution. Northeastern States; Alberta (Map 13).

Prepops rubellicollis (Knight)

Map 14

Platytylellus rubellicollis Knight, 1923*b*:555.
Prepops rubellicollis: Carvalho, 1959:341.

Length 6.65–7.77 mm; width 2.38–2.94 mm. Head black, base often reddish. Pronotum black, collar and side margins red, longitudinal median line often red; pubescence on red areas pale. Scutellum black, often red. Hemelytra black. Ventral surface black and red; legs black.

Remarks. This is the largest species encountered with completely black hemelytra. It is similar to *nigripilus* in appearance, but the hairs on the red markings of the pronotum are pale.

Habitat. Collected on herbaceous plants.

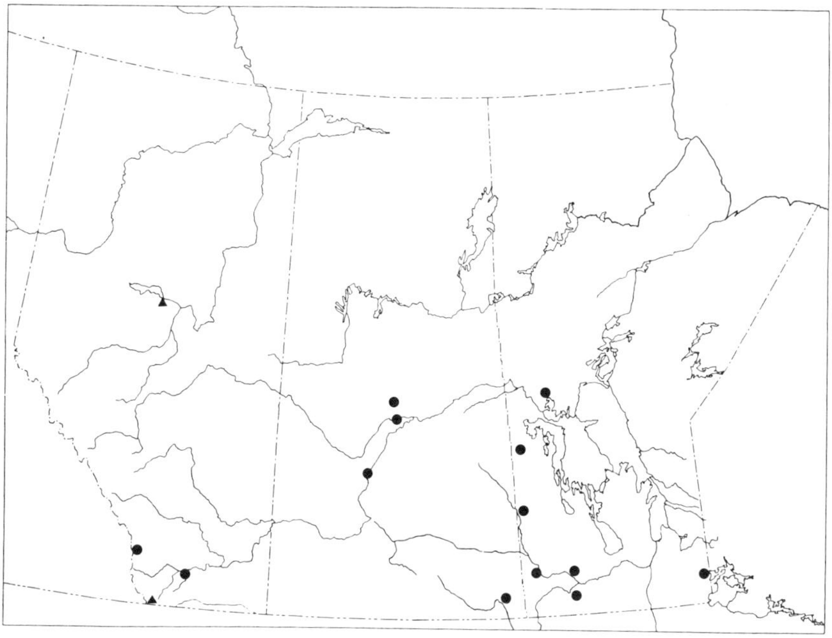

Map 14. Collection localities for *Prepops nigripilus* (▲) and *P. rubellicollis* (●).

Distribution. Northeastern States, Nebraska, British Columbia; now known to occur in the Prairie Provinces (Map 14).

Prepops borealis (Knight)

Map 14

Platytylellus borealis Knight, 1923*b*:551.
Prepops borealis: Carvalho, 1959:332.

Length 5.60–6.65 mm; width 2.10–2.73 mm. Head black, base often red. Pronotum black, collar and side margins often red. Scutellum and hemelytra black. Ventral surface black with red; legs black.

Remarks. This species is distinguished by the black scutellum and black hemelytra.

Habitat. Collected on herbaceous plants.

Distribution. Northeastern States, Ontario, Manitoba; Alberta, now known to occur in Saskatchewan (Map 14).

Prepops bivittis (Stål)

Map 15

Resthenia bivittis Stål, 1862:318.
Platytylellus basivittis Van Duzee, 1914:25.
Prepops bivittis: Carvalho, 1959:332.

Length 5.95–6.30 mm; width 2.10–2.80 mm. Head black, base often red. Pronotum black, collar, calli, side margins, and longitudinal median line often red. Scutellum red, basal angles black. Hemelytra black. Ventral surface black and red.

Remarks. This species is distinguished by the red markings on the pronotum and scutellum. The smaller size and shorter second antennal segments separate it from *rubellicollis* and *nigripilus*.

Habitat. Collected on herbaceous plants.

Distribution. Mexico, California, southeastern and northwestern States; Alberta, now known to occur in Saskatchewan and Manitoba (Map 15).

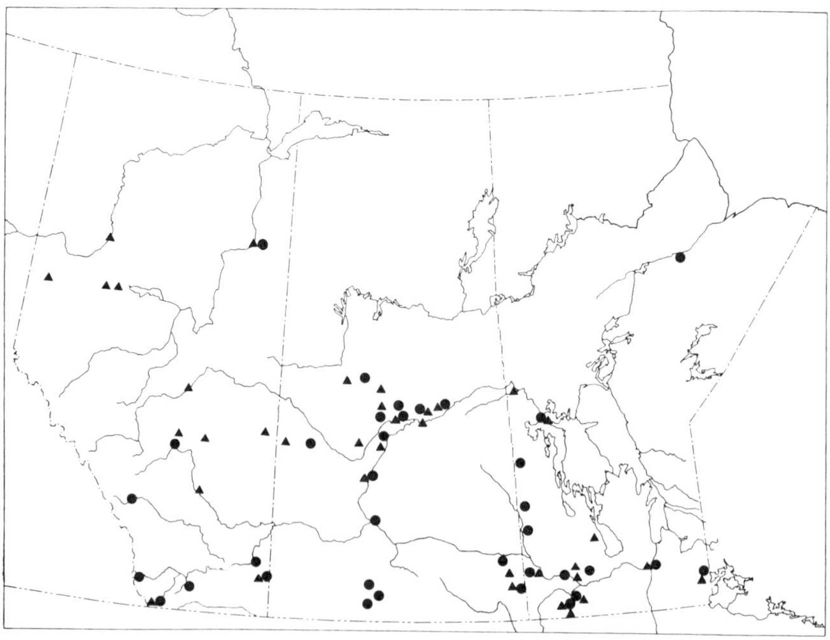

Map 15. Collection localities for *Prepops borealis* (▲) and *P. bivittis* (●).

Tribe Mirini

In the Prairie Provinces the tribe is represented by 23 genera and 103 species.

Key to genera of Mirini

1. First antennal segment with numerous flattened hairs (Fig. 41) .... ***Neurocolpus*** **Reuter** (p. 56)
   First antennal segment without flattened hairs 2
2. Pronotum with depressed black spot behind each callus (Fig. 42) .... ***Taedia*** **Distant** (p. 57)
   Pronotum without depressed black spots 3
3. Pronotum punctate between callus and collar (Fig. 43) 4
   Pronotum not punctate between callus and collar 5
4. Frons smooth .... ***Tropidosteptes*** **Uhler** (p. 60)
   Frons striate or grooved (Fig. 44) .... ***Neoborella*** **Knight** (p. 76)
5. Species with four longitudinal black lines on dorsal surface (Fig. 53) .... ***Poecilocapsus*** **Reuter** (p. 78)
   Species without four black lines 6
6. Second antennal segment shorter than head width .... ***Agnocoris*** **Reuter** (p. 80)
   Second antennal segment longer than head width 7
7. Species black and densely pubescent 8
   Species not black, if black nearly glabrous 9
8. Second antennal segment clavate; carina absent .... ***Capsus*** **Fabricius** (p. 82)
   Second antennal segment linear; carina present .... ***Irbisia*** **Reuter** (p. 83)
9. Carina between eyes present 10
   Carina between eyes absent 19
10. Eyes nearly spherical, ventral margin of eye not extending below antennal socket (Fig. 45) .... ***Lygidea*** **Reuter** (p. 86)
    Eyes elliptical, ventral margin extending below antennal socket (Fig. 46) 11
11. Species with woolly pubescence (Fig. 54) .... ***Polymerus*** **Hahn** (p. 88)
    Species without woolly pubescence 12
12. Pronotum roughly rugose; tarsal claws sharply angled .... ***Plesiocoris*** **Fieber** (p. 97)
    Pronotum punctate, or finely rugose; tarsal claws rounded 13
13. Frons with median longitudinal groove (Fig. 55) .... ***Salignus*** **Kelton** (p. 99)
    Frons without median groove 14
14. Head nearly vertical (Fig. 46) .... ***Orthops*** **Fieber** (p. 99)
    Head oblique (Fig. 47) 15
15. Pronotum coarsely punctate, punctures deep and wide apart 16
    Pronotum finely punctate, punctures shallow and close together 17
16. Scutellum deeply punctate .... ***Lygus*** **Hahn** (p. 100)
    Scutellum not punctate .... ***Platylygus*** **Van Duzee** (p. 135)
17. Reddish brown species with contrasting white collar .. ***Pinalitus*** **Kelton** (p. 139)
    Greenish species without contrasting collar 18
18. Base of head concave, eyes overlapping anterior angles of pronotum; (Fig. 64) on conifers .... ***Dichrooscytus*** **Fieber** (p. 141)

Base of head nearly straight, mostly on deciduous trees .... *Lygocoris* **Reuter** (p. 146)
19. Dorsal surface highly polished; glabrous .... *Metriorrhynchomiris* **Kirkaldy** (p. 162)
Dorsal surface not highly polished; pubescent 20
20. First segment of hind tarsus longer than second .... *Stenotus* **Jakovlev** (p. 163)
First segment of hind tarsus shorter than second 21
21. Jugal suture extending well below antennal socket; lorum tumid .... *Phytocoris* **Fallén** (p. 165)
Jugal suture extending directly to antennal socket; lorum not tumid 22
22. Vertex with shallow longitudinal groove .... *Adelphocoris* **Reuter** (p. 184)
Vertex without longitudinal groove, often with rounded depression .... *Calocoris* **Fieber** (p. 187)

Genus *Neurocolpus* Reuter

Elongate, robust species. Head oblique, frons elevated and separated from clypeus by deep notch; eyes large, carina between them absent. First antennal segment stout with flattened hairs. Pronotum subcampanulate, smooth. Hemelytra smooth; pubescence golden, long, dense. Legs strongly pilose.

Approximately eight species occur in North America, one occurs in the Prairie Provinces.

Neurocolpus nubilus (Say)

Fig. 41; Map 16

Capsus nubilus Say, 1832:22.
Neurocolpus nubilus: Reuter, 1875*d*:70.

Length 7.00–7.70 mm; width 2.52–2.80 mm. Head light brown, frons often marked with oblique, black bars. Rostrum extending to hind coxae. Pronotum yellowish brown with tufts of black, erect hairs, intermixed with golden, erect hairs; side margins rounded. Hemelytra mottled beige marked with dark brown.

Remarks. The flattened, black hairs on the first antennal segment are distinctive (Fig. 41).

Habitat. Collected on *Rhus typhina.*

Distribution. Mexico, widespread in USA, Ontario, Quebec; now known to occur in Manitoba (Map 16).

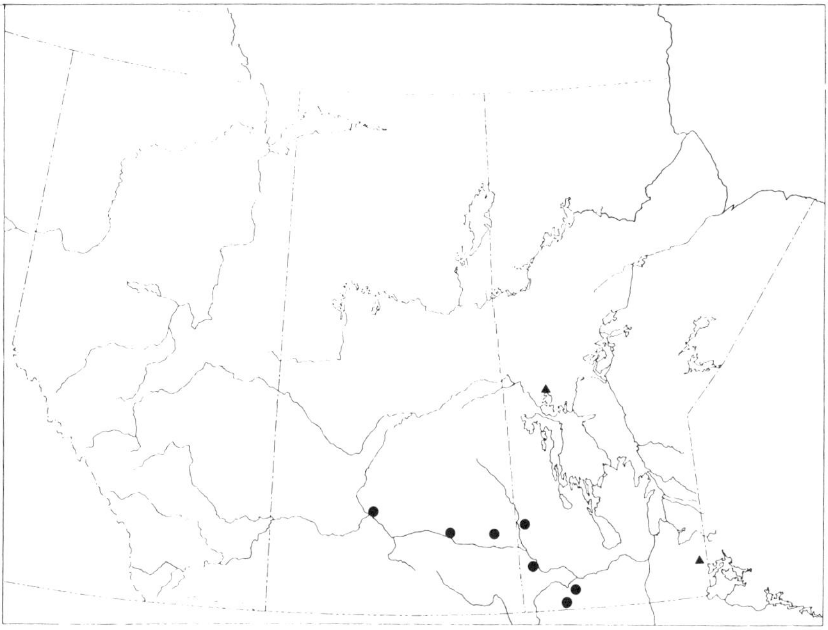

Map 16. Collection localities for *Neurocolpus nubilus* (▲) and *Taedia pallidula* (●).

Genus *Taedia* Distant

Elongate, robust species. Head oblique, eyes large, carina between them absent. Pronotum subcampanulate with black spot behind each callus. Hemelytra finely punctate; pubescence long, dense. Legs long, slender.

Approximately 30 species occur in North America, one is found in the Prairie Provinces. The genus was formerly known as *Paracalocoris*.

Taedia pallidula (McAtee)

Fig. 42; Map 16

Paracalocoris hawleyi var. *pallidulus* McAtee, 1916:380.
Paracalocoris pallidulus: Knight, 1930*c*:822.
Taedia pallidulus: Carvalho, 1959:262.

Length 6.30–7.00 mm; width 2.38–2.66 mm. Head brown, clypeus darker, frons often marked with oblique black bars. Rostrum 2.38–2.52 mm long. Pronotum mottled brown, velvety spots behind calli black; lateral

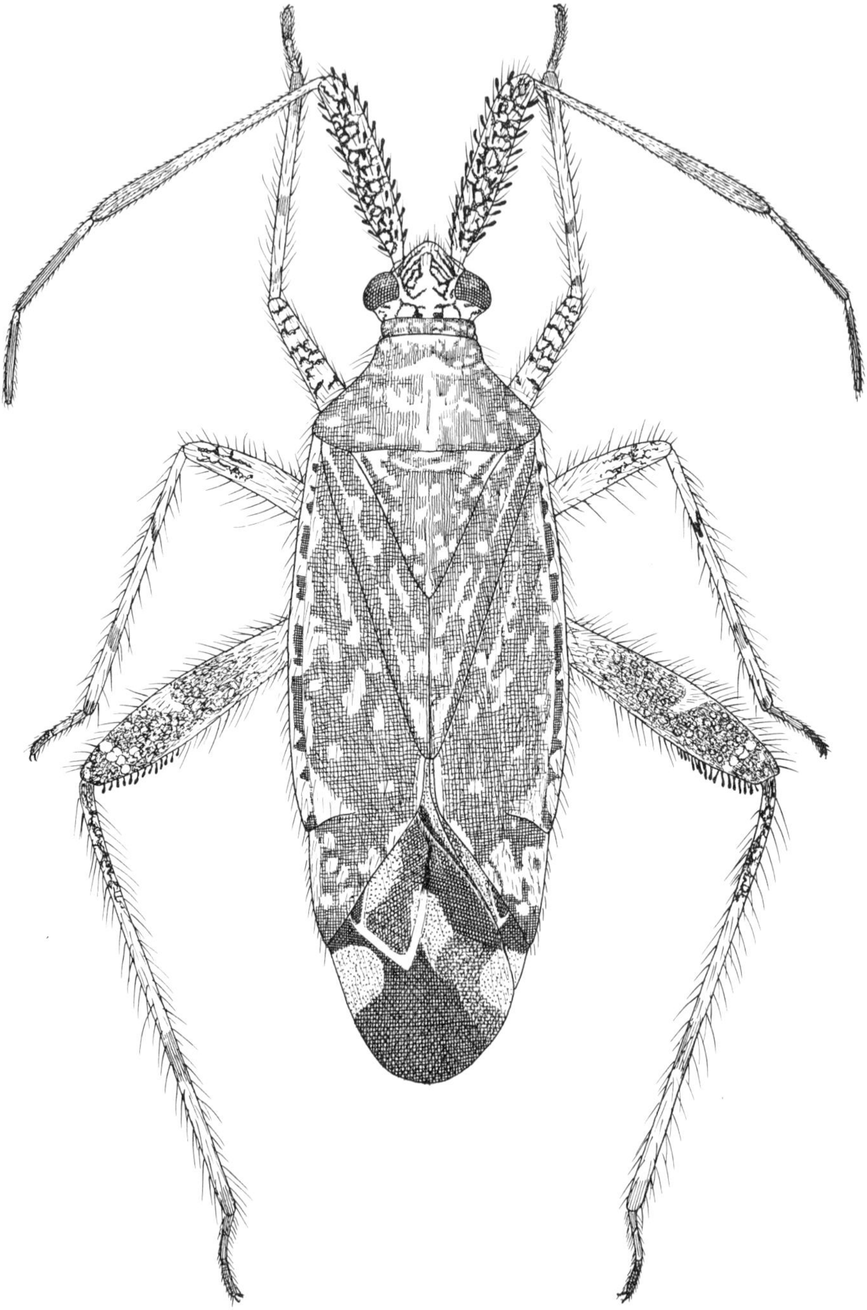

Fig. 41. *Neurocolpus nubilus*

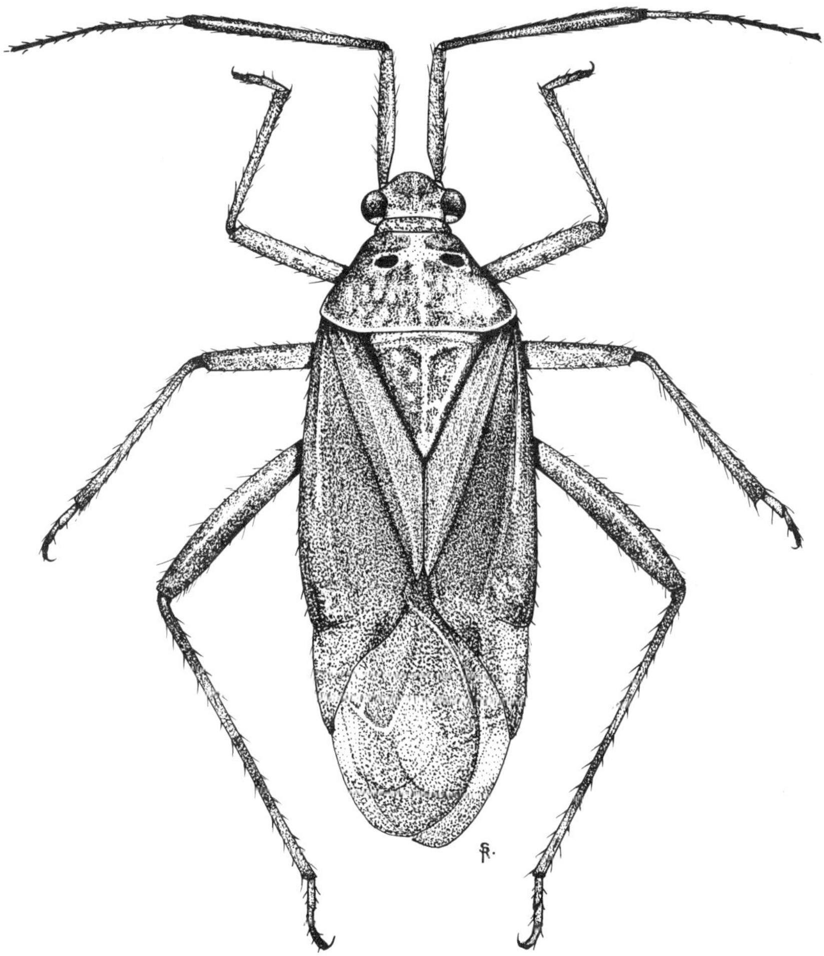

Fig. 42. *Taedia pallidula*

margins rounded. Scutellum brown, longitudinal median line yellow. Hemelytra mottled brown and yellow. Legs yellow marked with red, tibiae banded.

Remarks. The black, velvety spots behind the calli readily distinguish this species (Fig. 42).

Habitat. Collected on *Cornus stolonifera*.

Distribution. New York, North Central States, Ohio, Ontario; now known to occur in Manitoba and Saskatchewan (Map 16).

Genus *Tropidosteptes* Uhler

Pale green, green and red, or black and brown species. Head vertical, short, frons smooth or punctate, carina between eyes distinct. Pronotum trapeziform, strongly punctate, lateral margins rounded or carinate; collar prominent, calli smooth, area between calli and collar punctate. Hemelytra glabrous or pubescent. Osteolar peritreme large.

There are approximately 36 species of *Tropidosteptes* in North America, nine species occur in the Prairie Provinces. Species formerly included in the genera *Neoborus* Distant and *Xenoborus* Reuter are now placed in the genus *Tropidosteptes* (see Carvalho 1954, Akingbohungbe et al. 1972, and Kelton 1978*a*).

Key to species of *Tropidosteptes*

1. First antennal segment black 2
   First antennal segment pale 5
2. Pronotum all black .... ***pettiti* Reuter** (p. 60)
   Pronotum not all black 3
3. Head and pronotum without black markings .... ***commissuralis* (Reuter)** (p. 61)
   Head and pronotum with black markings 4
4. First antennal segment as long as width of vertex .... ***canadensis* Van Duzee** (p. 64)
   First antennal segment as long as width of vertex plus an eye .... ***plagifer* Reuter** (p. 68)
5. Rostrum 1.54 mm or longer .... ***palmeri* (Reuter)** (p. 70)
   Rostrum shorter 6
6. Scutellum black each side of pale median line .... ***glaber* (Knight)** (p. 71)
   Scutellum yellowish green 7
7. Clypeus and jugum green .... ***brooksi* Kelton** (p. 71)
   Clypeus and jugum marked with red or black 8
8. Pronotum and hemelytra densely pubescent .... ***pubescens* (Knight)** (p. 72)
   Pronotum and hemelytra almost glabrous .... ***amoenus* Reuter** (p. 74)

Tropidosteptes pettiti Reuter

Fig. 56; Map 17

Trichia punctulata Provancher, 1887:133.
Tropidosteptes pettiti Reuter, 1909:50.
Xenoborus pettiti: Knight, 1917*c*:82.

Length 5.46–5.74 mm; width 1.96–2.17 mm. Head and antennae black. Pronotum black, lateral margins carinate on anterior half. Hemelytra black. Scutellum pale. Ventral surface black, osteolar peritreme and legs pale.

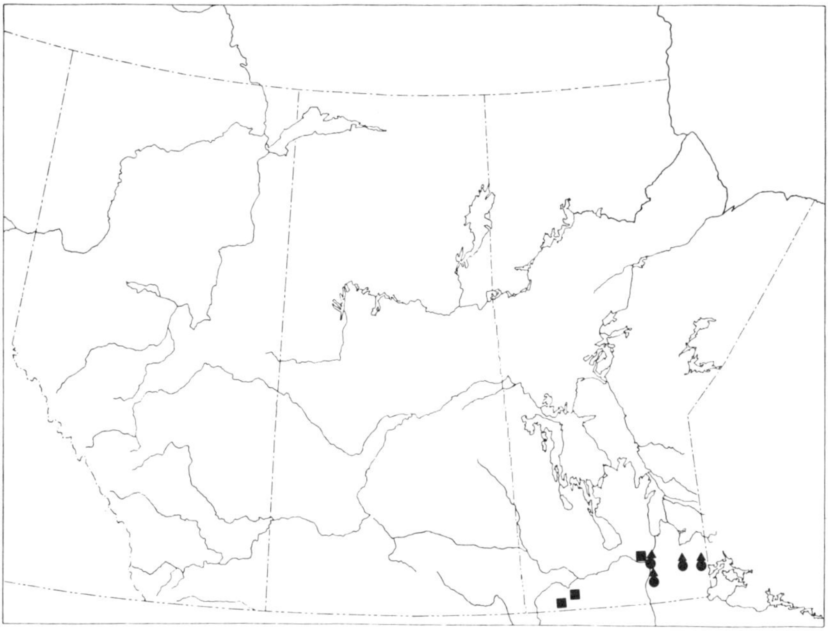

Map 17. Collection localities for *Tropidosteptes pettiti* (●), *T. commissuralis* (▲), and *T. canadensis* (■).

Remarks. The black hemelytra and the pale scutellum readily distinguish this species (Fig. 56).

Habitat. Collected on *Fraxinus pennsylvanica.*

Distribution. Northeastern States, Kansas, Missouri, Ontario, Quebec; now known to occur in Manitoba (Map 17).

Tropidosteptes commissuralis (Reuter)

Fig. 57; Map 17

Sthenarops chloris Provancher, 1887:134.
Neoborus (Xenoborus) commissuralis Reuter, 1908:112.
Tropidosteptes commissuralis Reuter, 1909:51.
Xenoborus commissuralis: Knight, 1917*c*:82.

Length 6.02–6.51 mm; width 2.10–2.40 mm. Head pale green; antennae black; pronotum pale green, lateral margins rounded. Hemelytra pale green, commissure black. Ventral surface and legs pale green.

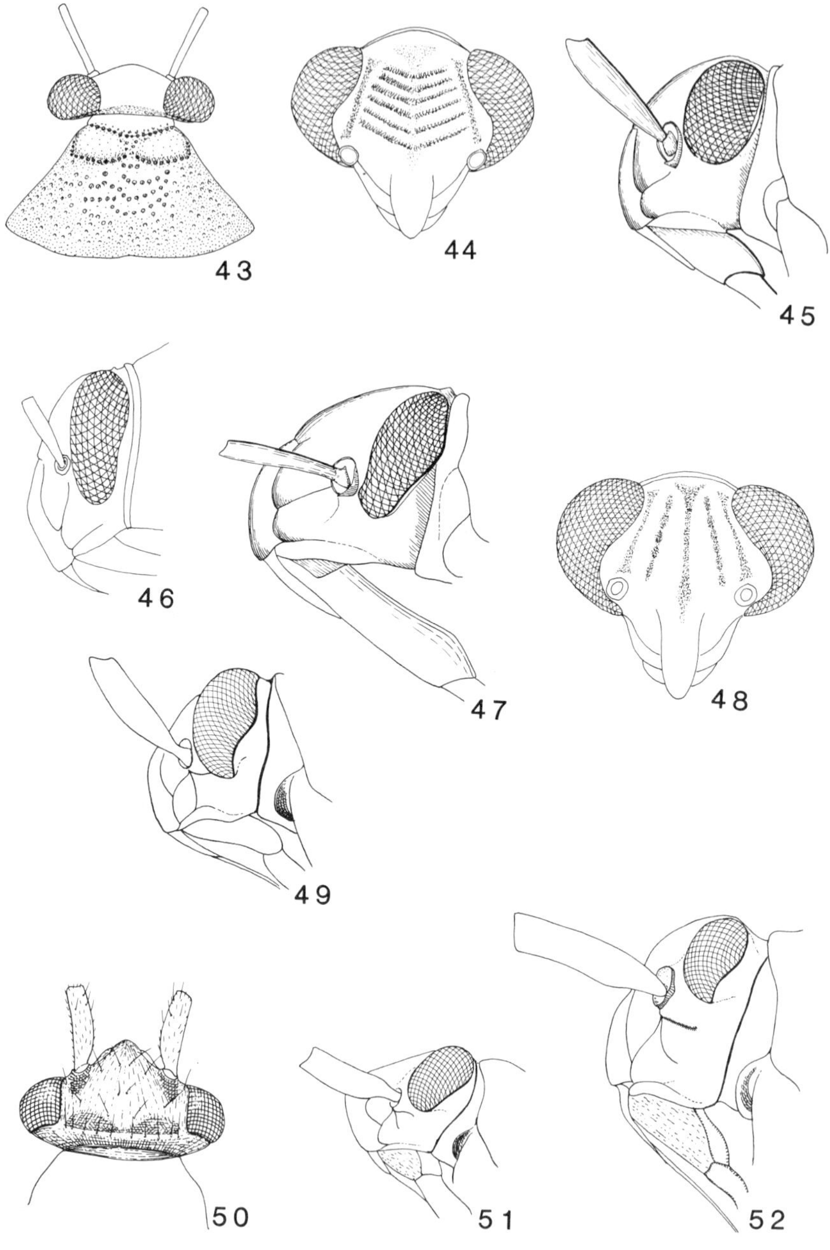

Figs. 43–52. Pronotum and heads of Mirini and Orthotylini. 43, *Tropidosteptes* and *Neoborella* spp.; 44, *Neoborella* spp.; 45, *Lygidea* spp.; 46, *Orthops* sp.; 47, *Platylygus* spp.; 48, *Lygus lineolaris*; 49, 50, *Hadronema* spp.; 51, *Lopidea* spp.; 52, *Labopidea* spp.

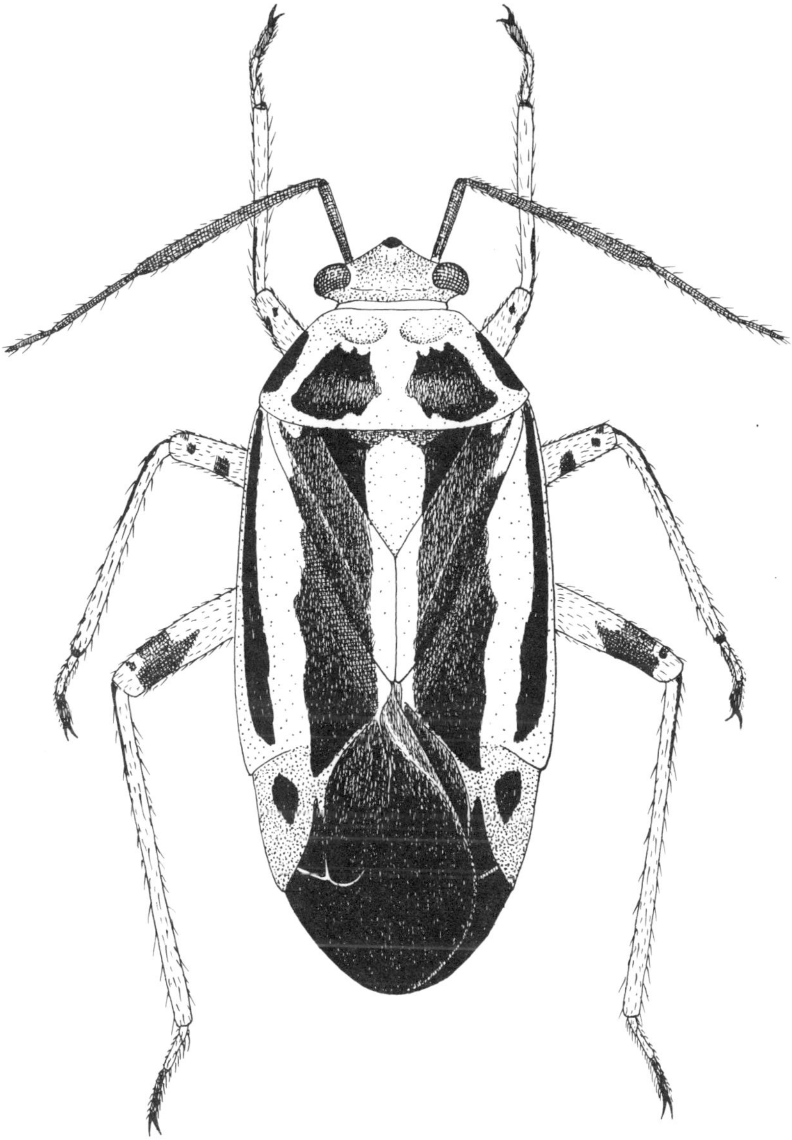

Fig. 53. *Poecilocapsus lineatus*

Remarks. The long and slender form, the pale green color, the black antennae, and the black commissure readily distinguish the species (Fig. 57).

Habitat. Collected on *Fraxinus pennsylvanica.*

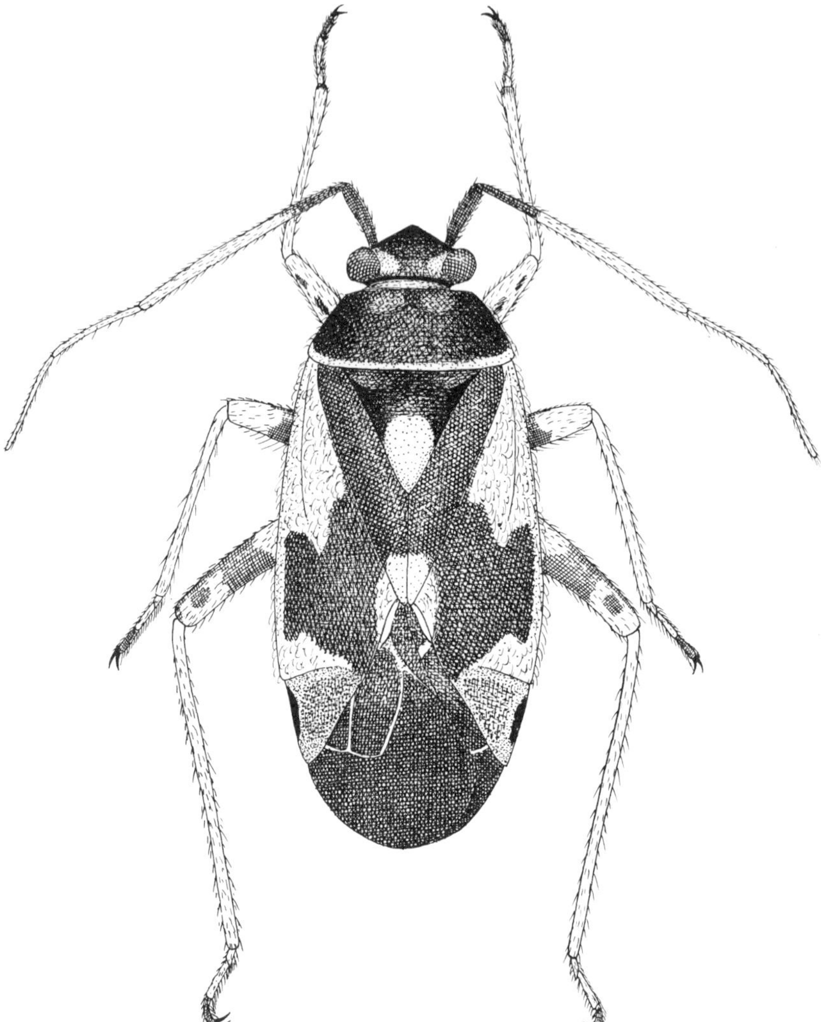

Fig. 54. *Polymerus unifasciatus*

Distribution. New York, North Central States, Ontario, Quebec, Nova Scotia; now known to occur in Manitoba (Map 17).

Tropidosteptes canadensis Van Duzee

Map 17

Tropidosteptes canadensis Van Duzee, 1912:486.
Neoborus canadensis: Van Duzee, 1917:351.

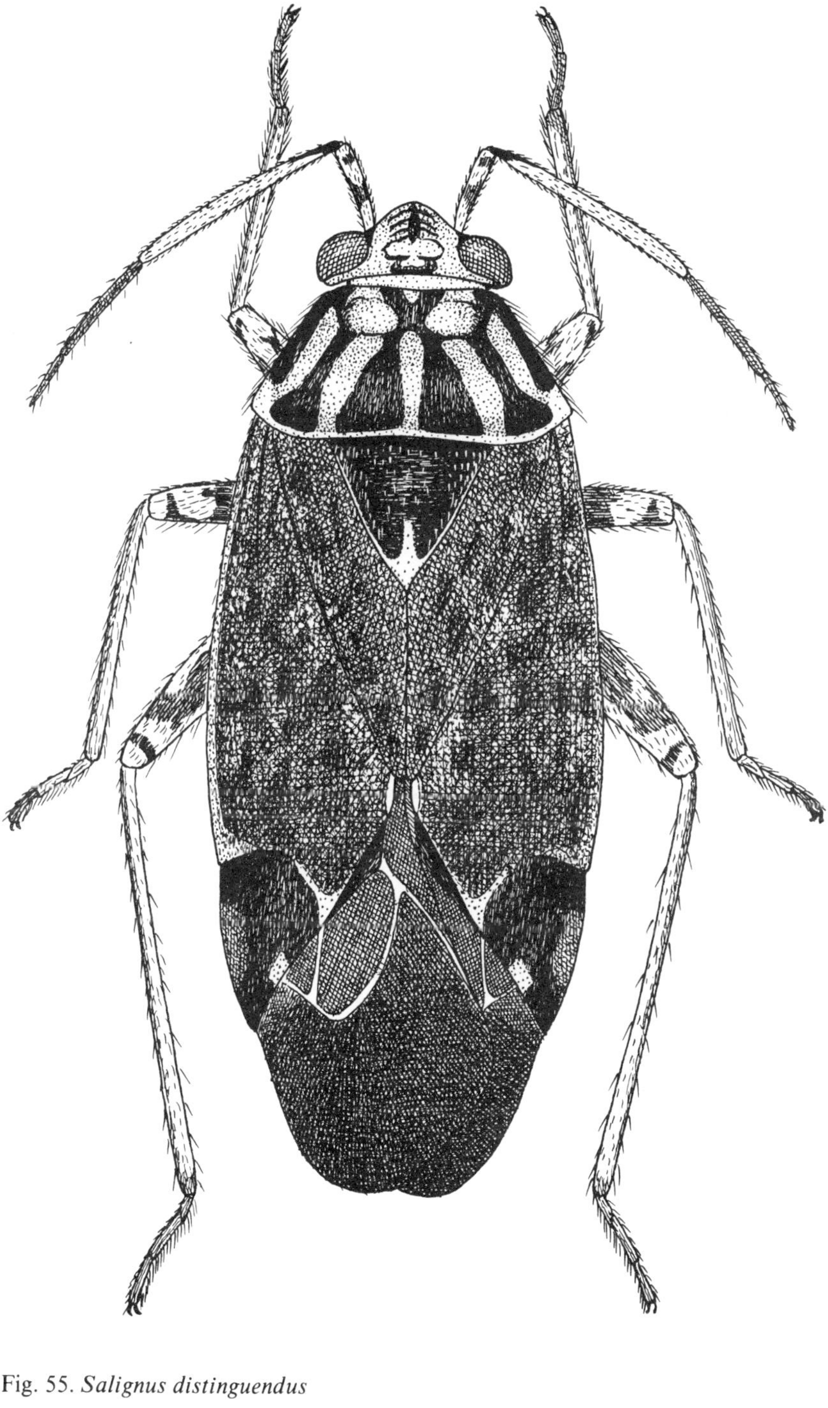

Fig. 55. *Salignus distinguendus*

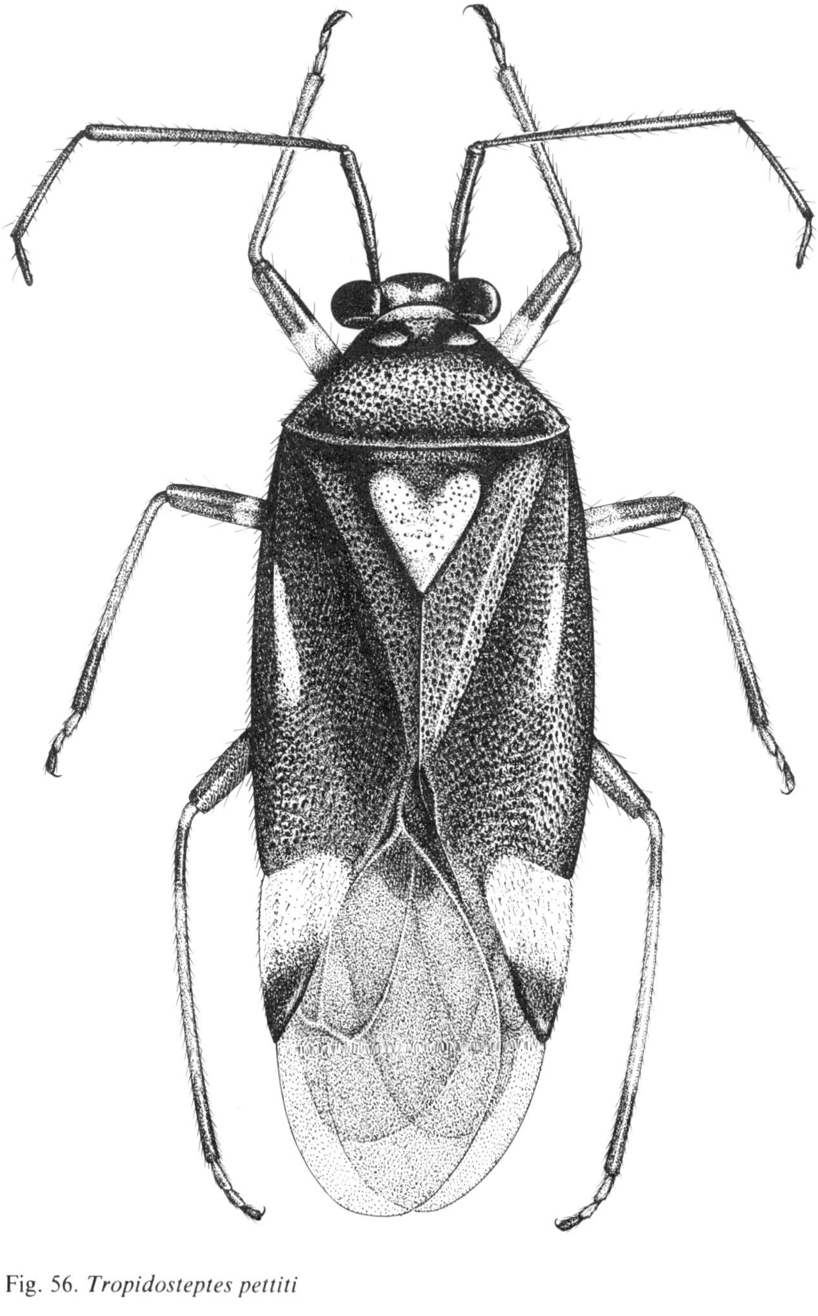

Fig. 56. *Tropidosteptes pettiti*

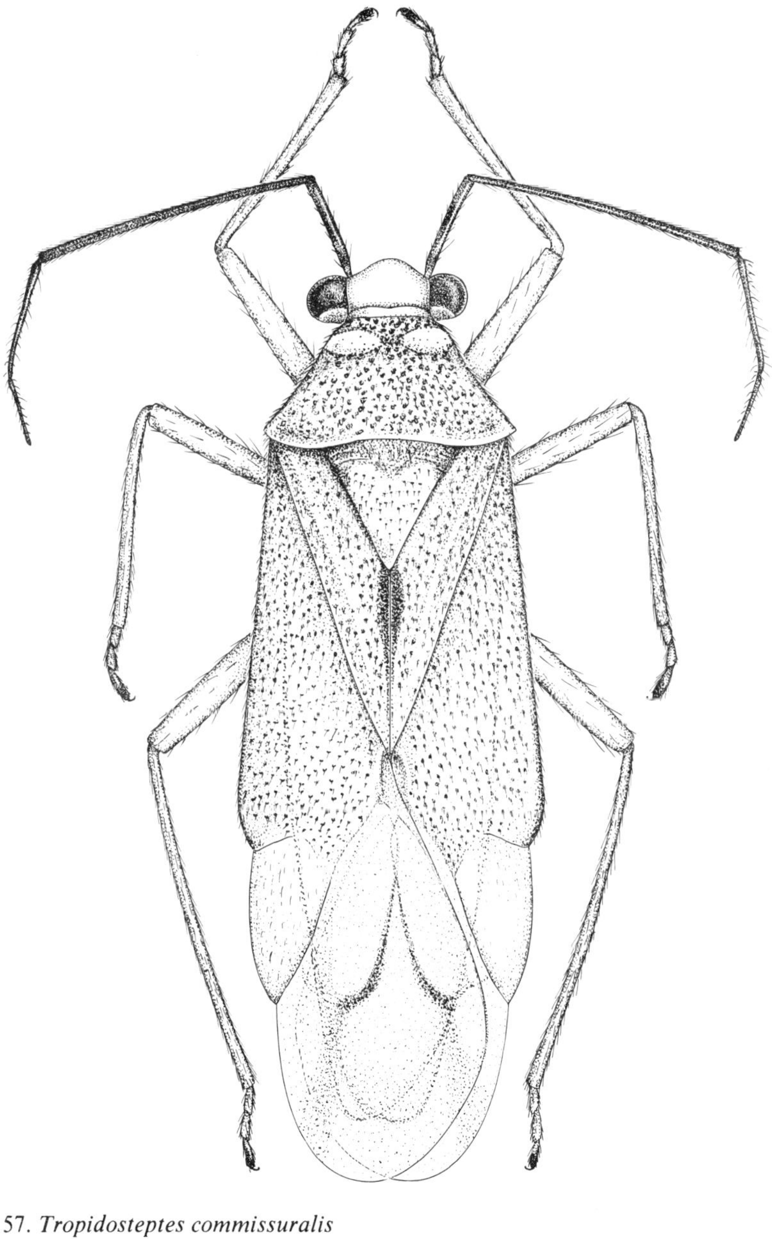

Fig. 57. *Tropidosteptes commissuralis*

Length 4.69–5.46 mm; width 1.96–2.38 mm. Head brown, clypeus black; first antennal segment black. Pronotum light brown, lateral margins angular. Scutellum light brown. Hemelytra light brown, clavus and apical portion of corium dark brown; pubescence long, dense. Ventral surface yellowish green; legs pale.

Remarks. This species is distinguished by the black clypeus, by the black first antennal segment, and by the pubescent hemelytra.

Habitat. Collected on *Fraxinus pennsylvanica.*

Distribution. Northeastern and Central States, Ontario, Quebec; now known to occur in Manitoba (Map 17).

Tropidosteptes plagifer Reuter

Fig. 58; Map 18

Tropidosteptes plagifer Reuter, 1909:51.
Xenoborus plagifer: Knight, 1917*c*:82.

Length 5.04–5.95 mm; width 1.82–2.24 mm. Head greenish yellow, clypeus and jugum black, frons marked with reddish tinge; antennae black. Pronotum pale green, callus and usually large triangular spot behind black; lateral margins rounded. Scutellum pale green. Hemelytra pale green, clavus and large spot on apical half of corium black. Ventral surface green; pleura and side of abdomen often black; legs pale.

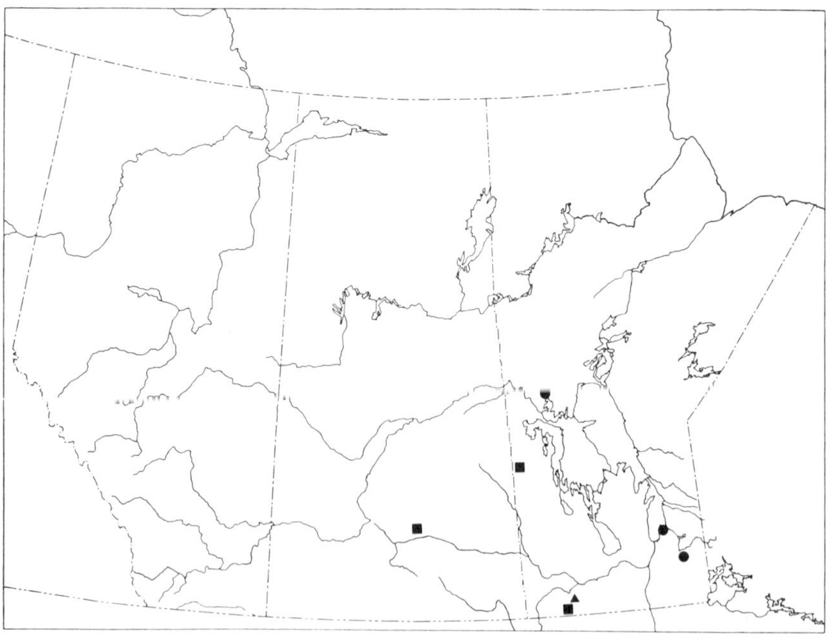

Map 18. Collection localities for *Tropidosteptes plagifer* (●), *T. palmeri* (■), and *T. glaber* (▲).

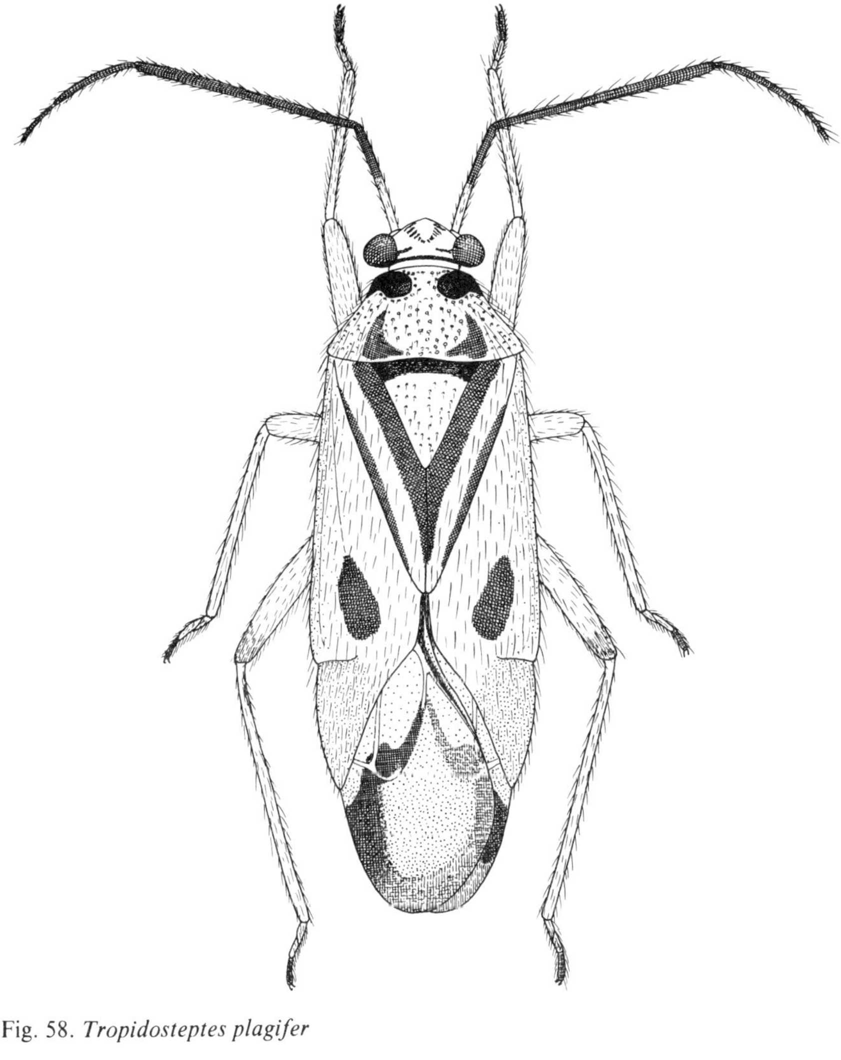

Fig. 58. *Tropidosteptes plagifer*

Remarks. This species is distinguished by the black antennal segments, by the black calli, and by the rounded lateral margins of the pronotum (Fig. 58).

Habitat. Collected on *Fraxinus nigra*.

Distribution. New York, North Central States, Ontario; now known to occur in Manitoba (Map 18).

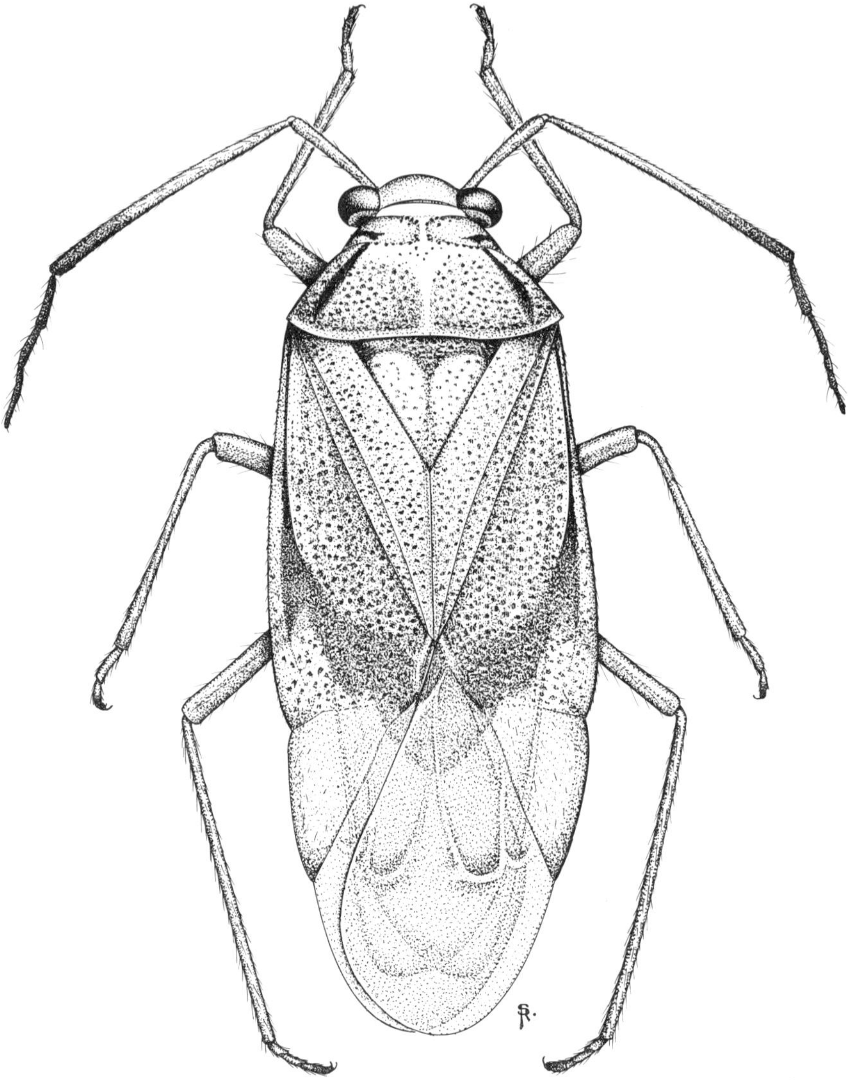

Fig. 59. *Tropidosteptes palmeri*

Tropidosteptes palmeri (Reuter)

Fig. 59; Map 18

Neoborus amoenus var. *palmeri* Reuter, 1908:112.
Tropidosteptes palmeri: Reuter, 1909:49.

Length 5.04–6.30 mm; width 1.33–2.80 mm. Head yellowish brown, clypeus and jugum often black; first antennal segment pale. Pronotum light yellowish brown, line above lateral margin, spot on callus, and line behind callus black in male; black markings absent in female; lateral margins carinate. Scutellum pale yellow. Hemelytra light yellowish brown, line along radial vein black or brown; glabrous. Ventral surface light brown with pleura darker; legs light yellowish brown.

Remarks. This species is distinguished by the glabrous hemelytra (Fig. 59).

Habitat. Collected on *Fraxinus pennsylvanica.*

Distribution. Northeastern and North Central States, Quebec; now known to occur in Manitoba and Saskatchewan (Map 18).

Tropidosteptes glaber (Knight)

Map 18

Neoborus glaber Knight, 1923:563.
Tropidosteptes glaber: Carvalho, 1959:271.

Length 4.48–4.68 mm; width 1.89–2.24 mm. Head light yellowish brown, clypeus and diagonal bars on frons black. Pronotum yellowish brown, callus and wide ray extending to posterior margin black. Scutellum yellowish green, longitudinal ray each side of median line black. Hemelytra light yellowish brown; glabrous clavus, costal margin, large spot on apical half of corium, and inner margin of cuneus dark brown. Ventral surface pale yellowish green; pleura and lateral line on abdomen dark brown; legs pale green.

Remarks. This species is distinguished by the black lines on the scutellum and by the glabrous hemelytra.

Habitat. Collected on *Fraxinus pennsylvanica.*

Distribution. Northeastern and Central States, Texas, Ontario; now known to occur in Manitoba and Saskatchewan (Map 18).

Tropidosteptes brooksi Kelton

Fig. 60; Map 19

Tropidosteptes brooksi Kelton, 1978:471.

Length 4.90–5.60 mm; width 1.96–2.24 mm. Head pale green, often red on vertex, clypeus and jugum pale green; eyes large in male, extending above vertex. Pronotum pale green, lateral margins carinate; callus and

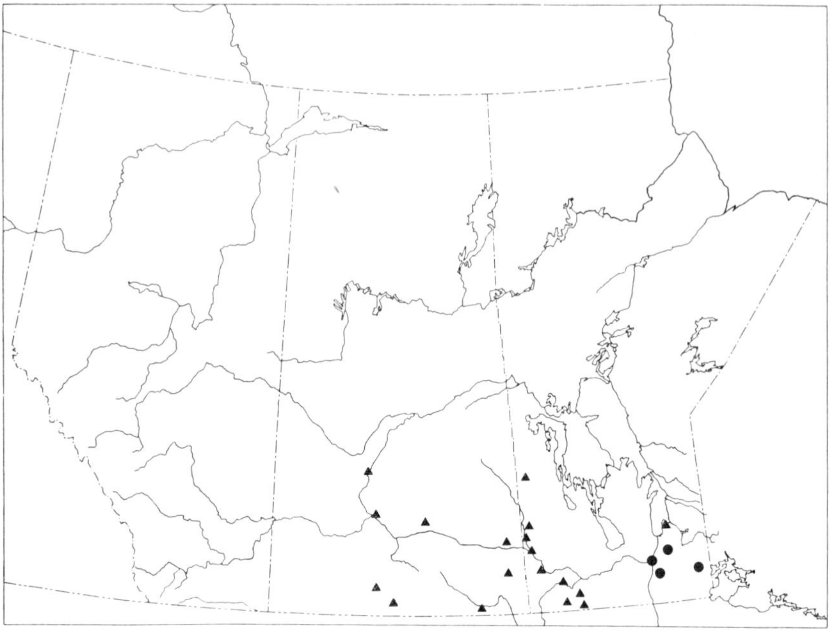

Map 19. Collection localities for *Tropidosteptes brooksi* (▲) and *T. pubescens* (●).

broad area behind often red. Scutellum pale green. Hemelytra pale green, apical area of clavus and corium often red in male. Ventral surface and legs pale green.

Remarks. This species is distinguished by the pale green clypeus and jugum, and by the overall green or green and red color (Fig. 60).

Habitat. Collected on *Fraxinus pennsylvanica.*

Distribution. Ontario, Quebec; Saskatchewan, Manitoba (Map 19).

Tropidosteptes pubescens (Knight)

Fig. 61; Map 19

Neoborus pubescens Knight, 1917*b*:81.
Tropidosteptes pubescens: Carvalho, 1959:272.

Length 4.55–4.83 mm; width 1.68–2.03 mm. Head pale green marked with reddish brown. Pronotum pale green, lateral margins carinate on anterior half; calli marked with black, two rays behind callus and ray along lateral margin black. Scutellum pale green, black at middle of base.

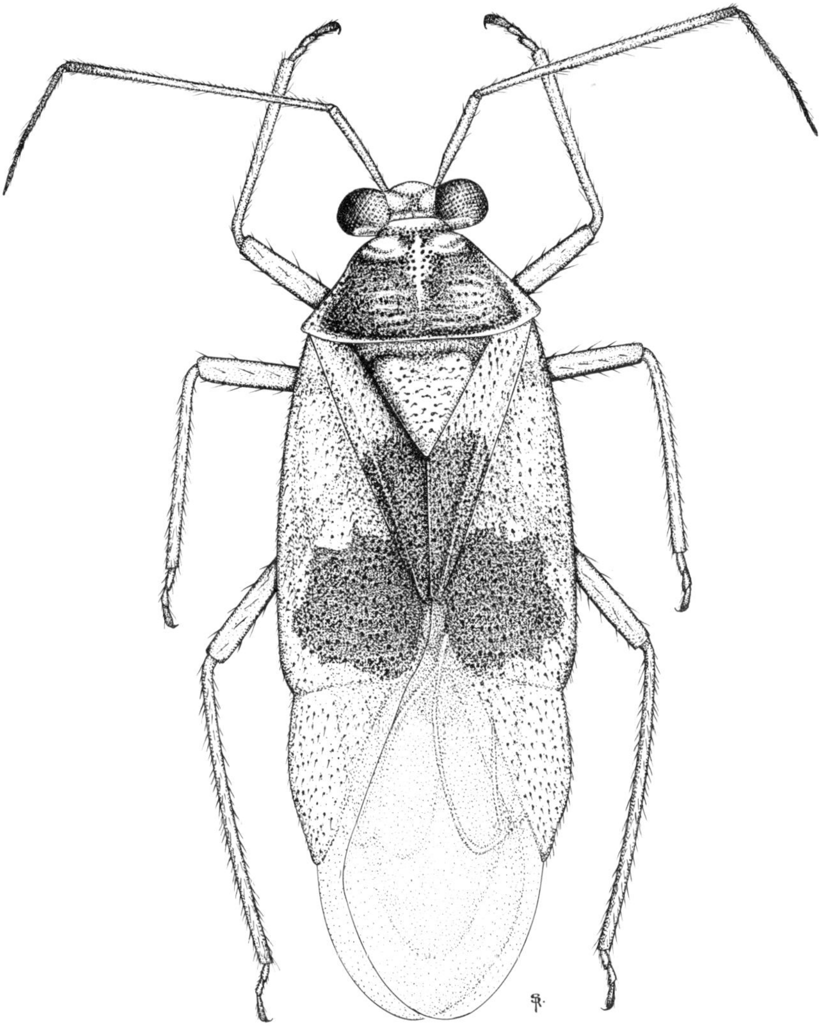

Fig. 60. *Tropidosteptes brooksi*

Hemelytra pale green, inner clavus, claval suture, costal margin, and apical corium black; pubescence long, dense. Ventral surface green marked with black; legs pale.

Remarks. This species is distinguished by the dark color markings and by the dense pubescence on the hemelytra (Fig. 61).

Habitat. Collected on *Fraxinus pennsylvanica.*

Distribution. Northeastern States, Ontario, Quebec; now known to occur in Manitoba (Map 19).

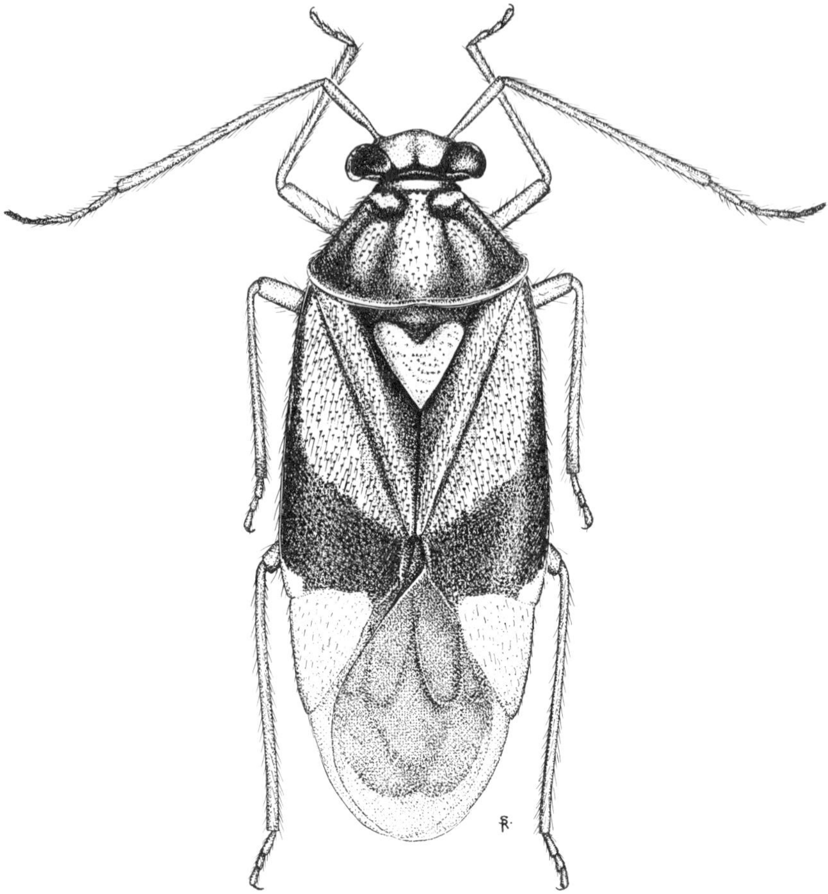

Fig. 61. *Tropidosteptes pubescens*

Tropidosteptes amoenus Reuter

Fig. 62; Map 20

Neoborus saxeus Uhler, 1894:264.
Tropidosteptes amoenus Reuter, 1909:48.

Length 4.20–5.04 mm; width 1.68–2.24 mm. Head yellow, clypeus and jugum marked with red or black, frons often marked with diagonal red bars. Pronotum green marked with red, or almost completely black. Hemelytra green with diagonal red bar near apex on corium, or nearly all black; glabrous. Ventral surface green or black; legs pale.

Remarks. This species varies greatly in color from green with red markings to almost black. The green frons resemble *brooksi,* but are separated from it by the black or red markings on the head and pronotum (Fig. 62).

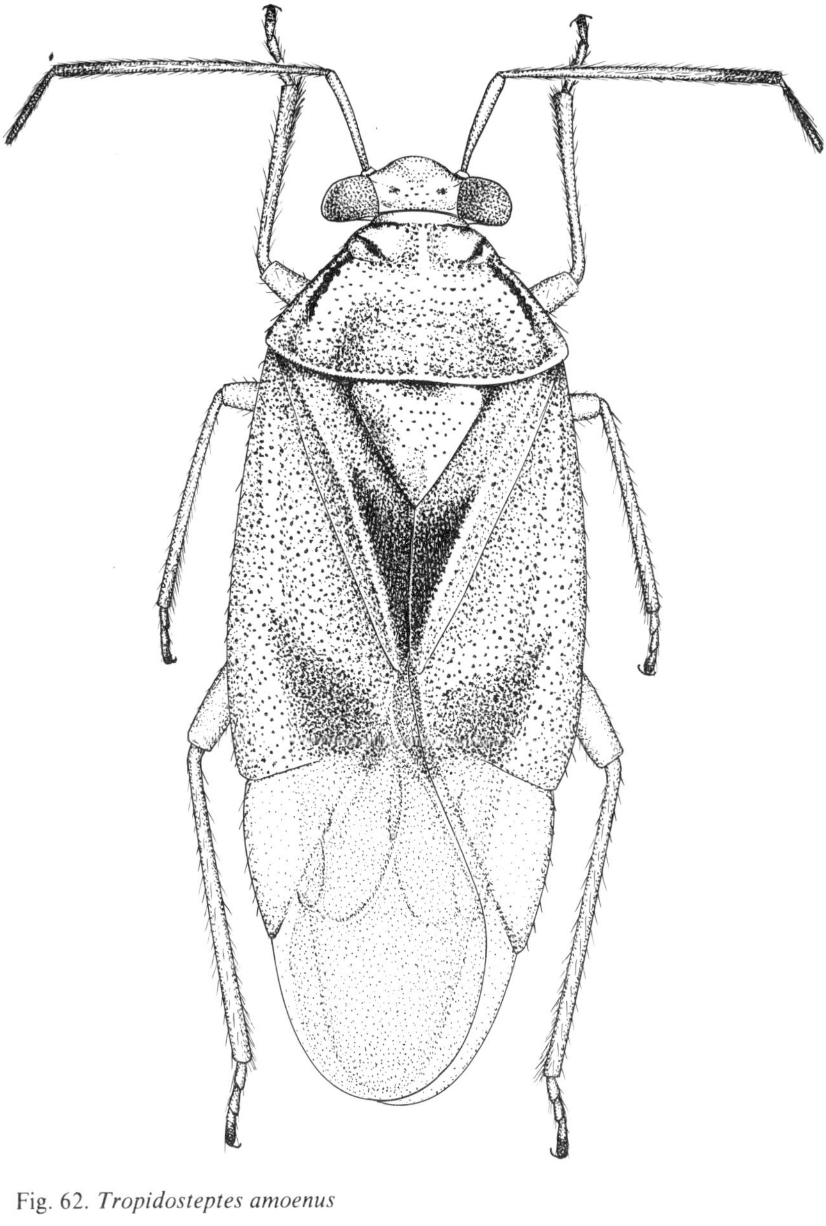

Fig. 62. *Tropidosteptes amoenus*

Habitat. Collected on *Fraxinus pennsylvanica.*

Distribution. Northeastern and Central States, Texas, British Columbia, Ontario, Quebec; now known to occur in the Prairie Provinces (Map 20).

Genus *Neoborella* Knight

Small, oblong, brown species. Head vertical, short, frons grooved and transversely striate; carina between eyes distinct. Pronotum trapeziform, strongly punctate, calli smooth, lateral margins rounded; area between calli and collar punctate. Scutellum tumid. Hemelytra finely pubescent. Osteolar peritreme large.

Kelton and Herring (1978) published a key to North American species of *Neoborella*; one species occurs in the Prairie Provinces.

Neoborella canadensis Kelton & Herring

Fig. 63; Map 20

Neoborella canadensis Kelton & Herring, 1978:779.

Length 4.20–4.76 mm; width 1.96–2.24 mm. Head light brown, punctate; frons with longitudinal median groove and several transverse grooves; first antennal segment light brown. Pronotum light brown, lateral margins

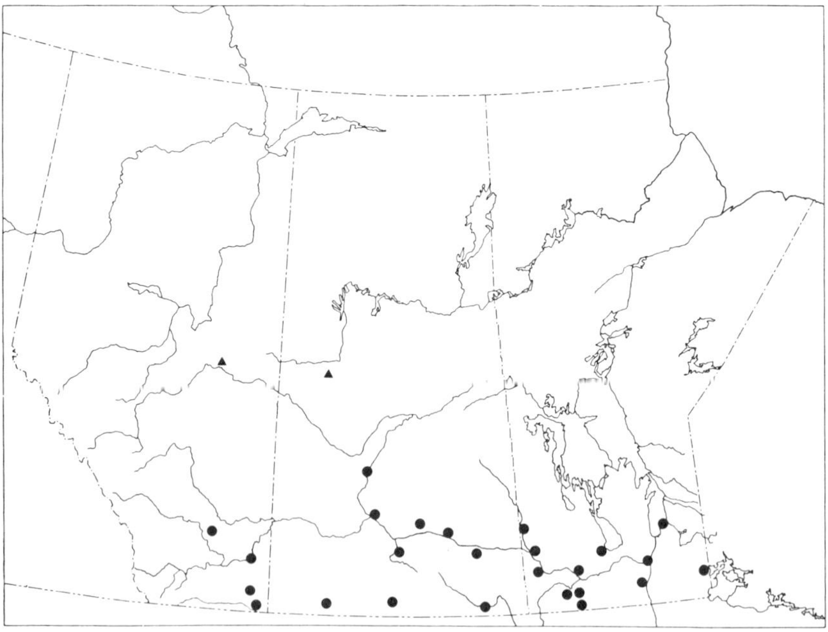

Map 20. Collection localities for *Tropidosteptes amoenus* (●) and *Neoborella canadensis* (▲).

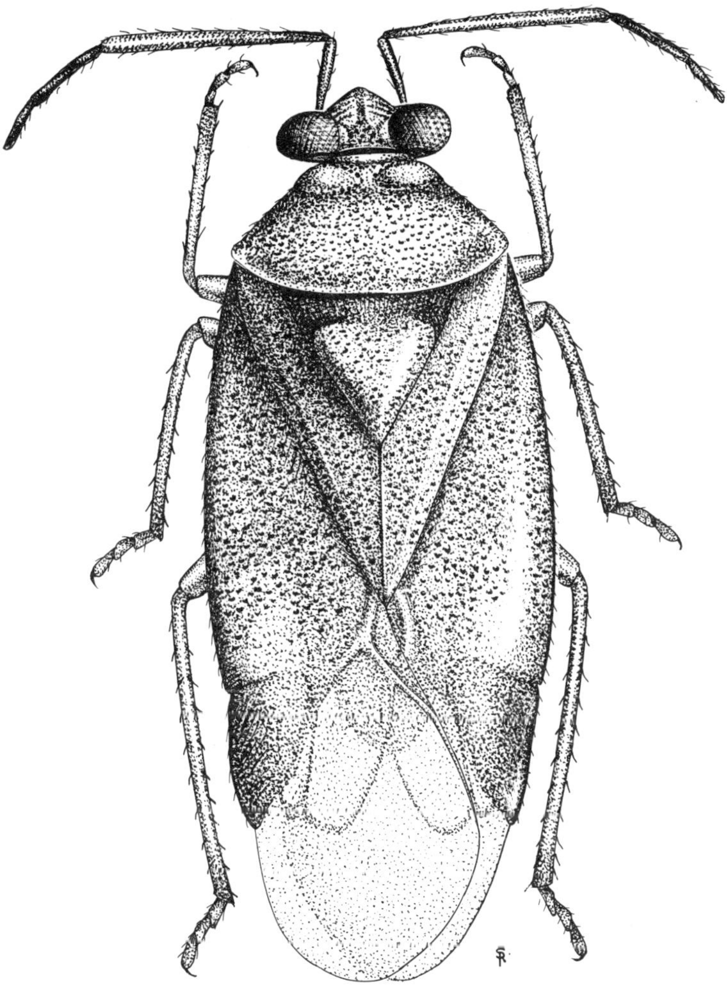

Fig. 63. *Neoborella canadensis*

rounded; callus often marked with transverse black bar. Scutellum brown. Hemelytra brown, shiny with bluish sheen; practically glabrous (Fig. 59). Ventral surface yellowish green; pleura and side margin of abdomen brown; legs mostly pale.

Remarks. This species resembles *xanthenes* in color and pubescence, but is larger and the eyes in the male are bulging.

Habitat. Collected on mistletoe growing on *Pinus banksiana*.

Distribution Alberta, Saskatchewan (Map 20).

Genus *Poecilocapsus* Reuter

Glabrous, green with longitudinal black lines. Head vertical, short, carina between eyes absent. Pronotum impunctate, lateral margins angular. Hemelytra impunctate, shiny.

Five species are known from North America, one occurs in the Prairie Provinces.

Poecilocapsus lineatus (Fabricius)

Fig. 53; Map 21

Lygaeus lineatus Fabricius, 1798:451.
Capsus quadrivittatus Say, 1832:20.
Phytocoris bellus Emmons, 1854:30.
Poecilocapsus lineatus: Reuter, 1875*d*:74.

Length 7.00–7.50 mm; width 2.80–3.50 mm. Head brown, clypeus and antennae black. Pronotum and hemelytra yellowish green with four black lines. Legs green.

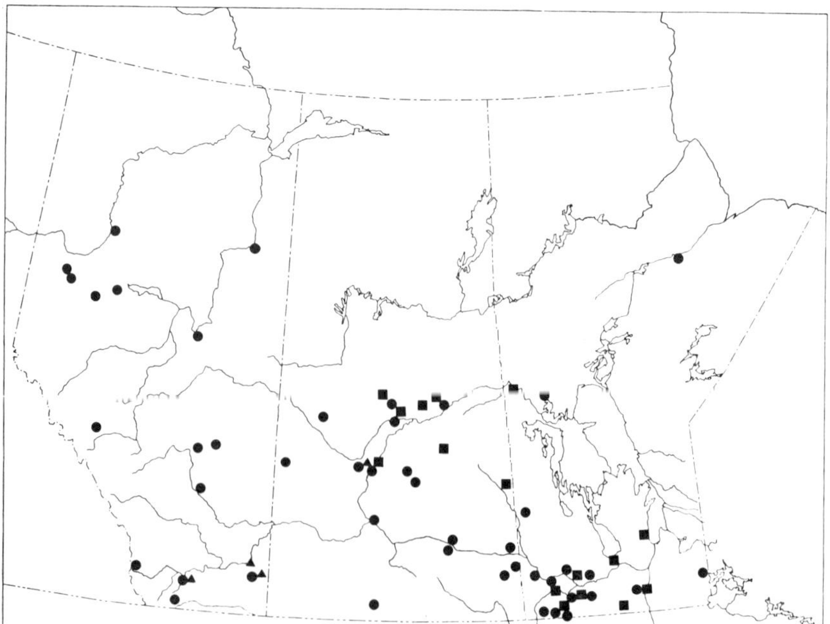

Map 21. Collection localities for *Poecilocapsus lineatus* (■), *Agnocoris rubicundus* (●), and *A. pulverulentus* (▲).

Remarks. This species is distinguished by the four black lines on the dorsum (Fig. 53).

Habitat. Collected on *Mentha arvensis.*

Distribution. Widespread in USA, Ontario, Quebec; now known to occur in Manitoba and Saskatchewan (Map 21).

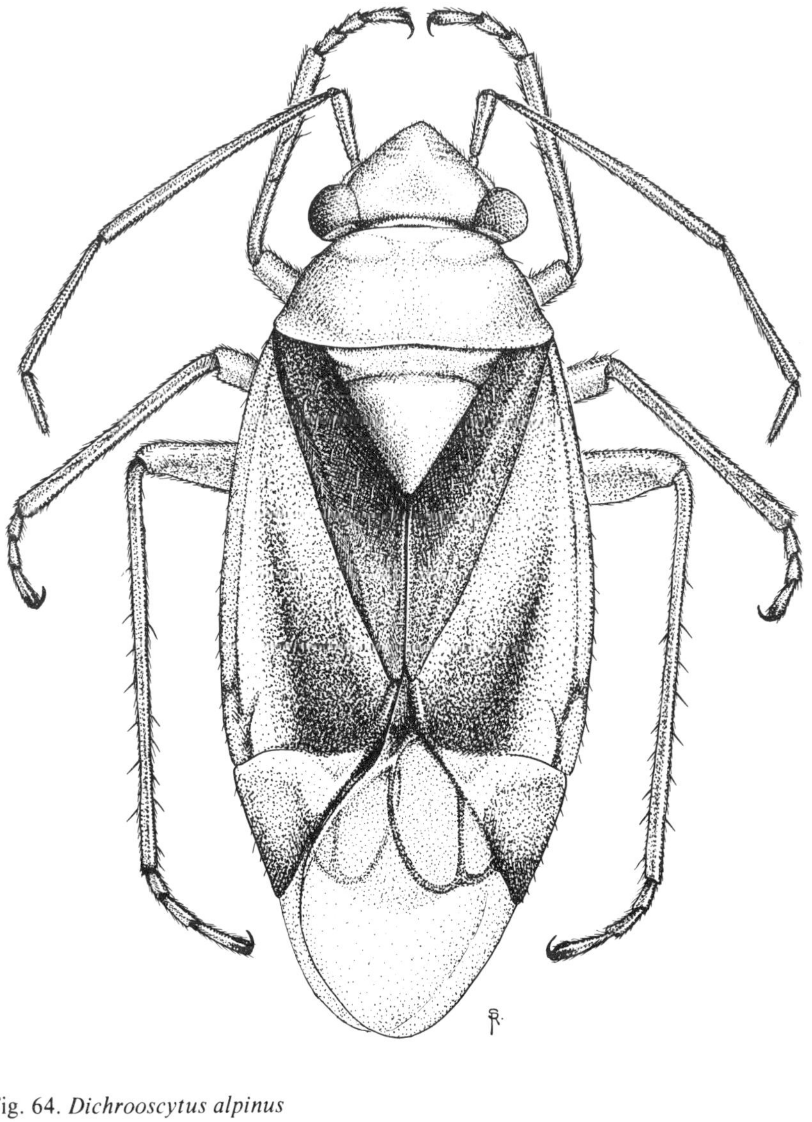

Fig. 64. *Dichrooscytus alpinus*

Genus *Agnocoris* Reuter

Oval, small, brown or reddish brown species. Head oblique, short, wide; carina between eyes distinct; second antennal segment shorter than head width; rostrum extending to hind coxae. Pronotum finely punctate, lateral margins rounded, calli smooth, pubescent. Hemelytra shiny, densely pubescent.

Four species are known from North America; two are found in the Prairie Provinces, one is Holarctic in distribution.

Key to species of *Agnocoris*

1. Reddish or reddish brown species; pubescence short, appressed; male claspers (Fig. 65) .... ***rubicundus* (Fallén)** (p. 80)
   Brown or fawn species, pubescence long, slanting, hemelytra appears mottled; male claspers (Fig. 66) .... ***pulverulentus* (Uhler)** (p. 80)

Agnocoris rubicundus (Fallén)

Fig. 65; Map 21

Lygaeus rubicundus Fallén, 1807:84.
Agnocoris rubicundus: Reuter, 1875*a*:63.

Length 4.20–4.69 mm; width 2.10–2.38 mm. Head orange marked with red. Rostrum 1.64–1.69 mm long. Pronotum reddish brown, calli orange, spots behind calli often black. Scutellum red, two median dashes often black. Hemelytra red. Ventral surface mostly red; legs tinged with red.

Remarks. This species varies from reddish brown to dark red. Wagner and Slater (1952), and Moore (1955) suggested that *rubicundus* did not occur in North America. Kelton (1955) referred to the red adults of *rubicundus* as *pulverulentus*, and to the reddish brown adults (in part) as *utahensis*. Moore (1956) concluded that *rubicundus* occurred in North America.

Habitat. Collected on *Salix candida* and *S. interior*.

Distribution. Idaho, Ontario; Prairie Provinces (Map 21).

Agnocoris pulverulentus (Uhler)

Fig. 66; Map 21

Hadronema pulverulentus Uhler, 1892:183.
Agnocoris pulverulentus: Kelton, 1955*a*:282.

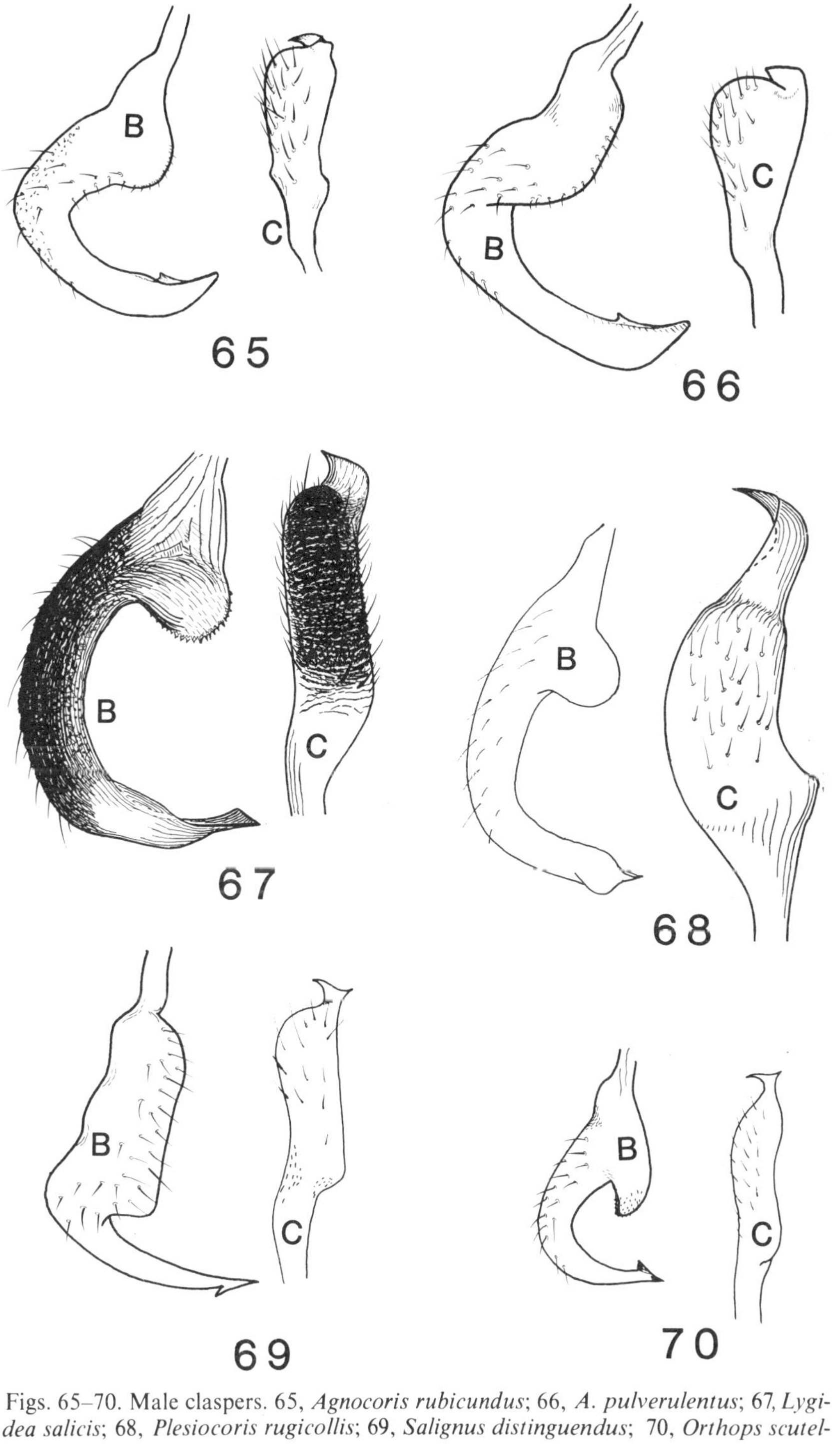

Figs. 65–70. Male claspers. 65, *Agnocoris rubicundus*; 66, *A. pulverulentus*; 67, *Lygidea salicis*; 68, *Plesiocoris rugicollis*; 69, *Salignus distinguendus*; 70, *Orthops scutellatus*

Length 4.62–5.32 mm; width 2.10–2.52 mm. Head light yellowish brown, base of clypeus, middle of frons, area above antennal socket, and vertex often marked with dark brown; frons weakly striate; rostrum 1.61–1.68 mm long. Pronotum yellowish brown, longitudinal dash along outer callus dark reddish brown. Scutellum dark brown, basal angles, apex, and median line often pale. Hemelytra light brown, inner clavus and apical half of corium often dark brown; pubescence silvery, appears mottled with brown. Ventral surface marked with dark brown; legs mostly pale.

Remarks. This species is generally larger than *rubicundus* and is mostly marked with brown. The pubescence on the hemelytra is longer and appears mottled. The right clasper is broader than in *rubicundus*. *A. utahensis* does not occur in the Prairie Provinces and reference to it by Kelton (1955*b*) in part should refer to *pulverulentus*.

Habitat. Collected on *Salix amygdaloides*.

Distribution. Northeastern and Central States, Idaho; Alberta (Map 21).

Genus *Capsus* Fabricius

Black, shiny species. Head oblique, carina between eyes absent; second antennal segment clavate. Pronotum and hemelytra punctate. Ventral surface and femora black, tibiae pale.

Two species occur in North America, both introduced from Europe; one is found in the Prairie Provinces.

Capsus simulans (Stål)

Map 22

Deraeocoris simulans Stål, 1858:186.
Capsus simulans: Walker, 1873:89.

Length 5.60–6.02 mm; width 2.52–3.08 mm. Head black, area between eyes often pale; second antennal segment moderately clavate. Pubescence on hemelytra golden brown, short, appressed.

Remarks. This species may be confused with those of *Irbisia*, but the clavate second antennal segment will readily distinguish *simulans*.

Habitat. Collected on *Bromus* spp.

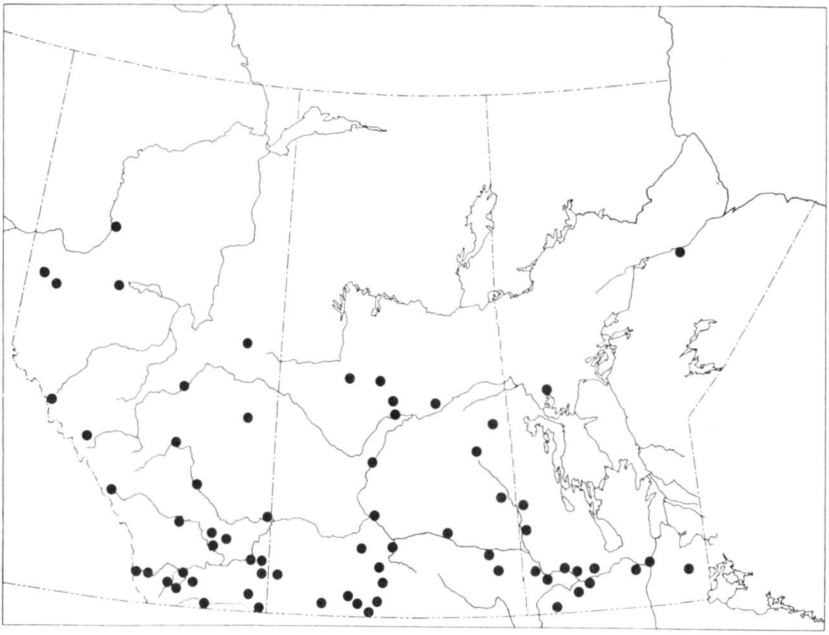

Map 22. Collection localities for *Capsus simulans*.

Distribution. Alaska, North Central States; Alberta, now known to occur in Saskatchewan and Manitoba (Map 22).

Genus *Irbisia* Reuter

Elongate-oval, black, shiny, pubescent species. Head oblique, carina between eyes distinct, second antennal segment linear. Pronotum and hemelytra punctate; pubescence black or silvery. Ventral surface mostly black; legs pale or black; osteolar peritreme pale.

Approximately 34 species are known from North America, four occur in the Prairie Provinces.

Key to species of *Irbisia*

1. Pubescence on pronotum and hemelytra black .... ***fuscipubescens* Knight** (p. 84)
   Pubescence on pronotum and hemelytra silvery 2
2. Legs black .... ***nigripes* Knight** (p. 84)
   Legs partly pale 3
3. Costal margin of hemelytra convex .... ***brachycera* (Uhler)** (p. 85)
   Costal margin of hemelytra straight .... ***elongata* Knight** (p. 86)

Irbisia fuscipubescens Knight

Map 23

Irbisia fuscipubescens Knight, 1941*a*:76.

Length 5.53–5.95 mm; width 2.38–2.73 mm. Pubescence on pronotum and hemelytra black, short, appressed. Femora black, apices pale; tibiae pale, basal one third often blackish.

Remarks. This species is distinguished by the black pubescence.

Habitat. Collected on herbaceous plants.

Distribution. Washington; Alberta (Map 23).

Irbisia nigripes Knight

Map 23

Irbisia nigripes Knight, 1925*a*:94.

Length 5.74–6.30 mm; width 2.52–2.80 mm. Pubescence on pronotum and hemelytra silvery and black, semiappressed. Femora and tibiae black.

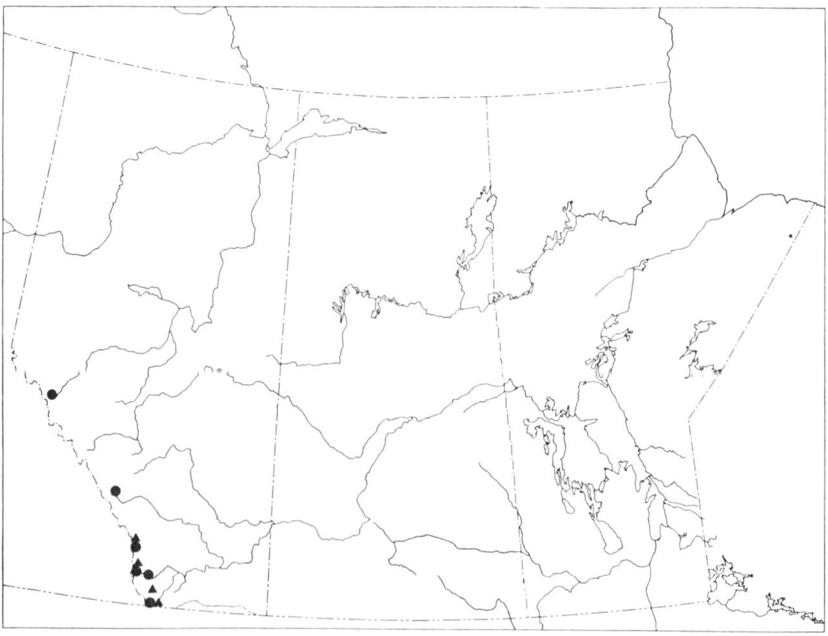

Map 23. Collection localities for *Irbisia fuscipubescens* (●) and *I. nigripes* (▲).

Remarks. This species is distinguished by the black legs.

Habitat. Collected on herbaceous plants.

Distribution. Idaho, Montana, British Columbia; Alberta (Map 23).

Irbisia brachycera (Uhler)

Map 24

Rhophalotomus brachycerus Uhler, 1872:416.
Irbisia brachycera: Cockerell, 1910:370.
Capsus solani Heidemann, 1910:200.
Irbisia arcuata Van Duzee, 1921:148.

Length 5.60–5.88 mm; width 2.24–2.66 mm. Pubescence on head, pronotum, and hemelytra silvery, appressed. Femora with pale areas, tibiae pale.

Remarks. This species is distinguished by the appressed, silvery pubescence.

Habitat. Collected on prairie grasses.

Distribution. Mexico, western States; Alberta, now known to occur in Saskatchewan (Map 24).

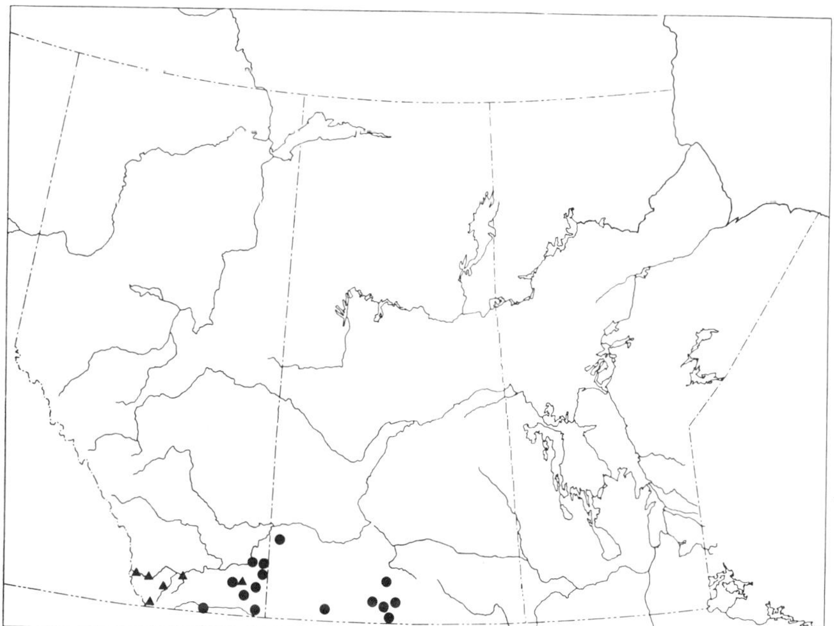

Map 24. Collection localities for *Irbisia brachycera* (●) and *I. elongata* (▲).

Irbisia elongata Knight

Map 24

Irbisia elongata Knight, 1941*a*:77.

Length 5.46–6.65 mm; width 2.10–2.52 mm. Pubescence on pronotum and hemelytra silvery, long, dense, woolly. Femora mostly pale, often with black bars or spots; tibiae pale.

Remarks. This species resembles *brachycera*, but is more elongate, and the pubescence is longer and denser.

Habitat. Collected on *Agropyron repens*.

Distribution. Northwestern States, British Columbia; now known to occur in Alberta (Map 24).

Genus *Lygidea* Reuter

Elongate, reddish brown pubescent species. Head nearly vertical, frons smooth, eyes nearly spherical, positioned above antennal socket, carina between them distinct. Rostrum extending to middle coxae. Pronotum punctate, pubescent; calli glabrous. Hemelytra punctate, pubescence long, dense. Ventral surface green, lateral margins black; legs mostly green.

Nine species are known from North America, two occur in the Prairie Provinces.

Key to species of *Lygidea*

1. Larger species 6.0–7.0 mm; pilosity on second antennal segment longer than thickness of segment .... ***rosacea* Reuter** (p. 86)
   Smaller species 5.6–6.0 mm; pilosity on second antennal segment shorter than thickness of segment .... ***salicis* Knight** (p. 87)

Lygidea rosacea Reuter

Map 25

Lygidea rubecula var. *rosacea* Reuter, 1909:46.
Lygidea rosacea: Knight, 1923*b*:569.

Length 6.02–7.00 mm; width 2.38–2.87 mm. Head yellow, clypeus and line above antennal socket black. Second antennal segment strongly pilose. Pronotum brown to dark brown, median longitudinal line pale. Hemelytra dark brown, cuneus often red; pubescence golden, long, dense.

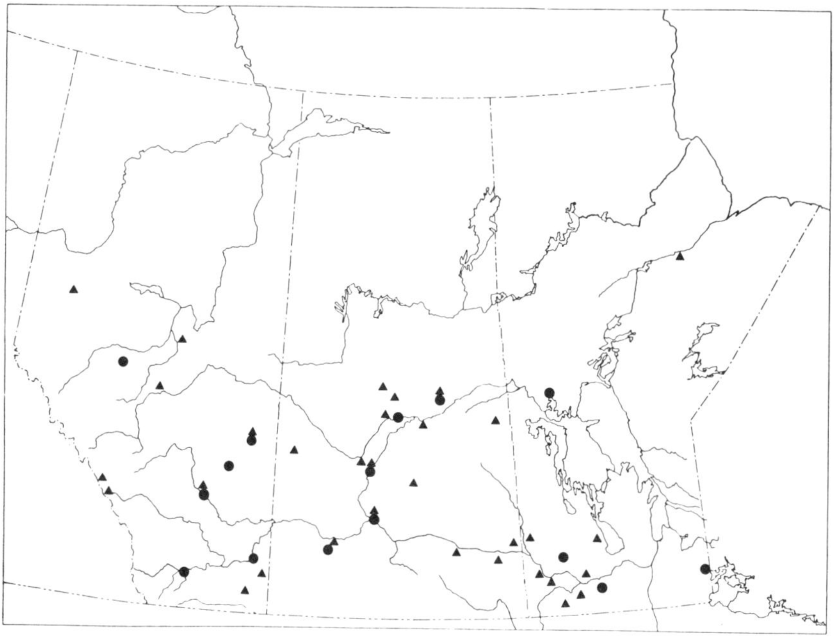

Map 25. Collection localities for *Lygidea rosacea* (●) and *L. salicis* (▲).

Remarks. This species is distinguished by the large size and by the pilose second antennal segment.

Habitat. Collected on *Salix interior*.

Distribution. North Central States, Ontario; Prairie Provinces (Map 25).

Lygidea salicis Knight

Figs. 45, 67; Map 25

Lygidea salicis Knight, 1939*a*:22.

Length 5.60–6.02 mm; width 2.24–2.52 mm. Much like *rosacea* in color and appearance, but smaller in size.

Remarks. This species is separated from *rosacea* by the shorter pilosity on the second antennal segment. The genital claspers (Fig. 63) are similar.

Habitat. Collected on *Salix* spp.

Distribution. New York, Michigan, Illinois, Ontario; now known to occur in the Prairie Provinces (Map 25).

Genus *Polymerus* Hahn

Elongate-oval, pubescent species. Head oblique, short; eyes large, prominent, carina between them distinct. Pronotum trapeziform, rugose, calli inconspicuous. Hemelytra finely punctate; pubescence woolly, shiny, often intermixed with erect hairs. Males elongate, females oval, more robust.

Approximately 43 species are known from North America, 13 occur in the Prairie Provinces.

Key to species of *Polymerus*

1. Rostrum extending to hind coxae or beyond 2
   Rostrum not extending to hind coxae 4
2. Dorsal surface green or stained with reddish brown; legs mostly pale, tibiae red in males .... ***rufipes*** **Knight** (p. 88)
   Dorsal surface black with costal margin and cuneus pale; legs green or black .. 3
3. Pubescence on dorsum silvery, appressed, without erect, black hairs .... ***sculleni*** **Knight** (p. 89)
   Pubescence on dorsum silvery, appressed, intermixed with erect, black hairs .... ***hirtus*** **Knight** (p. 90)
4. Rostrum more than 1.40 mm 5
   Rostrum less than 1.40 mm 7
5. Cuneus with extensive red markings .... ***unifasciatus*** **(Fabricius)** (p. 90)
   Cuneus mostly pale or black 6
6. Pronotum with extensive pale green markings .... ***balli*** **Knight** (p. 91)
   Pronotum black without pale markings, only basal margin pale .... ***basivittis*** **(Reuter)** (p. 92)
7. Cuneus red 8
   Cuneus mostly black, outer margin pale 10
8. Femora black .... ***rubrocuneatus*** **Knight** (p. 93)
   Femora with extensive pale areas 9
9. Tibiae black .... ***chrysopsis*** **Knight** (p. 94)
   Tibiae mostly pale .... ***venaticus*** **(Uhler)** (p. 94)
10. Tibiae black, femora red .... ***brevirostris*** **Knight** (p. 95)
    Tibiae yellow or marked with pale 11
11. Legs yellow .... ***severini*** **Knight** (p. 96)
    Legs strongly marked with brown or black 12
12. Frons strongly tumid .... ***tumidifrons*** **Knight** (p. 97)
    Frons not strongly tumid .... ***diffusus*** **(Uhler)** (p. 97)

Polymerus rufipes Knight

Map 26

Polymerus basalis var. *rufipes* Knight, 1926*b*:167.

Length 4.30–5.32 mm; width 2.03–2.52 mm. Head yellow, clypeus and submedian area on frons often black. Rostrum 2.10–2.38 mm long, extend-

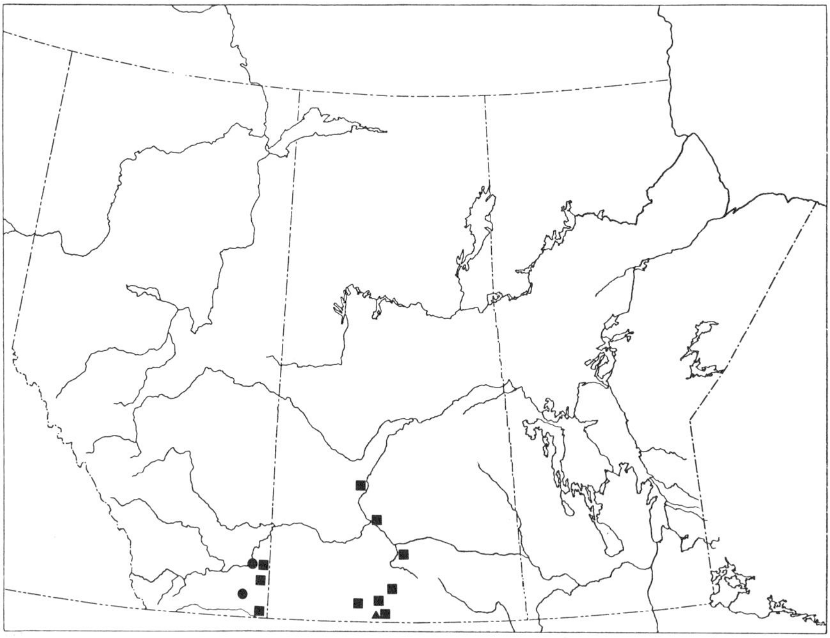

Map 26. Collection localities for *Polymerus rufipes* (●), *P. sculleni* (▲), and *P. hirtus* (■).

ing to hind coxae. Pronotum yellowish green, calli and rays behind often black. Scutellum yellowish green, median area at base often black. Hemelytra yellowish green, clavus and corium often marked with reddish tinge; cuneus reddish, often green in females; pubescence golden, appressed. Legs yellow, femora often annulate with reddish tinge or black; tibiae red in males, often green in females.

Remarks. This species was originally described as a variety of *basalis*. However, it differs from *basalis* in the shorter rostrum, longer second antennal segment, and in the markings on the scutellum. The reddish tibiae in the males are distinctive.

Habitat. Collected on *Erigeron* spp.

Distribution. Wyoming, Colorado; now known to occur in Alberta (Map 26).

Polymerus sculleni Knight

Map 26

Polymerus sculleni Knight, 1943*b*:180.

Length 4.34–5.01 mm; width 1.75–2.03 mm. Head black, spot on vertex next to eye and carina between eye yellow. Rostrum 1.96–2.10 mm long, extending beyond hind coxae. Pronotum black, collar and median line between calli yellow. Scutellum black, median line or apex yellow. Hemelytra black, costal margin and cuneus pale; pubescence silvery, recumbent. Legs marked with fuscous, black.

Remarks. This species is distinguished by the long rostrum, by the yellow median line on pronotum and scutellum, and by the pale costal margin.

Habitat. Unknown.

Distribution. Oregon; now known to occur in Saskatchewan (Map 26).

Polymerus hirtus Knight

Map 26

Polymerus hirtus Knight, 1943*b*:180.

Length 4.20–4.62 mm; width 1.68–1.82 mm. Head black, lorum and jugum often paler, carina brown. Rostrum 1.75–1.89 mm long, extending to hind coxae. Pronotum black. Scutellum black, tip often pale. Hemelytra mostly black, costal margin and cuneus yellowish white; pubescence sericeous, intermixed with erect, black hairs. Ventral surface and femora black, tibiae reddish brown.

Remarks. The females are unknown. Originally the species was described from a single male, collected 12 July. The prairie specimens were collected from June to the middle of July with no females in the series. Probably they appear much later in the season.

Habitat. Collected on mixed vegetation.

Distribution. Idaho; now known to occur in Alberta and Saskatchewan (Map 26).

Polymerus unifasciatus (Fabricius)

Fig. 54; Map 27

Lygaeus unifasciatus Fabricius, 1794:178.
Polymerus unifasciatus: Wagner, 1939:42.

Length 4.48–5.60 mm; width 1.82–2.66 mm. Head black, spot on top next to each eye pale. Rostrum 1.54–1.68 mm long, extending to middle coxae. Pronotum black, basal margin pale. Scutellum black, apical half pale. Hemelytra generally black, basal half of corium, area adjacent to cuneus, and area at tip of clavus pale; cuneus pale at base and apex, middle portion

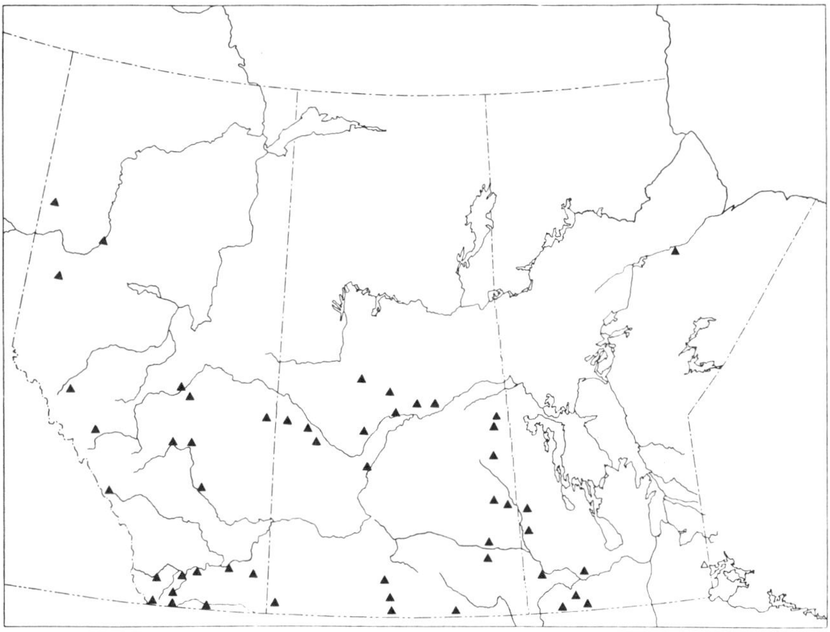

Map 27. Collection localities for *Polymerus unifasciatus.*

black and red; pubescence mostly golden, appressed. Ventral surface mostly black; legs yellow marked with black.

Remarks. All North American specimens of this Holarctic species refer to *lateralis* Hahn, a variety of *unifasciatus*. It is distinguished from *balli*, a closely related species, by the red cuneus and by the black pronotum (Fig. 54).

Habitat. Collected on *Galium boreale*.

Distribution. Alaska, northeastern and North Central States, California, Ontario, Quebec; Alberta, now known to occur in Saskatchewan and Manitoba (Map 27).

Polymerus balli Knight

Map 28

Polymerus balli Knight, 1925*b*:250.

Length 4.20–4.97 mm; width 1.68–2.03 mm. Head yellow, clypeus, median line on frons, and carina black. Rostrum 1.40–1.54 mm long, extending to middle coxae. Pronotum green with four longitudinal black

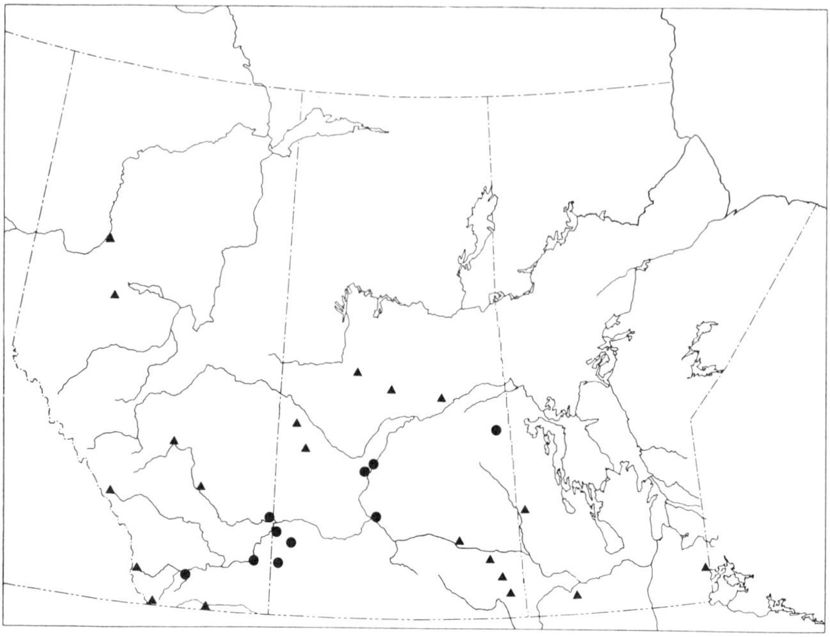

Map 28. Collection localities for *Polymerus balli* (●) and *P. basivittis* (▲).

rays. Scutellum black, apical half yellow. Hemelytra yellow, middle of clavus and subdivided patch on apical corium black; cuneus fuscous at middle; pubescence golden, appressed. Ventral surface mostly green, legs green with several black spots.

Remarks. This species is closely related to *unifasciatus*, separated from it by the green and black pronotum, and the absence of red on the cuneus.

Habitat. Collected on *Psoralea lanceolata.*

Distribution. Idaho, Colorado, Nebraska; Alberta, Saskatchewan, now known to occur in Manitoba (Map 28).

Polymerus basivittis (Reuter)

Map 28

Poeciloscytus basivittis Reuter, 1909:61.
Polymerus basivittis: Knight, 1925*a*:95.

Length 4.34–5.05 mm; width 2.24–2.66 mm. Head black, spot on vertex next to each eye pale. Rostrum 1.54–1.82 mm long, extending to middle coxae. Pronotum black, basal margin pale. Scutellum black, apex pale. Hemelytra generally black, tip of clavus, anterior half of inner corium, area adjacent to cuneus, and outer margin of embolium pale; cuneus pale, central spot black; pubescence golden, appressed. Ventral surface black; legs marked with red or black.

Remarks. The species is variable in color. Females generally have less black on the head, pronotum, and hemelytra, and the ventral surface is partly pale. The reddish tinge on the legs is more pronounced in the female.

Habitat. Collected on *Galium boreale*.

Distribution. Montana, Wyoming, Colorado; Alberta, now known to occur in Saskatchewan and Manitoba (Map 28).

Polymerus rubrocuneatus Knight

Map 29

Polymerus rubrocuneatus Knight, 1925*b*:247.

Length 5.04–5.74 mm; width 2.38–2.66 mm. Head black, spot at vertex next to eye yellow. Rostrum 1.12–1.26 mm long, extending just beyond front

Map 29. Collection localities for *Polymerus rubrocuneatus* (●) and *P. chrysopsis* (■).

coxae. Pronotum black, basal margin pale. Scutellum black. Hemelytra black, apex of embolium and cuneus red; pubescence silvery, appressed. Ventral surface and legs black.

Remarks. This species is distinguished by the short rostrum, red cuneus, and black femora.

Habitat. Collected on mixed vegetation.

Distribution. North Dakota, South Dakota; Manitoba, now known to occur in Saskatchewan and Alberta (Map 29).

Polymerus chrysopsis Knight

Map 29

Polymerus chrysopsis Knight, 1925*b*:245.

Length 5.60–6.44 mm; width 2.38–2.80 mm. Head black, spot on top next to each eye yellow. Rostrum 1.19–1.26 mm long, extending beyond front coxae. Pronotum black, basal margin pale. Scutellum black. Hemelytra black, embolium pale, cuneus red; pubescence silvery, appressed. Ventral surface black, coxae and femora orange, tibiae black.

Remarks. This species is distinguished by the short rostrum, red cuneus, and black tibiae.

Habitat. Collected on *Chrysopsis villosa.*

Distribution. North Central States; Manitoba, now known to occur in Saskatchewan (Map 29).

Polymerus venaticus (Uhler)

Map 30

Poeciloscytus venaticus Uhler, 1872:414.
Rhopalotomus rubronotatus Provancher, 1872:105.
Polymerus venaticus: Tucker, 1907:60.

Length 4.90–6.30 mm; width 2.38–2.52 mm. Head black, spot on top next to each eye yellow. Rostrum 0.98–1.12 mm long, extending to front coxae. Pronotum black, basal margin feebly pale. Scutellum black. Hemelytra black, cuneus orange; pubescence silvery and black, appressed. Ventral surface black; legs partly pale.

Remarks. This species has a short rostrum and an orange cuneus like *rubrocuneatus* and *chrysopsis,* but differs in the bicolored femur and tibia.

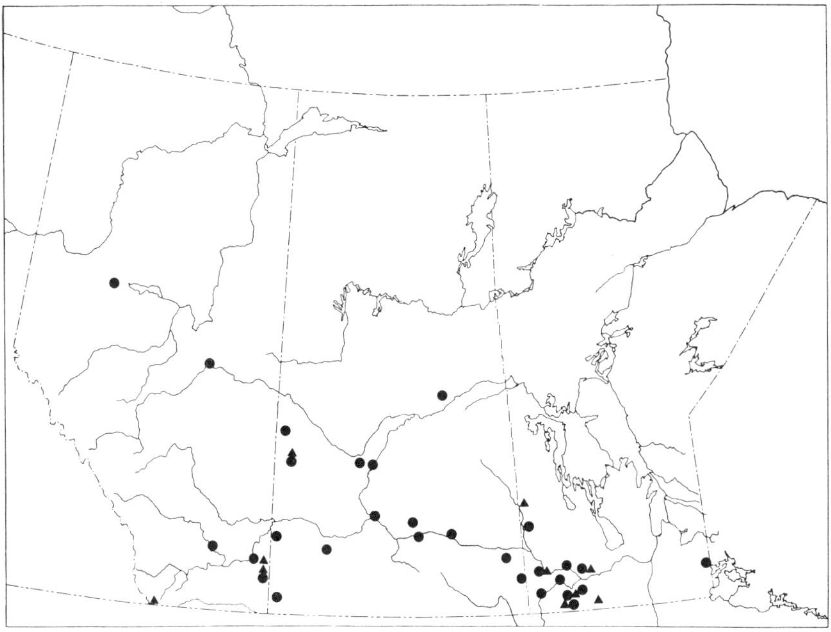

Map 30. Collection localities for *Polymerus venaticus* (●) and *P. brevirostris* (▲).

Habitat. Collected on *Solidago* spp. and *Medicago sativa.*

Distribution. Widespread in USA, Ontario, Quebec, British Columbia; Alberta, Manitoba, now known to occur in Saskatchewan (Map 30).

Polymerus brevirostris Knight

Map 30

Polymerus brevirostris Knight, 1925*b*:246.

Length 4.69–4.90 mm; width 2.17–2.52 mm. Head black, spot on top next to each eye reddish, lorum and jugum often red. Rostrum 0.84–0.98 mm long, extending to front coxae. Pronotum black, basal margin pale. Scutellum black. Hemelytra black, narrow costal margin pale, cuneus black; pubescence golden, appressed. Ventral surface black, coxae and femora reddish, tibiae black.

Remarks. This species is distinguished by the short rostrum, black cuneus, reddish femora, and black tibiae.

Habitat. Collected on mixed vegetation.

Distribution. North Central States; Manitoba, now known to occur in Saskatchewan and Alberta (Map 30).

Polymerus severini Knight

Map 31

Polymerus severini Knight, 1925*b*:247.

Length 5.40–6.30 mm; width 2.50–2.80 mm. Head black, spot on top next to each eye pale. Rostrum 1.12–1.18 mm long, extending just beyond front coxae. Pronotum black, basal margin pale. Scutellum black. Hemelytra black, embolium and outer cuneus pale; pubescence silvery, appressed. Ventral surface black, legs mostly yellow.

Remarks. This species is distinguished by the short rostrum and by the yellow legs.

Habitat. Unknown.

Distribution. Minnesota, South Dakota; Alberta (Map 31).

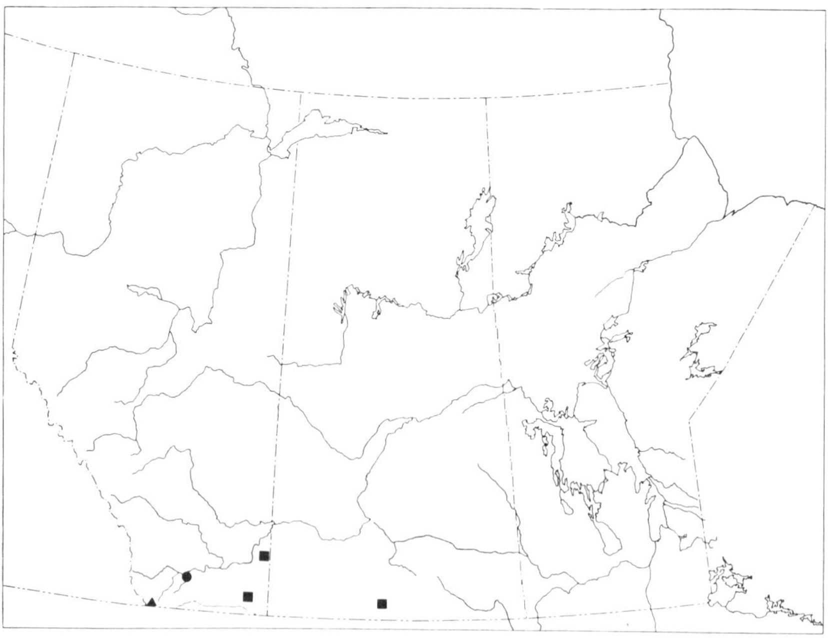

Map 31. Collection localities for *Polymerus severini* (●), *P. tumidifrons* (▲), and *P. diffusus* (■).

Polymerus tumidifrons Knight

Map 31

Polymerus tumidifrons Knight, 1925*b*:248.

Length 5.60–6.30 mm; width 2.24–2.94 mm. Head black, spot on top next to each eye yellow, frons tumid. Rostrum 1.19–1.40 mm long, extending to margin of mesosternum. Pronotum black, basal margin pale. Scutellum black. Hemelytra black, embolium and narrow adjacent area on corium and outer margins of cuneus white; pubescence silvery, appressed. Ventral surface black; legs mostly black.

Remarks. This species is distinguished by the short rostrum and by the tumid frons.

Habitat. Collected on *Helianthus* spp.

Distribution. Montana, Utah; now known to occur in Alberta (Map 31).

Polymerus diffusus (Uhler)

Map 31

Poeciloscytus diffusus Uhler, 1872:415.
Poeciloscytus divergens Parshley, 1914:141.
Polymerus diffusus: Knight, 1926*d*:165.

Length 4.90–6.44 mm; width 2.10–2.80 mm. Head black, spot on top next to each eye pale. Rostrum 1.26–1.40 mm long, extending to apex of mesosternum. Pronotum black. Scutellum black. Hemelytra black, embolium, adjacent corium, and outer margin of cuneus pale; pubescence silvery, appressed. Ventral surface black; legs yellow marked with black.

Remarks. The males are more slender than the females, and the females may also have the jugum, scutellum, coxae, and abdomen partly pale. Similar to *tumidifrons* in appearance, but frons not tumid.

Habitat. Collected on *Artemisia frigida* and *A. tridentata.*

Distribution. Idaho, Utah, Colorado; now known to occur in Alberta and Saskatchewan (Map 31).

Genus *Plesiocoris* Fieber

Elongate, pale green, glabrous species. Head oblique, frons striate, carina between eyes distinct. Pronotum strongly rugose, calli smooth.

Hemelytra glabrous, pubescence short, sparse. Ventral surface and legs green.

One Holarctic species occurs in North America.

Plesiocoris rugicollis (Fallén)

Fig. 68; Map 32

Lygaeus rugicollis Fallén, 1807:76.
Plesiocoris rugicollis: Fieber, 1861:272.

Length 5.74–6.02 mm; width 2.10–2.28 mm. Head, pronotum, and hemelytra green, embolium pale.

Remarks. This species is distinguished by the overall green color, by the rugose pronotum, and by the genital clasper (Fig. 68).

Habitat. Collected on *Salix* spp.

Distribution. Alaska, British Columbia; Alberta (Map 32).

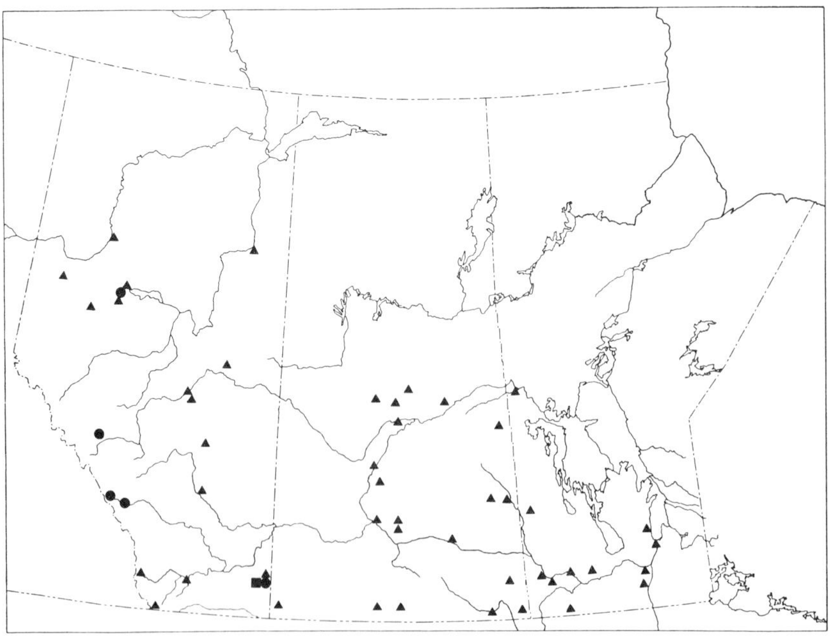

Map 32. Collection localities for *Plesiocoris rugicollis* (■), *Salignus distinguendus* (●), and *Orthops scutellatus* (▲).

Genus *Salignus* Kelton

Oval, reddish brown, mottled species. Head short, blunt, frons transversely striate, carina between eyes distinct. Rostrum extending to hind coxae. Pronotum finely punctate, lateral margins angulate, calli raised, pubescent. Hemelytra pruinose, densely pubescent, appears mottled.

This Nearctic genus is monobasic.

Salignus distinguendus (Reuter)

Figs. 55, 69; Map 32

Lygus distinguendus Reuter, 1875:544.
Salignus distinguendus: Kelton, 1955*a*:283.

Length 5.04–5.46 mm; width 2.10–2.38 mm. Head strongly marked with black. Rostrum 1.82–1.96 mm long. Pronotum brown, strongly marked with black, basal margin and collar white. Scutellum black, lateral margins and apex pale. Hemelytra mottled brown; pubescence long, dense. Ventral surface mostly dark brown, femora brown, tibiae pale; male claspers (Fig. 65).

Remarks. This species is distinguished by the mottled appearance of the hemelytra, by the striate frons (Fig. 55), and by the distinctive male claspers (Fig. 69).

Habitat. Collected on *Salix bebbiana.*

Distribution. Alaska, western States, Yukon, British Columbia; Prairie Provinces (Map 32).

Genus *Orthops* Fieber

Small, oblong, green species. Head vertical, short; frons smooth, carina between eyes distinct. Rostrum extending to middle coxae. Pronotum deeply punctate, lateral margins rounded, calli raised, smooth, pubescent. Hemelytra shiny, pubescence long, dense.

Only one species occurs in North America. Wagner and Slater (1952) showed that *Orthops campestris* (L) does not occur in the Nearctic region. Henry and Wheeler (1974) reported *Orthops rubricatus* (Fallén) from Pennsylvania; Kelton (1977) transferred the species to *Pinalitus* Kelton.

Orthops scutellatus Uhler

Figs. 46, 70; Map 32

Orthops scutellatus Uhler, 1877:420.

Length 4.20–4.90 mm; width 1.82–2.10 mm. Head mostly green, clypeus and frons often black. Rostrum 1.40–1.54 mm long. Pronotum green, calli black, rays behind often black. Scutellum green, often marked with black. Hemelytra mostly green, clavus and triangular area at apex of corium often black; punctate, shiny; pubescence long, dense. Ventral surface and legs mostly green.

Remarks. The species is distinguished by the nearly vertical head (Fig. 46), by the moderately large and deep punctures on the pronotum and hemelytra, and by the genital claspers (Fig. 70).

Habitat. Collected on members of the carrot family, often encountered in carrot seed production.

Distribution. Mexico, widespread in USA, British Columbia, Ontario, Quebec; Prairie Provinces (Map 32).

Genus *Lygus* Hahn

Elongate-oblong, medium to large species. Head oblique, frons smooth or striate, eyes large, conspicuous, carina between them prominent, or obsolete in one species. Rostrum variable in length. Pronotum deeply and coarsely punctate; anterior angles rounded or angular, calli smooth, glabrous or pubescent. Hemelytra green or reddish brown; pubescence short and sparse, or long and dense.

The genus was revised by Kelton (1975) where additional information on taxonomy, distribution, and habits is given.

Thirty-four species are known from North America, 22 occur in the Prairie Provinces.

Key to species of *Lygus*

1. Frons obliquely striate or grooved (Fig. 44) 2
   Frons smooth, not striate 6
2. Rostrum more than 3.0 mm in length .... ***convexicollis* Reuter** (p. 101)
   Rostrum less than 2.9 mm in length 3
3. Pubescence on hemelytra short, sparse (Fig. 72) ....
   .... ***nigropallidus* Knight** (p. 102)
   Pubescence on hemelytra long, dense 4

4. Anterior angles of pronotum rounded (Fig. 73); rostrum less than 2.2 mm .... ***nubilus*** **Van Duzee** (p. 104)
   Anterior angles of pronotum angular (Fig. 74); rostrum 2.2 mm or longer 5
5. Mesoscutum black with lateral areas red or yellow (Fig. 74) .... ***humeralis*** **Knight** (p. 104)
   Mesoscutum black throughout (Fig. 75) .... ***perplexus*** **Stanger** (p. 108)
6. Hind tibia black or reddish brown (Fig. 76) .... ***atritibialis*** **Knight** (p. 108)
   Hind tibia green or yellow 7
7. Carina between eyes almost obsolete; front and middle tibiae with reddish stripes (Fig. 77) .... ***vanduzeei*** **Knight** (p. 111)
   Carina between eyes complete; tibiae without reddish stripes 8
8. Rostrum 2.8 mm or longer .... ***potentillae*** **Kelton** (p. 111)
   Rostrum less than 2.8 mm 9
9. Frons with submedian lines (Fig. 48) 10
   Frons without submedian lines 13
10. Hemelytra with patches of silvery pubescence (Fig. 79) .... ***plagiatus*** **Uhler** (p. 113)
    Hemelytra without patches of silvery pubescence 11
11. Mesoscutum black (Fig. 80) .... ***unctuosus*** **(Kelton)** (p. 116)
    Mesoscutum with lateral areas red or yellow 12
12. Pubescence on hemelytra short, sparse (Fig. 81) .... ***rufidorsus*** **(Kelton)** (p. 118)
    Pubescence on hemelytra moderately long, dense (Fig. 82) .... ***lineolaris*** **(Palisot de Beauvois)** (p. 118)
13. Mesoscutum with lateral areas red or yellow (Fig. 83) 14
    Mesoscutum black 15
14. Green species .... ***desertinus*** **Knight** (p. 119)
    Yellow or reddish brown species .... ***rubroclarus*** **Knight** (p. 123)
15. Anterior angles of pronotum angular 16
    Anterior angles of pronotum rounded 17
16. Rostrum over 2.5 mm .... ***varius*** **Knight** (p. 125)
    Rostrum 2.5 mm or less .... ***ravus*** **Stanger** (p. 125)
17. Pubescence on hemelytra short, sparse 18
    Pubescence on hemelytra long, dense 20
18. Cuneus mostly unicolorous; second antennal segment 1.6 mm or shorter .... ***borealis*** **(Kelton)** (p. 125)
    Cuneus with black tip; second antennal segment longer than 1.6 mm 19
19. Hemelytra dark reddish brown .... ***columbiensis*** **Knight** (p. 129)
    Hemelytra greenish yellow .... ***shulli*** **Knight** (p. 131)
20. Corium with triangular brown or black pattern on apical half (Fig. 90) .... ***solidaginis*** **(Kelton)** (p. 133)
    Corium without triangular pattern 21
21. Pronotum and hemelytra reddish .... ***rubrosignatus*** **Knight** (p. 133)
    Pronotum and hemelytra green .... ***elisus*** **Van Duzee** (p. 133)

Lygus convexicollis Reuter

Fig. 71; Map 33

Lygus convexicollis Reuter, 1875*c*:72.
Lygus dolichorhynchus Stanger, 1942:166.

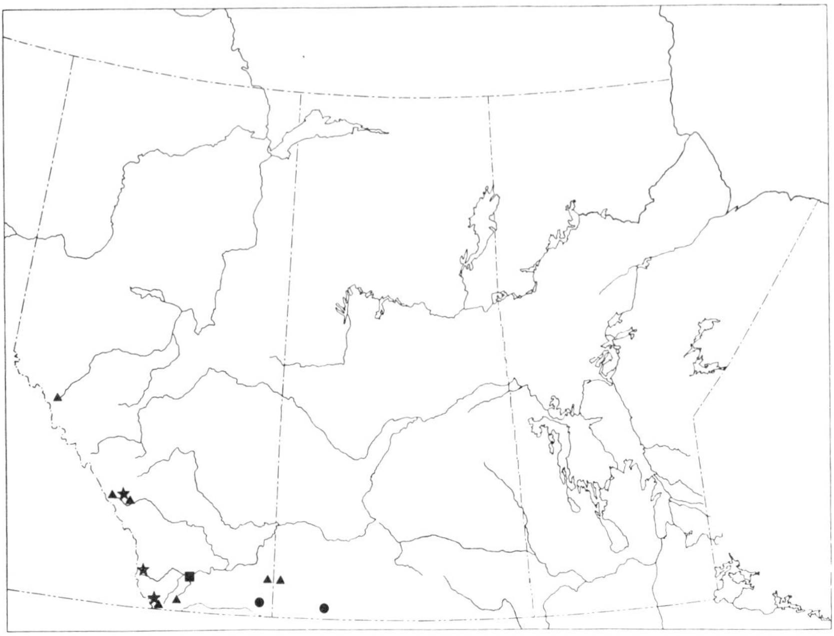

Map 33. Collection localities for *Lygus convexicollis* (●), *L. nigropallidus* (▲), *L. nubilus* (■), and *L. humeralis* (★).

Length 6.30–7.14 mm; width 2.80–3.29 mm. Head yellowish brown, frons striate. Rostrum 3.08–3.99 mm long. Pronotum reddish brown, anterior angles prominent. Mesoscutum black. Hemelytra reddish brown, pubescence long, dense (Fig. 71).

Remarks. This species has the longest rostrum.

Habitat. Collected on herbaceous plants.

Distribution. Western States, British Columbia; Alberta, Saskatchewan (Map 33).

Lygus nigropallidus Knight

Fig. 72; Map 33

Lygus nigropallidus Knight, 1917*a*:579.

Length 5.18–6.02 mm; width 2.52–2.94 mm. Head dull yellowish brown, frons striate. Rostrum 1.75–1.96 mm long. Pronotum dull green. Mesoscutum black. Hemelytra dull green, pubescence short, closely appressed (Fig. 72).

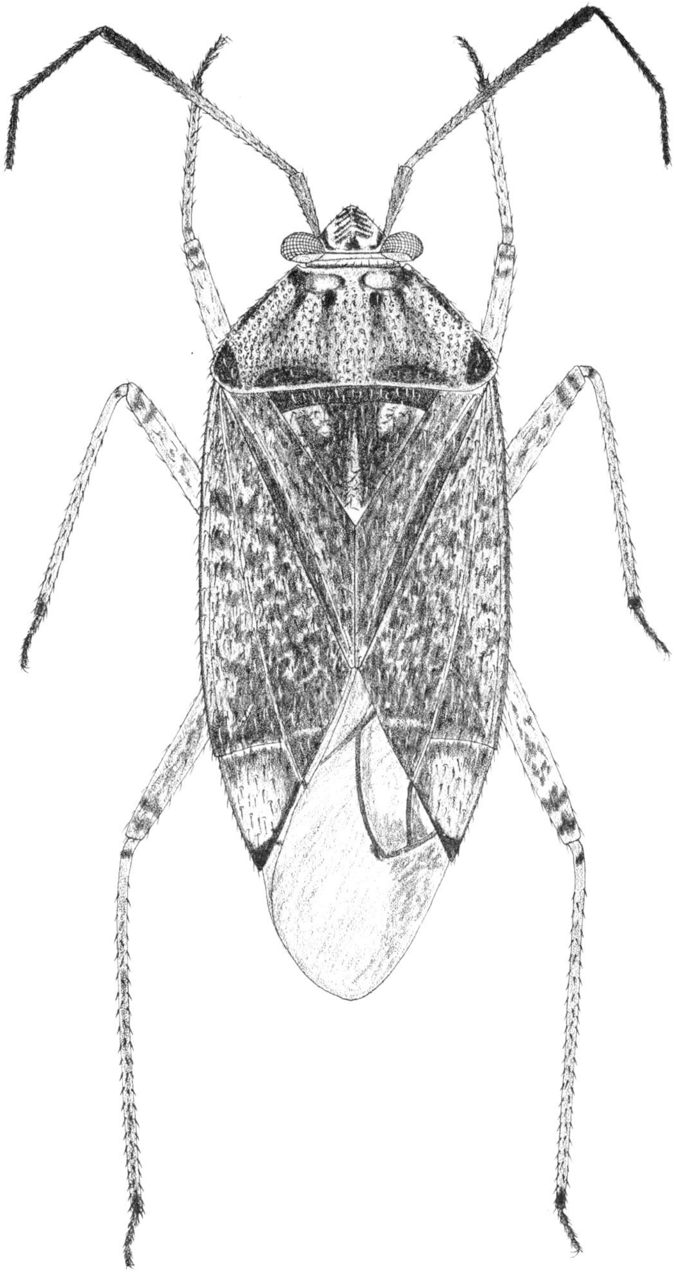

Fig. 71. *Lygus convexicollis*

Remarks. This species is distinguished by the striate frons, by the short rostrum, and by the short pubescence on the hemelytra.

Habitat. Collected on *Lupinus argenteus* and *Oenothera biennis* var. *canescens*.

Distribution. Western States, British Columbia; Alberta, Saskatchewan (Map 33).

Lygus nubilus Van Duzee

Fig. 73; Map 33

Lygus distinguendus var. *nubilus* Van Duzee, 1914:20.
Lygus nubilus Van Duzee, 1917:350.
Lygus ultranubilus Knight, 1917*b*:583.
Lygus epelys Hussey, 1954:196.

Length 4.34–5.50 mm; width 1.89–2.38 mm. Head yellowish brown, frons striate. Calli pubescent. Rostrum 1.54–2.10 mm long. Pronotum yellowish brown. Mesoscutum yellowish or light reddish. Hemelytra yellowish mottled with dark brown, pubescence long, dense (Fig. 73).

Remarks. This species is distinguished by the small size, short rostrum, pubescent calli, and striate frons.

Habitat. Collected on *Sambucus* spp.

Distribution. Western States, Connecticut, Michigan, British Columbia; Alberta (Map 33).

Lygus humeralis Knight

Fig. 74; Map 33

Lygus humeralis Knight, 1917*a*:570.
Lygus ceanothus var. *deleticus* Knight, 1941*b*:270.

Length 6.16–6.65 mm; width 2.66–3.95 mm. Head dark reddish brown, frons striate. Rostrum 2.59–2.83 mm long. Pronotum reddish brown to black, anterior angles prominent. Mesoscutum black, lateral areas red or yellow. Hemelytra dark reddish brown to sepia brown, pubescence yellowish, long, dense, hemelytra appears mottled (Fig. 74).

Remarks. This species is distinguished by the striate frons, the prominent anterior angles of pronotum, and the mottled appearance of the pubescence.

Habitat. Collected on *Ceanothus* spp.

Distribution. Western States, British Columbia; Alberta (Map 33).

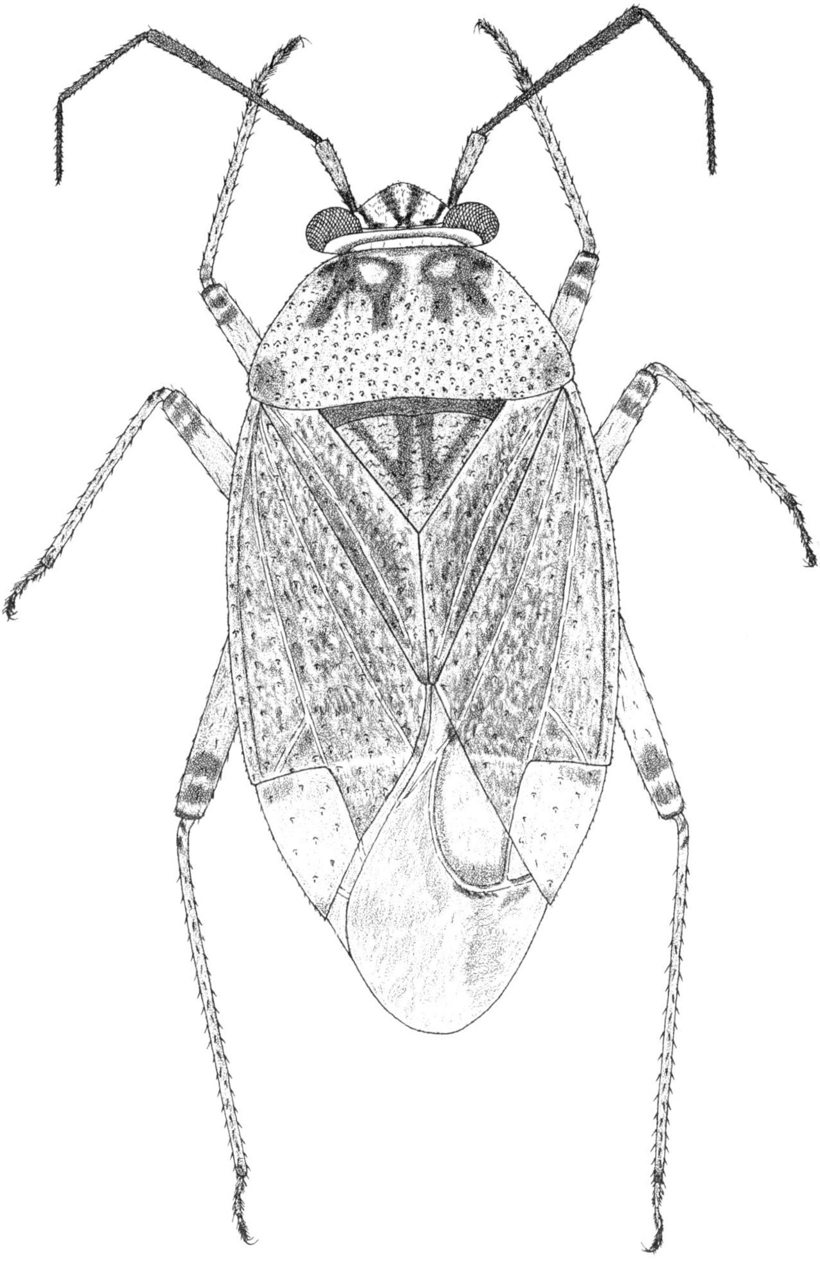

Fig. 72. *Lygus nigropallidus*

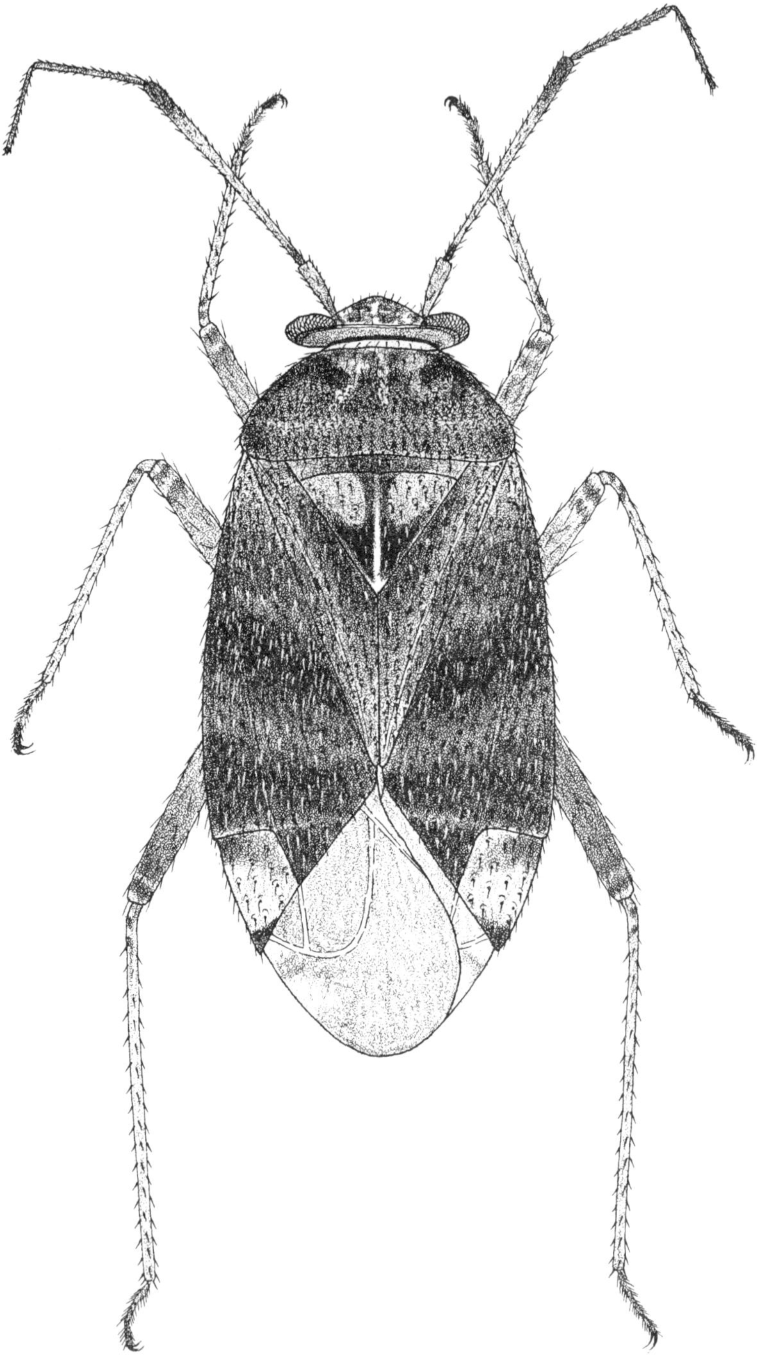

Fig. 73. *Lygus nubilus*

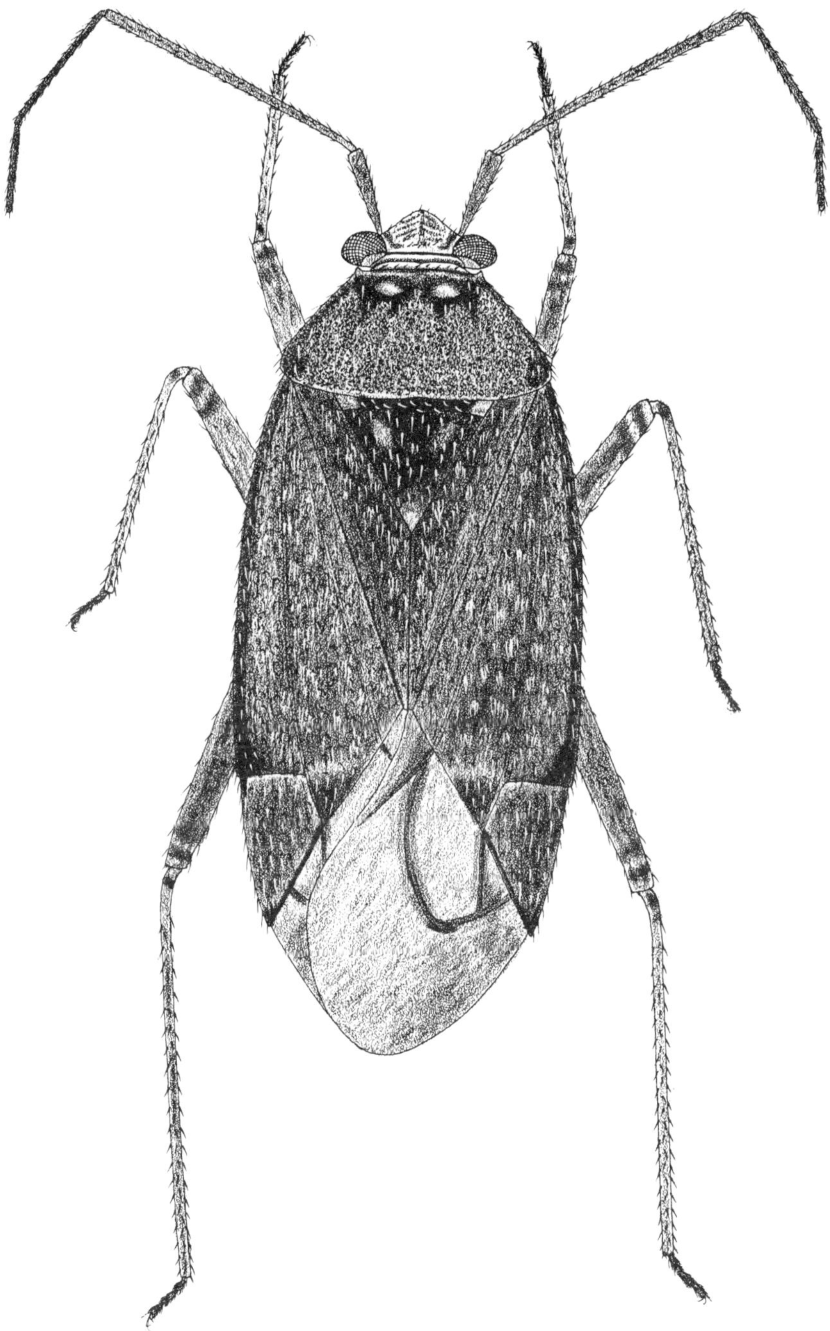

Fig. 74. *Lygus humeralis*

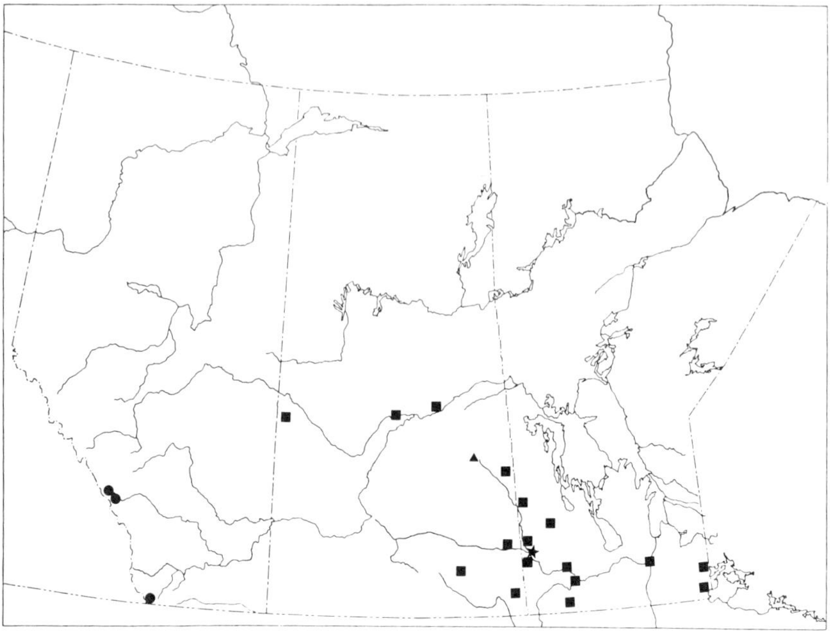

Map 34. Collection localities for *Lygus perplexus* (●), *L. atritibialis* (★), and *L. vanduzeei* (■).

Lygus perplexus Stanger

Fig. 75; Map 34

Lygus perplexus Stanger, 1942:163.

Length 5.04–5.60 mm; width 2.38–2.66 mm. Head yellowish brown, frons striate. Anterior angles prominent. Rostrum 2.31–2.80 mm long. Mesoscutum black. Hemelytra dull greenish brown, densely mottled with darker brown, pubescence moderately long, dense (Fig. 75).

Remarks. This species is distinguished by the striate frons, by the prominent anterior angles of pronotum, and by the black mesoscutum.

Habitat. Collected on herbaceous plants in alpine meadows.

Distribution. Alaska, Western States, Yukon, British Columbia; Alberta (Map 34).

Lygus atritibialis Knight

Fig. 76; Map 34

Lygus atritibialis Knight, 1941*c*:152.

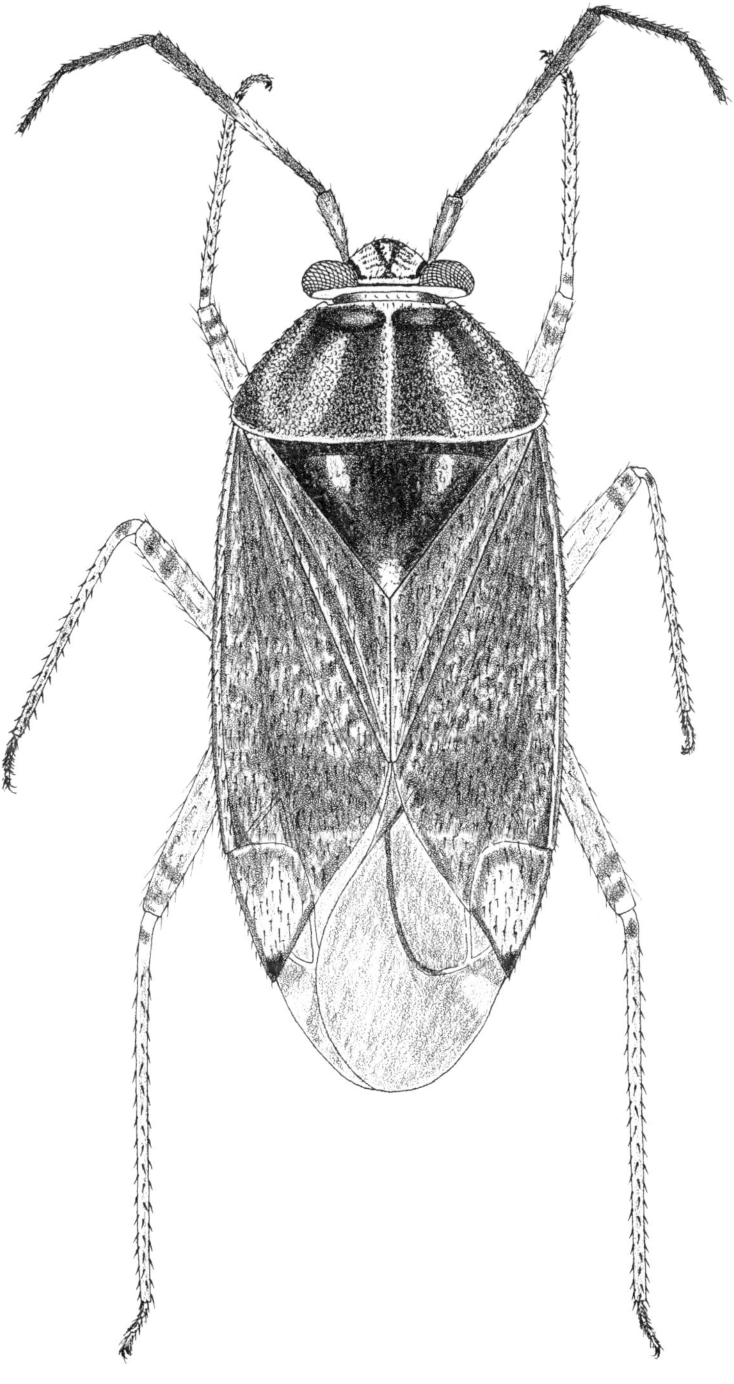

Fig. 75. *Lygus perplexus*

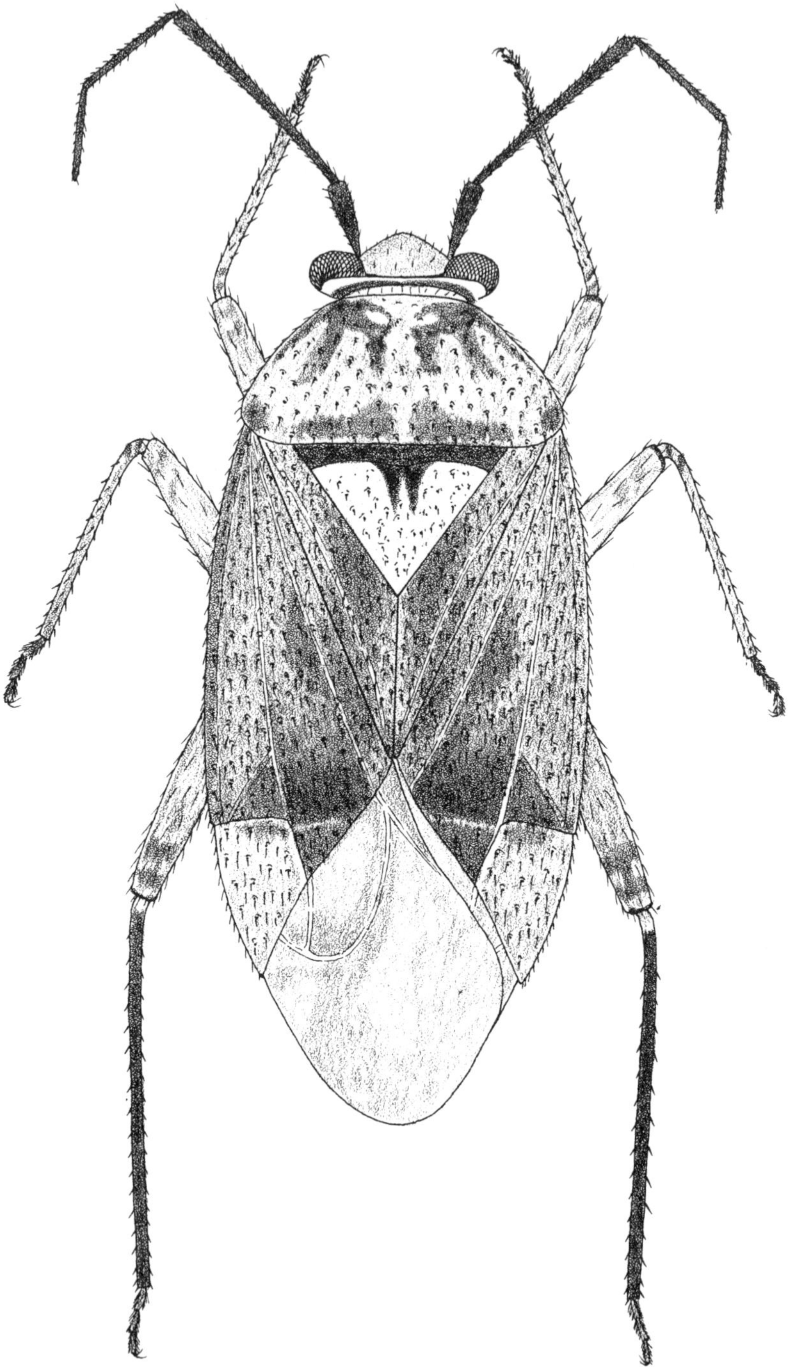

Fig. 76. *Lygus atritibialis*

Length 5.25–5.88 mm; width 2.45–2.73 mm. Head yellow, frons smooth. Rostrum 2.17–2.45 mm long. Pronotum yellow, anterior angles rounded. Mesoscutum black. Hemelytra yellowish, apical area of corium brown, pubescence long, dense. Hind tibia black in adults and reddish brown in teneral specimens (Fig. 76).

Remarks. This species is distinguished by the black, or in teneral specimens reddish brown hind tibia.

Habitat. Collected on *Artemisia frigida.*

Distribution. North Central States, British Columbia; Alberta, Saskatchewan (Map 34).

Lygus vanduzeei Knight

Fig. 77; Map 34

Lygus vanduzeei Knight, 1917*a*:565.

Length 5.74–7.14 mm; width 3.08–3.50 mm. Head brown or reddish, frons smooth, carina almost obsolete. Rostrum 2.66–2.94 mm long. Pronotum yellowish brown, anterior angles rounded. Mesoscutum yellowish brown. Hemelytra mostly brown, pubescence short, sparse. Front and middle tibiae usually with longitudinal reddish stripes (Fig. 77).

Remarks. This species is readily distinguished by the large size, by the reduced carina on the head, and by the long rostrum. The reddish stripes on the front and middle tibiae are pronounced.

Habitat. Collected on *Solidago* spp.

Distribution. North Central and eastern States, Eastern Canada; Manitoba, Saskatchewan (Map 34).

Lygus potentillae Kelton

Fig. 78; Map 35

Lygus potentillae Kelton, 1973:1545.

Length 5.60–5.95 mm; width 2.52–2.94 mm. Head reddish, frons smooth. Rostrum 2.87–3.15 mm long. Pronotum light reddish brown, anterior angles prominent. Mesoscutum black. Hemelytra reddish brown, pubescence short, sparse (Fig. 78).

Remarks. This is the only species with a smooth frons and prominent carina to have a long rostrum.

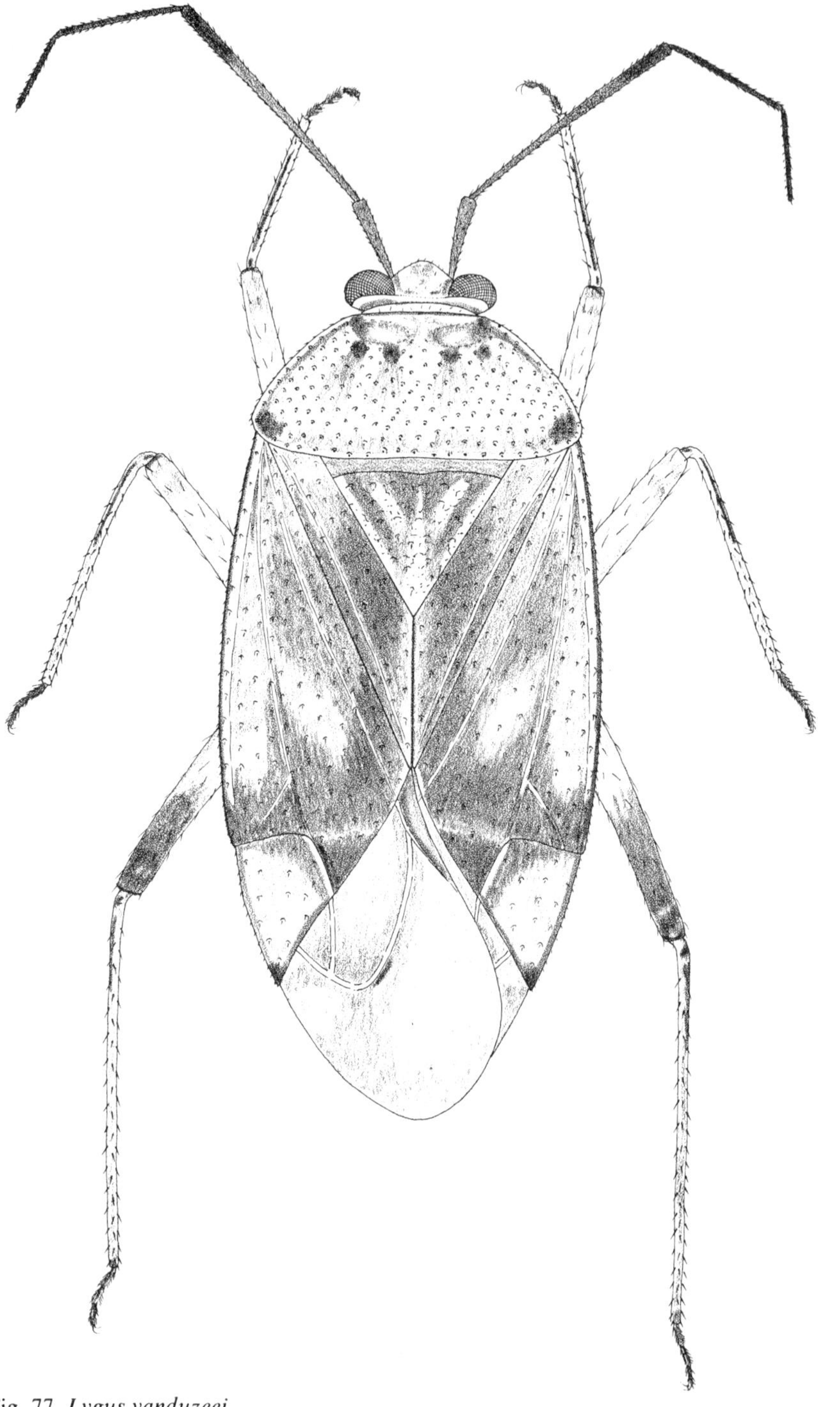

Fig. 77. *Lygus vanduzeei*

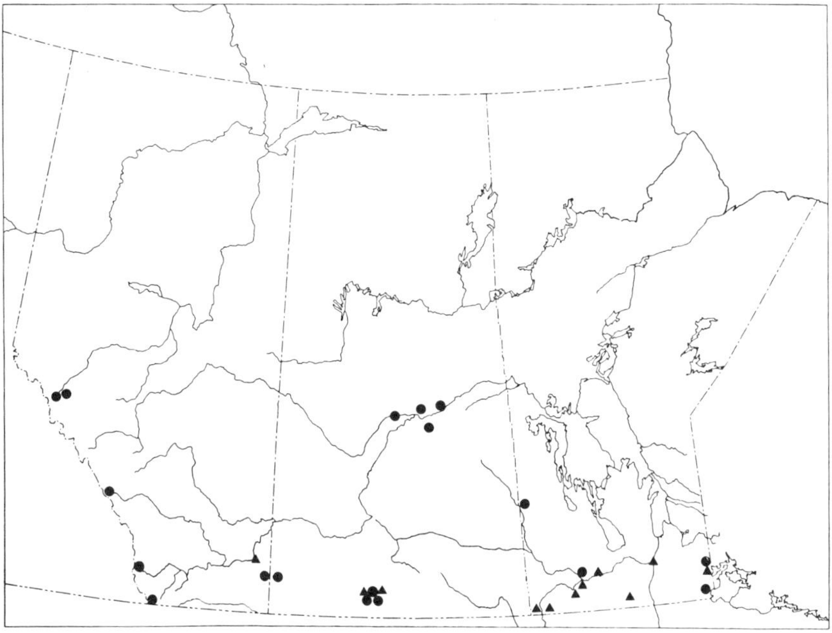

Map 35. Collection localities for *Lygus potentillae* (●) and *L. plagiatus* (▲).

Habitat. Collected on *Potentilla fruticosa.*

Distribution. Alaska, West Central States, northern and Eastern Canada, British Columbia; Prairie Provinces (Map 35).

Lygus plagiatus Uhler

Figs. 48, 79; Map 35

Lygus plagiatus Uhler, 1895:35.

Length 5.03–5.95 mm; width 2.52–2.94 mm. Head light brown, frons smooth with black submedian lines. Rostrum 1.96–2.17 mm long. Pronotum dull green to brown, anterior angles rounded. Mesoscutum black, lateral areas red. Hemelytra dull green mottled with brown or black, pubescence moderately long, dense, surface appears mottled (Fig. 79).

Remarks. This species is distinguished by the submedian lines on the smooth frons (Fig. 48), the dense pubescence, and the mottled appearance of the hemelytra.

Habitat. Collected on *Helianthus* spp.

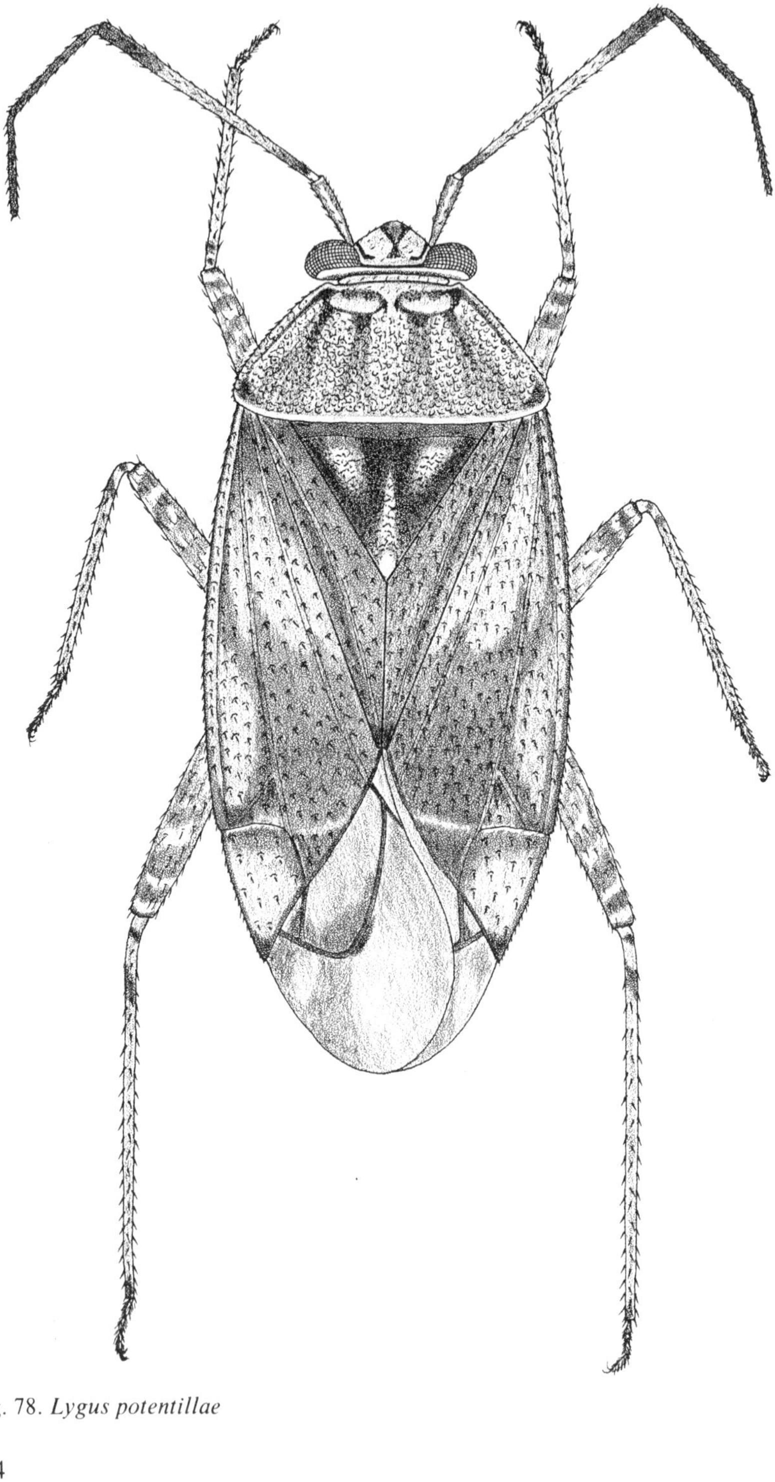

Fig. 78. *Lygus potentillae*

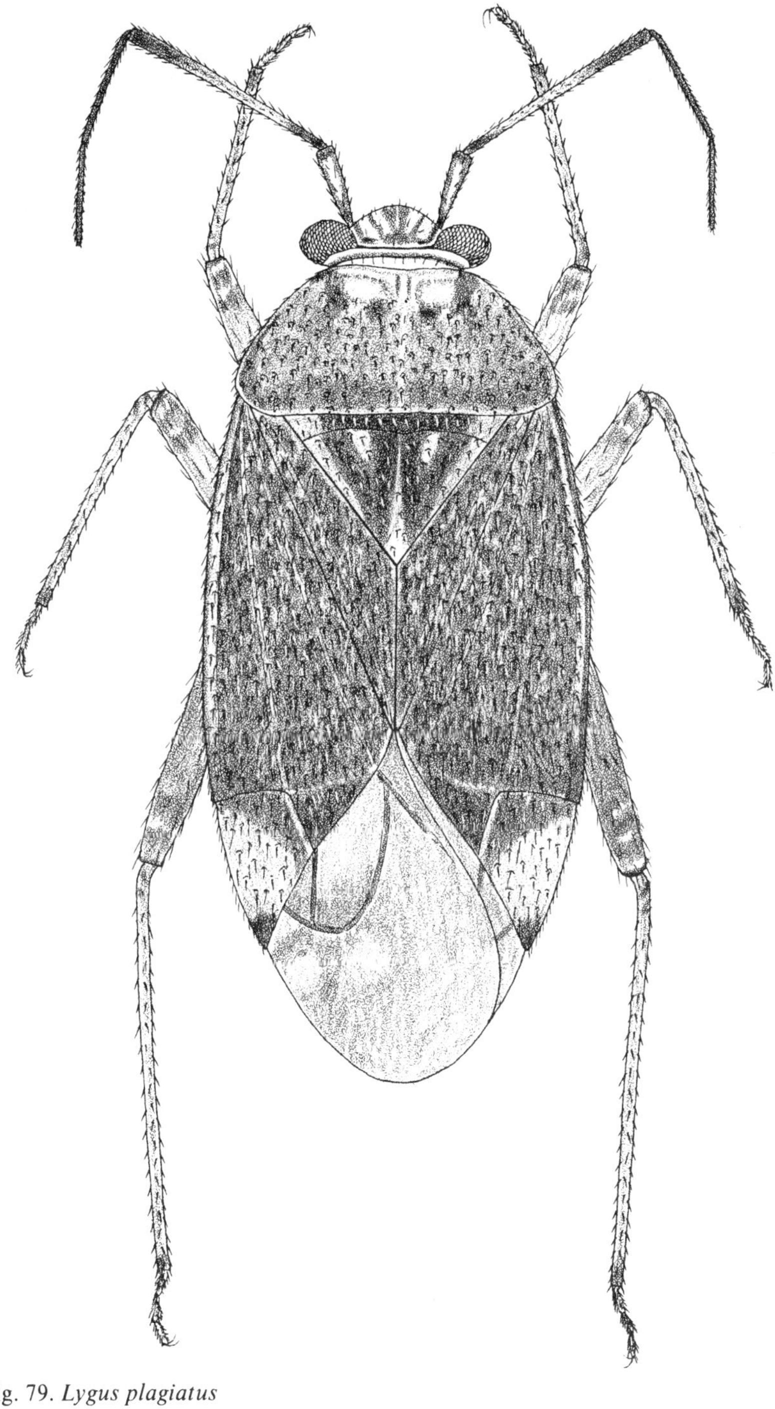

Fig. 79. *Lygus plagiatus*

Distribution. Northeastern and North Central States, Ontario, Quebec; Prairie Provinces (Map 35).

Lygus unctuosus (Kelton)

Fig. 80; Map 36

Liocoris unctuosus Kelton, 1955*b*:486.
Lygus unctuosus: Carvalho, 1959:156.

Length 5.25–5.95 mm; width 2.52–2.94 mm. Head yellowish brown, frons smooth with black submedian lines. Rostrum 2.10–2.24 mm long. Pronotum yellowish green, anterior angles rounded. Mesoscutum black. Hemelytra greenish brown irregularly mottled with fuscous or black, pubescence short, sparse (Fig. 80).

Remarks. This species is distinguished by the dull greenish color, the mottled appearance, the short pubescence, and the black mesoscutum.

Habitat. Collected on *Matricaria matricarioides, Medicago sativa*, and other herbaceous plants.

Distribution. Alaska, North Central States, Colorado, Northwest Territories, Ontario, Quebec, New Brunswick; Prairie Provinces (Map 36).

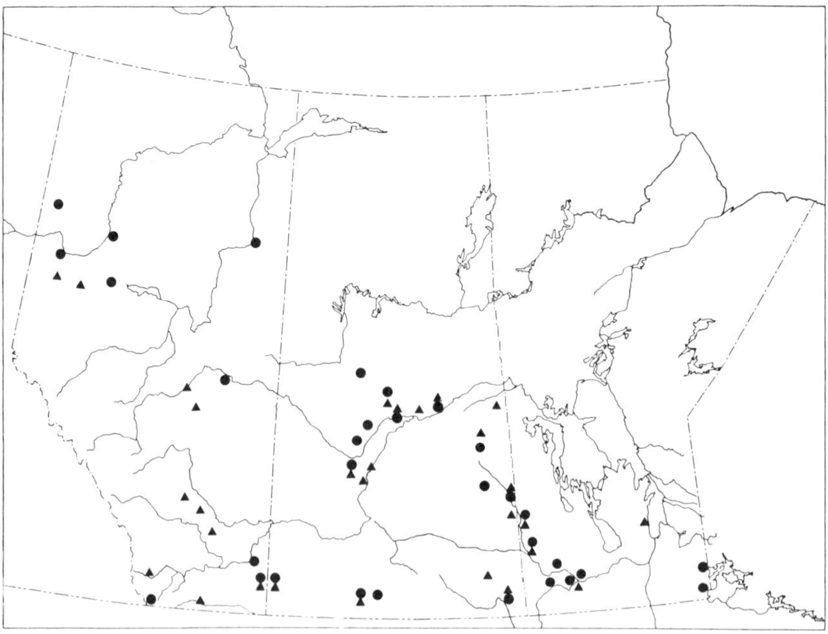

Map 36. Collection localities for *Lygus unctuosus* (●) and *L. rufidorsus* (▲).

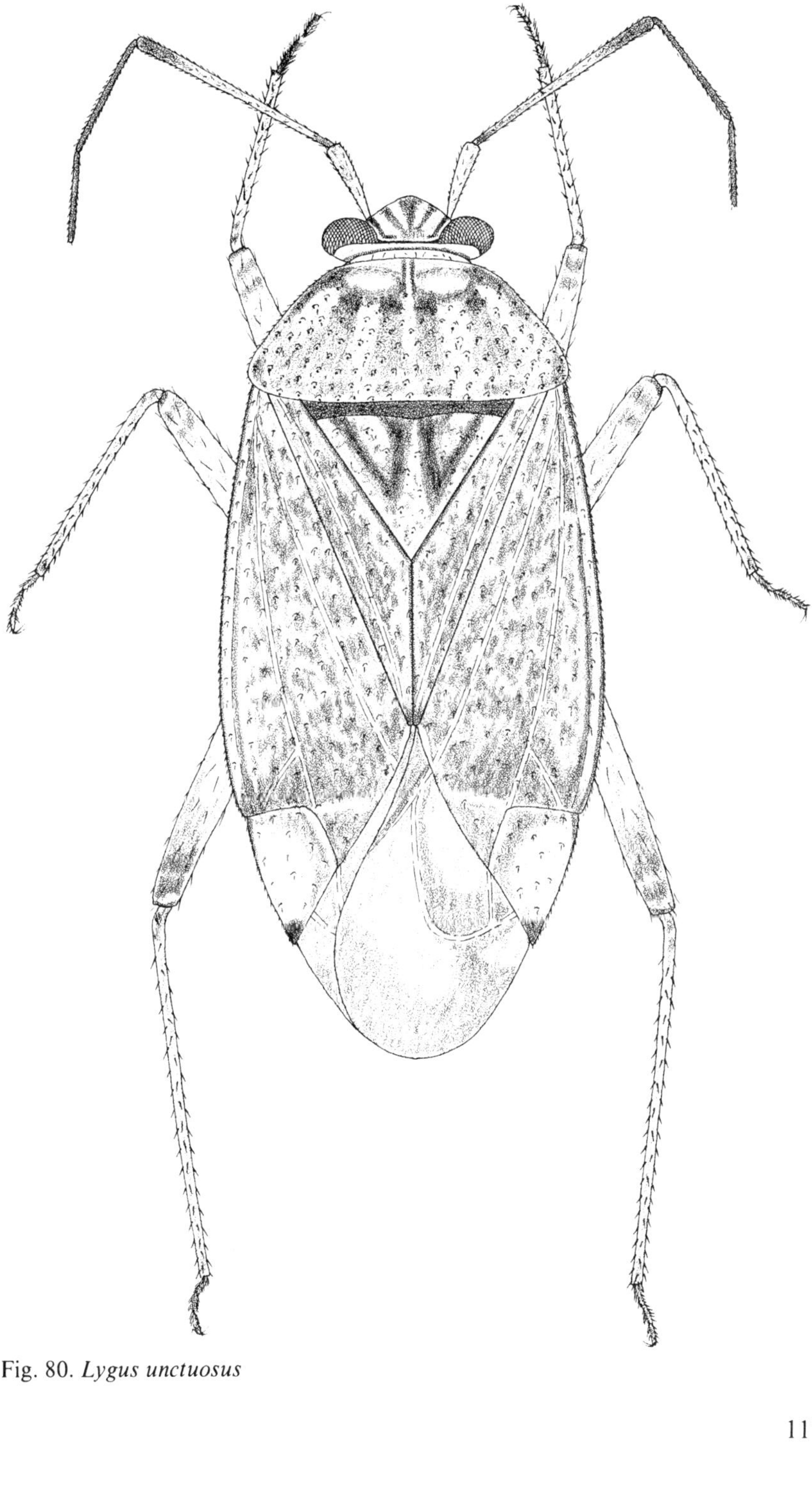

Fig. 80. *Lygus unctuosus*

Lygus rufidorsus (Kelton)

Fig. 81; Map 36

Liocoris rufidorsus Kelton, 1955*b*:484.
Lygus rufidorsus: Carvalho, 1959:155.

Length 5.60–6.30 mm; width 2.59–3.01 mm. Head yellowish brown, frons smooth with reddish brown submedian lines. Rostrum 2.10–2.38 mm long. Pronotum yellowish brown, anterior angles rounded. Mesoscutum black, lateral areas reddish. Hemelytra shiny, reddish brown, pubescence short, sparse (Fig. 81).

Remarks. This species is distinguished by the submedian lines on the frons, by the reddish lateral areas on the mesoscutum, by the shiny, reddish brown hemelytra, and by the short pubescence.

Habitat. Collected on a great variety of plants, and is a pest of alfalfa.

Distribution. West Central States, New Hampshire, Northwest Territories, Eastern Canada, British Columbia; Prairie Provinces (Map 36).

Lygus lineolaris (Palisot de Beauvois)

Fig. 82; Map 37

Capsus lineolaris Palisot de Beauvois, 1818:187.
Lygus oblineatus Say, 1832:21.
Lygus lineolaris: Uhler, 1872:413.
Capsus flavonotatus Provancher, 1872:103.

Length 4.90–5.95 mm; width 2.38–3.01 mm. Head yellowish brown, frons smooth with black submedian lines. Rostrum 2.17–2.52 mm long. Pronotum yellowish brown to reddish brown, anterior angles rounded. Mesoscutum black, lateral areas pale or reddish. Hemelytra reddish brown, pubescence moderately long, dense (Fig. 82).

Remarks. This species is distinguished by the submedian lines on the frons, by the pale or reddish lateral areas on the mesoscutum, and by the dense pubescence on the hemelytra that is not mottled in appearance.

Habitat. Collected on a great variety of plants, and is a pest of alfalfa.

Distribution. Widespread in North America; Prairie Provinces (Map 37).

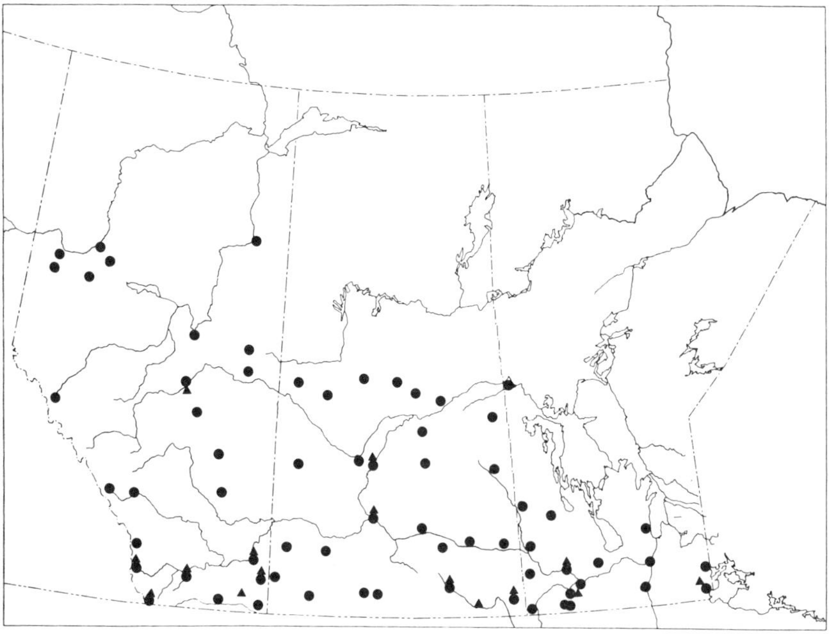

Map 37. Collection localities for *Lygus lineolaris* (●) and *L. desertinus* (▲).

Lygus desertinus Knight

Fig. 83; Map 37

Lygus desertus Knight, 1944:471.
Lygus desertinus Knight, 1968*a*:189.

Length 5.18–5.95 mm; width 2.24–2.66 mm. Head yellowish green, frons smooth. Rostrum 2.03–2.17 mm long. Pronotum yellowish green. Mesoscutum black, lateral margins yellow. Hemelytra pale green, pubescence moderately long, dense (Fig. 83).

Remarks. This species is distinguished by the green color and by the yellow margins on the mesoscutum.

Habitat. Collected on a great variety of plants, and is a pest of alfalfa.

Distribution. Western States, Northwest Territories, British Columbia; Prairie Provinces (Map 37).

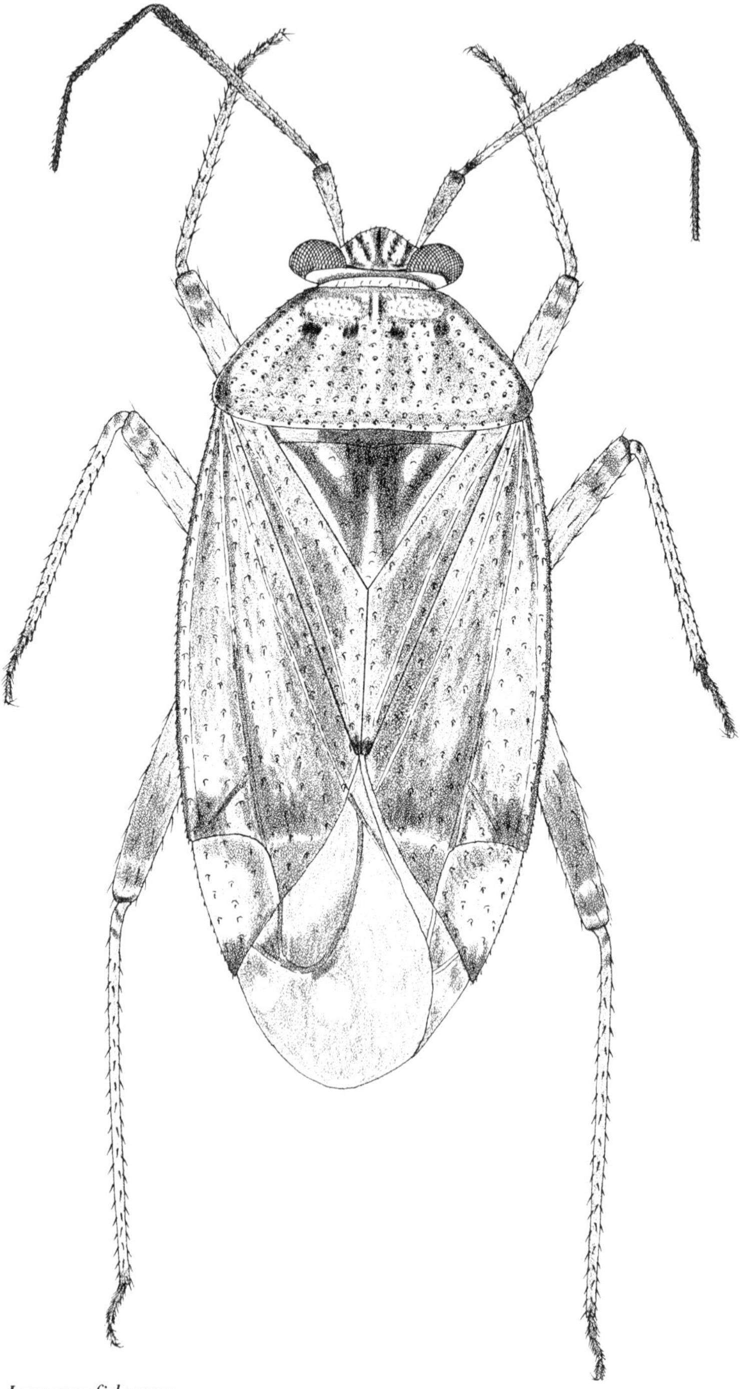

Fig. 81. *Lygus rufidorsus*

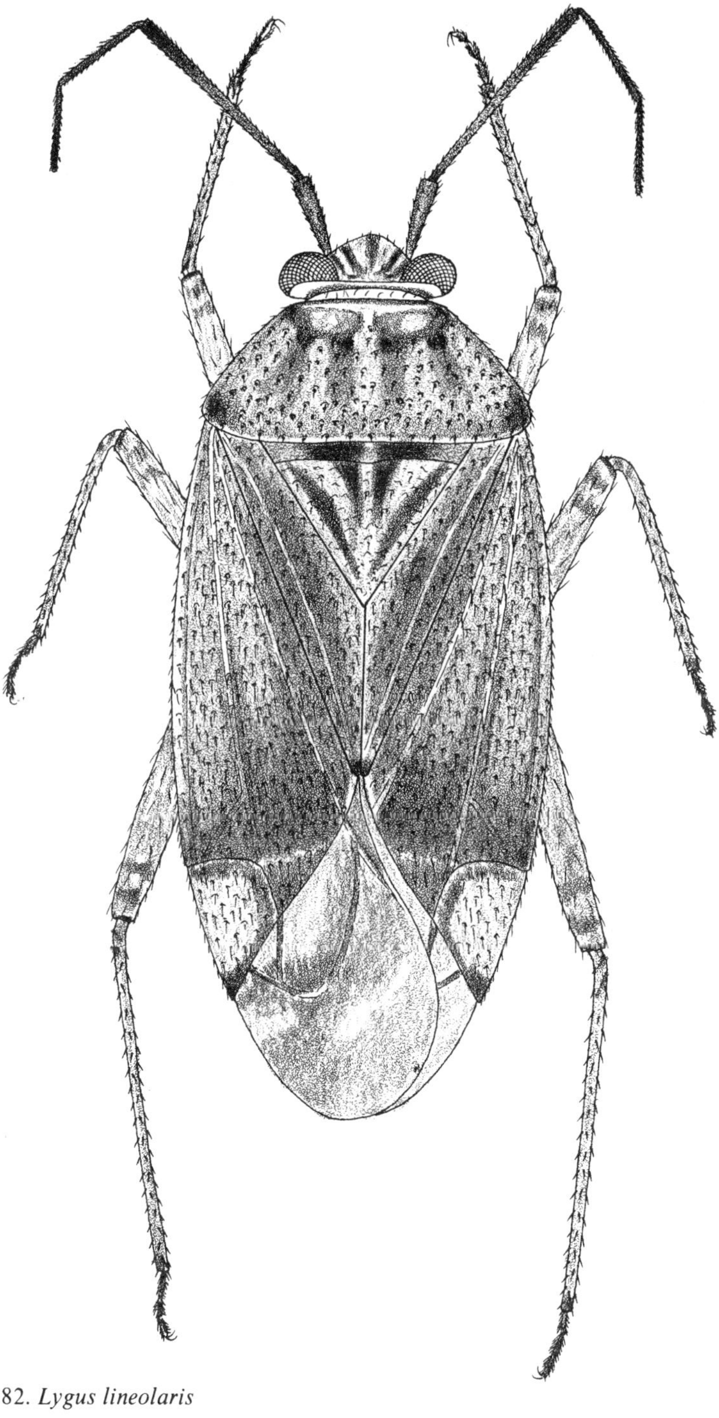

Fig. 82. *Lygus lineolaris*

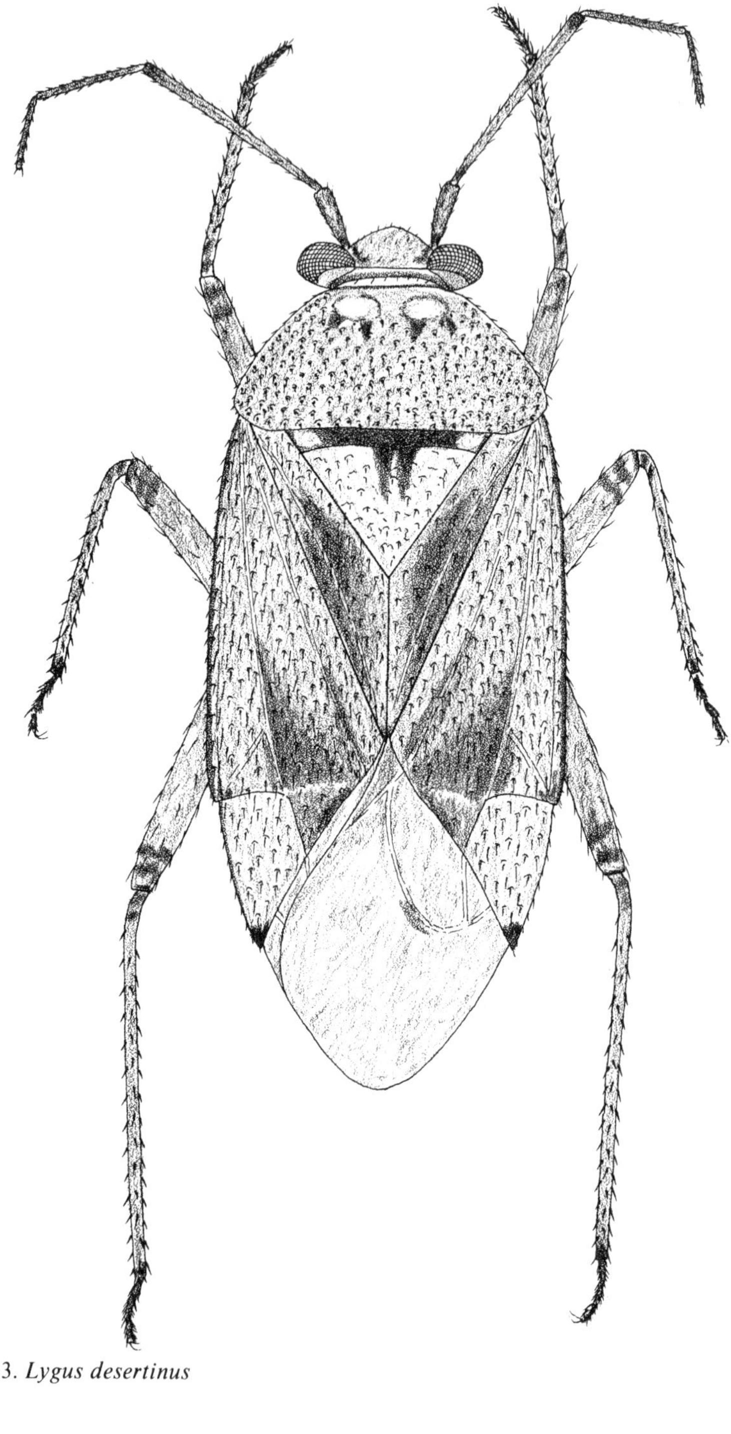

Fig. 83. *Lygus desertinus*

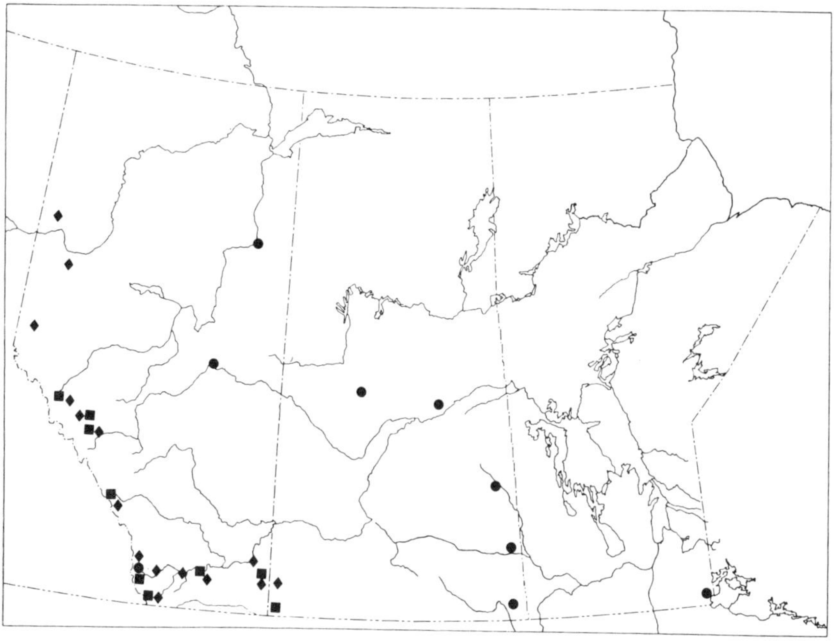

Map 38. Collection localities for *Lygus rubroclarus* (●), *L. varius* (◆), and *L. ravus* (■).

Lygus rubroclarus Knight

Fig. 84; Map 38

Lygus vanduzeei var. *rubroclarus* Knight, 1917*a*:567.
Lygus frisoni Knight, 1941*c*:151.
Lygus rubroclarus: Knight, 1953:518.

Length 5.50–6.60 mm; width 2.66–3.22 mm. Head light yellowish brown, frons smooth. Rostrum 2.38–2.66 mm long. Pronotum reddish brown, anterior angles rounded. Mesoscutum black, lateral margins yellow or red. Hemelytra reddish or yellowish brown, pubescence short, dense (Fig. 84).

Remarks. This species is distinguished by the robust size, reddish color, smooth frons, and the yellow or red lateral areas on the mesoscutum.

Habitat. Collected on a great variety of plants.

Distribution. Northeastern and North Central States, North Carolina, Eastern Canada, British Columbia; Prairie Provinces (Map 38).

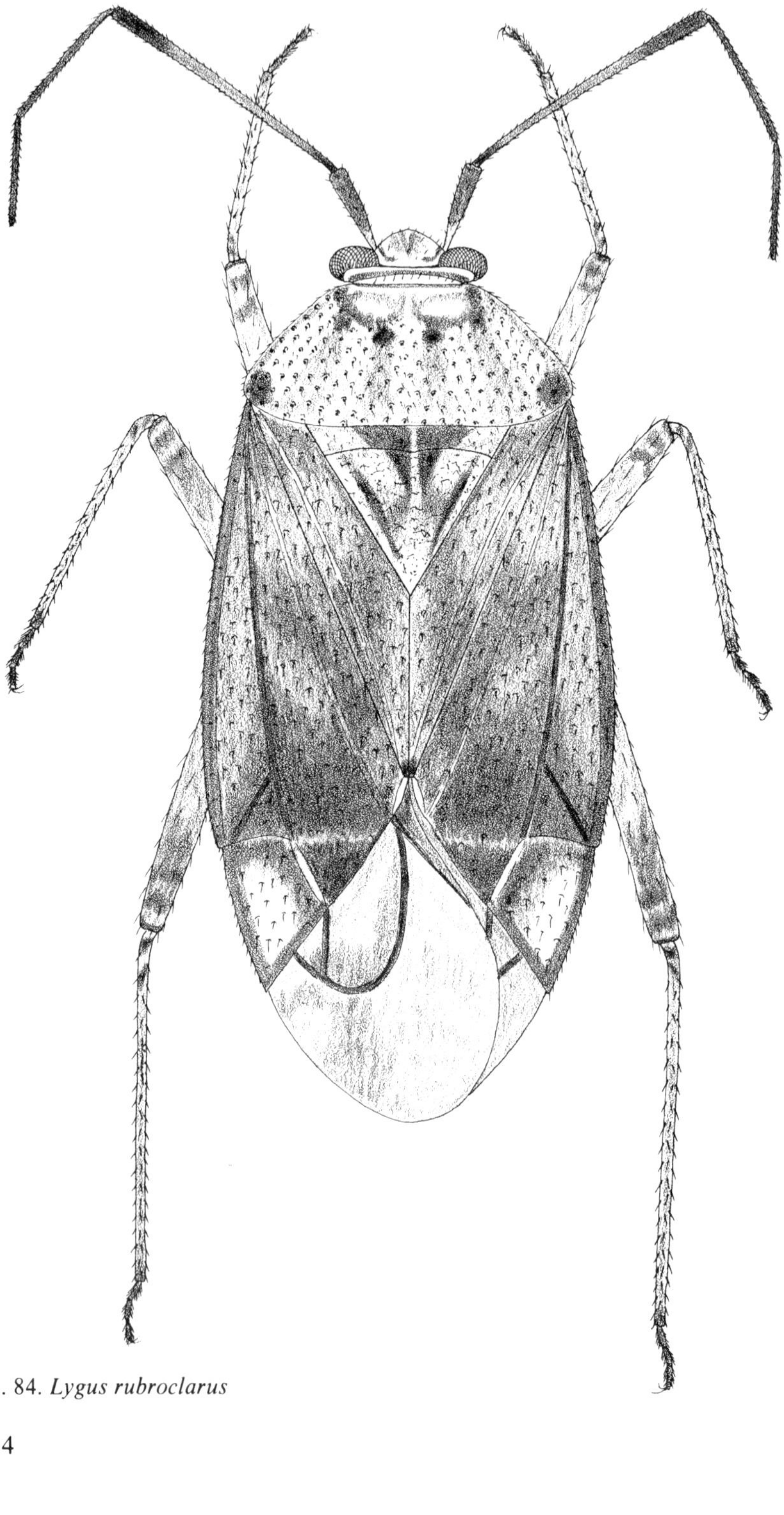

Fig. 84. *Lygus rubroclarus*

Lygus varius Knight

Fig. 85; Map 38

Lygus varius Knight, 1944:473.

Length 5.60–5.88 mm; width 2.73–3.08 mm. Head light yellowish brown, frons smooth. Rostrum 2.59–2.80 mm long. Pronotum dull green, anterior angles prominent. Mesoscutum black. Hemelytra dull green, often dark brown on apical half of corium, pubescence short, sparse (Fig. 85).

Remarks. This species may be confused with *potentillae*, because the two are similar in size and color, but is separated from it by the shorter rostrum.

Habitat. Collected on a great variety of plants.

Distribution. Alaska, western States, northern and Eastern Canada, British Columbia; Alberta and Saskatchewan (Map 38).

Lygus ravus Stanger

Fig. 86; Map 38

Lygus ravus Stanger, 1942:164.
Lygus nigritus Knight, 1944:472.

Length 5.25–5.74 mm; width 2.52–2.80 mm. Head yellowish brown, frons smooth. Rostrum 2.24–2.52 mm long. Pronotum dull green, anterior angles prominent. Mesoscutum black. Hemelytra dull green, strongly marked with brown or black, pubescence short, sparse (Fig. 86).

Remarks. This species resembles *varius* in color, markings, and pubescence. It is separated from *varius* by the shorter rostrum.

Habitat. Collected on a great variety of plants.

Distribution. Alaska, western States, northern and Eastern Canada, British Columbia; Prairie Provinces (Map 38).

Lygus borealis (Kelton)

Fig. 87; Map 39

Liocoris borealis Kelton, 1955*b*:488.
Lygus borealis: Carvalho, 1959:148.

Length 5.35–5.95 mm; width 2.52–2.94 mm. Head yellowish green, frons smooth. Rostrum 2.24–2.38 mm long. Pronotum yellowish green,

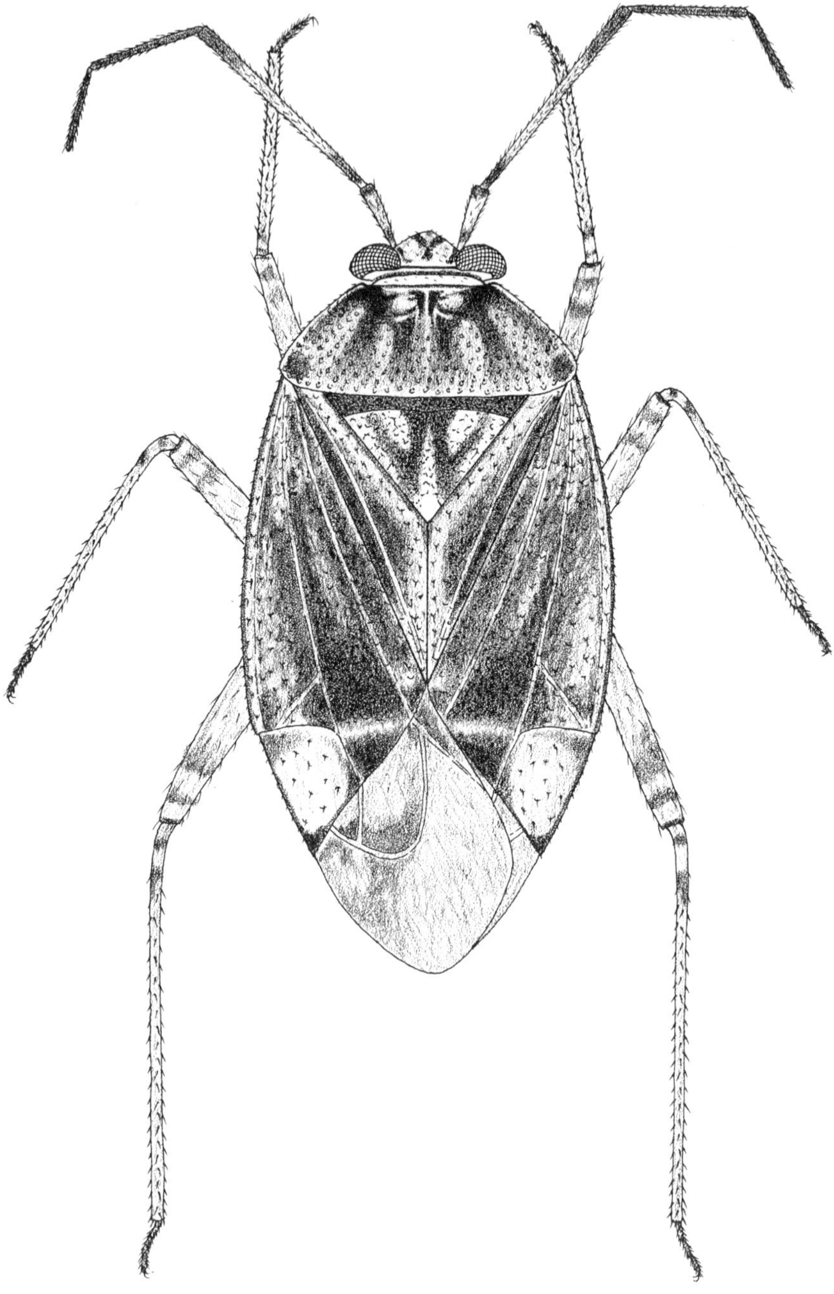

Fig. 85. *Lygus varius*

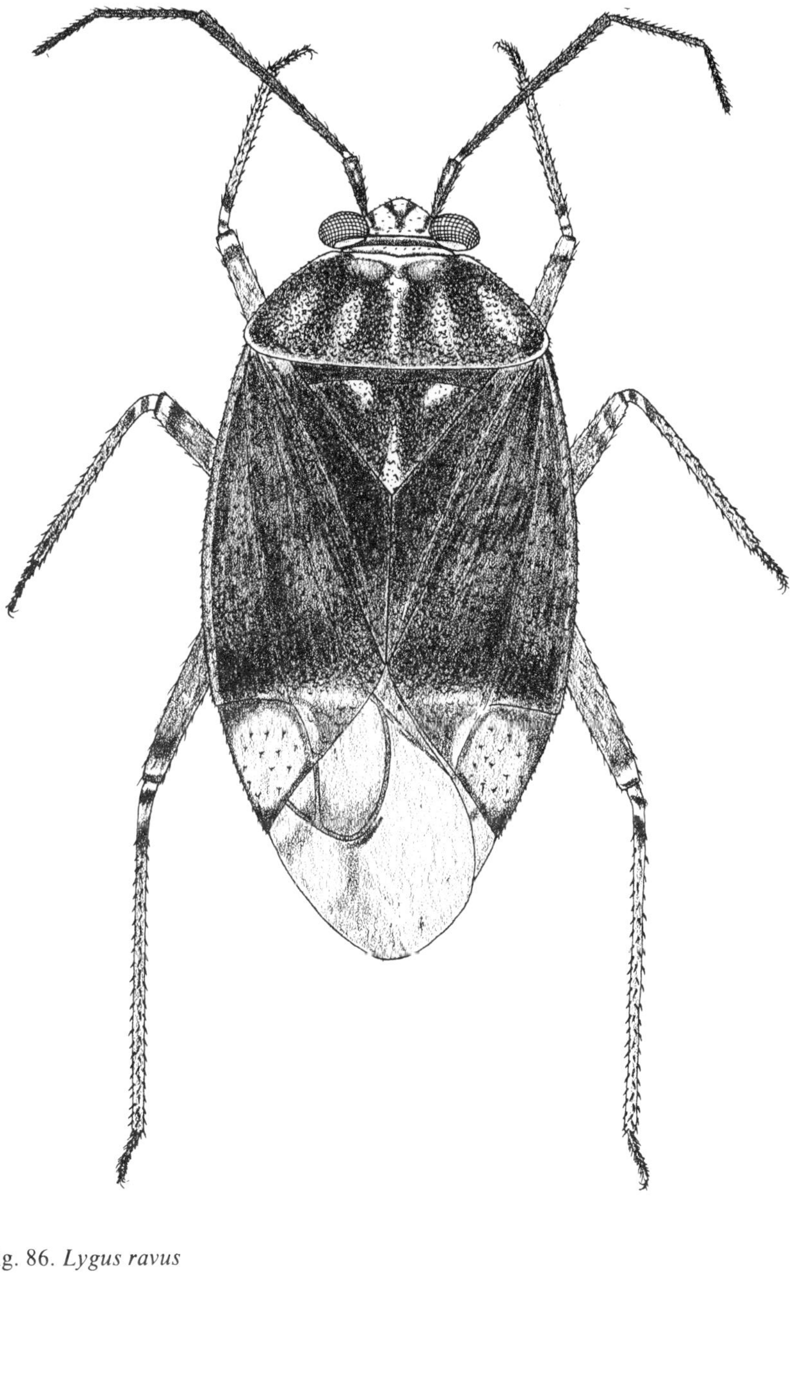

Fig. 86. *Lygus ravus*

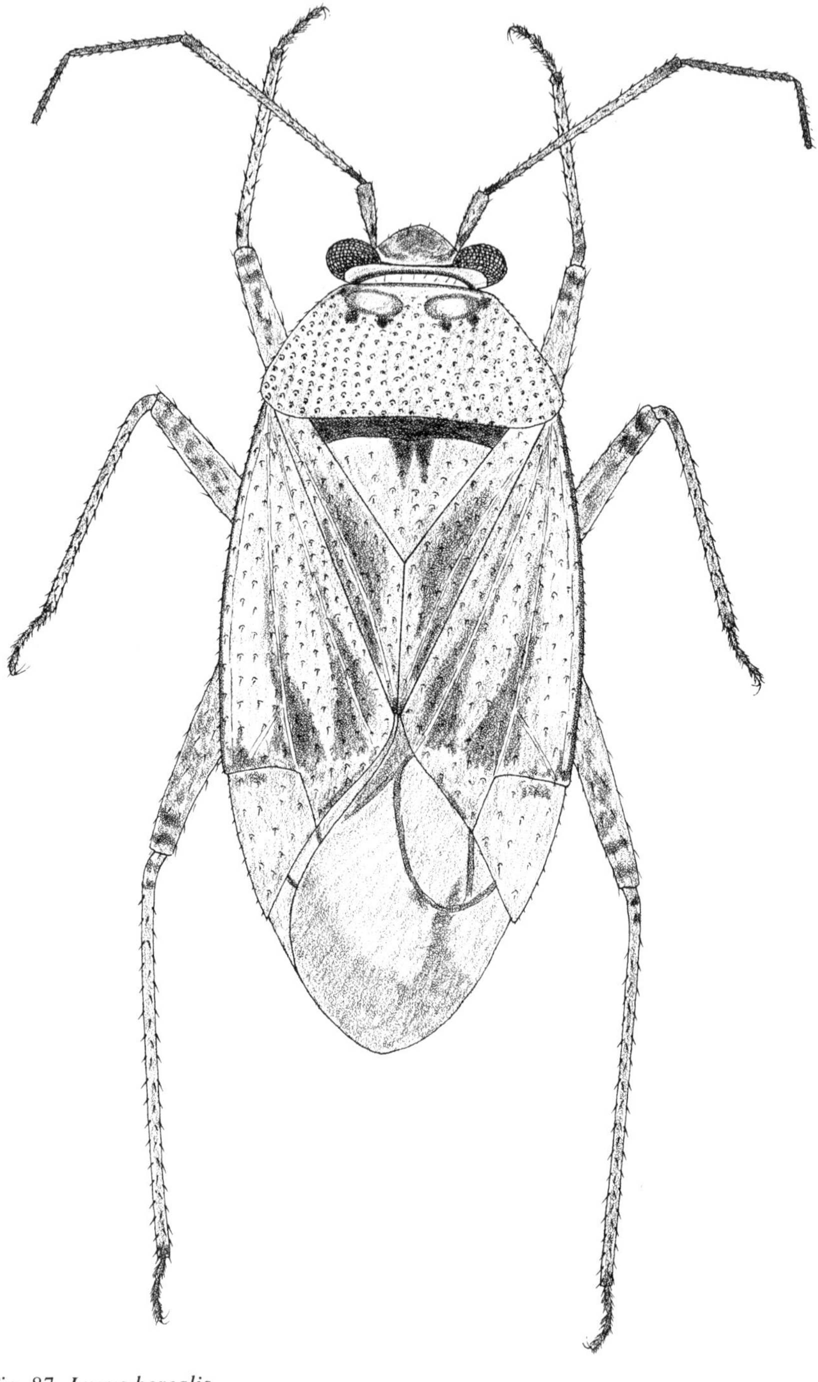

Fig. 87. *Lygus borealis*

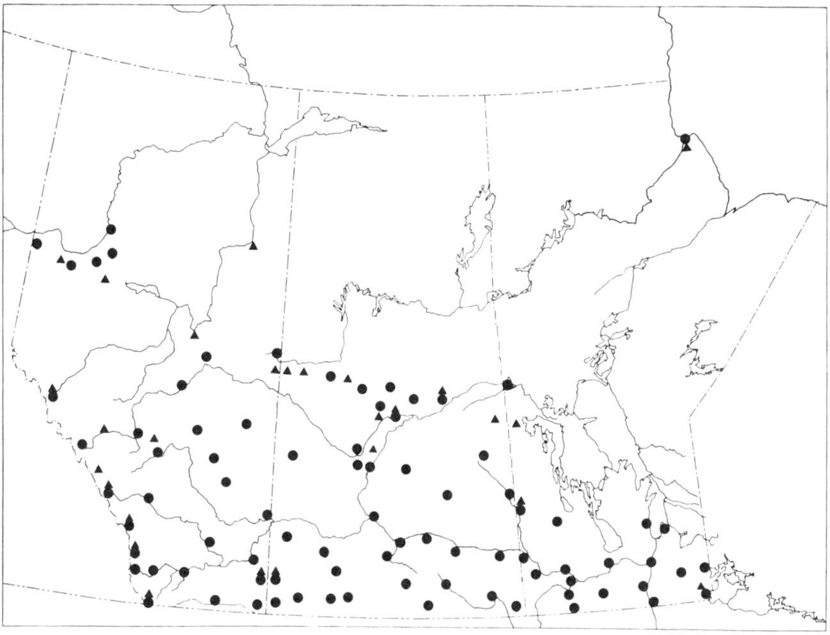

Map 39. Collection localities for *Lygus borealis* (●) and *L. columbiensis* (▲).

anterior angles rounded. Mesoscutum black. Hemelytra pale green, pubescence short, sparse; cuneus green (Fig. 187).

Remarks. This species is distinguished from the other green species by the short and sparse pubescence on the hemelytra, and usually by the uniformly green cuneus.

Habitat. Collected on *Medicago sativa* and other plants.

Distribution. Alaska, North Central and northwestern States, Eastern Canada, British Columbia; Prairie Provinces (Map 39).

Lygus columbiensis Knight

Fig. 88; Map 39

Lygus columbiensis Knight, 1917*a*:571.
Lygus superiorensis Knight, 1917*b*:572.

Length 4.75–6.72 mm; width 2.52–3.01 mm. Head reddish brown, frons smooth. Rostrum 2.38–2.66 mm long. Pronotum reddish brown, anterior angles rounded. Mesoscutum black. Hemelytra dark reddish brown, pubescence short, sparse (Fig. 88).

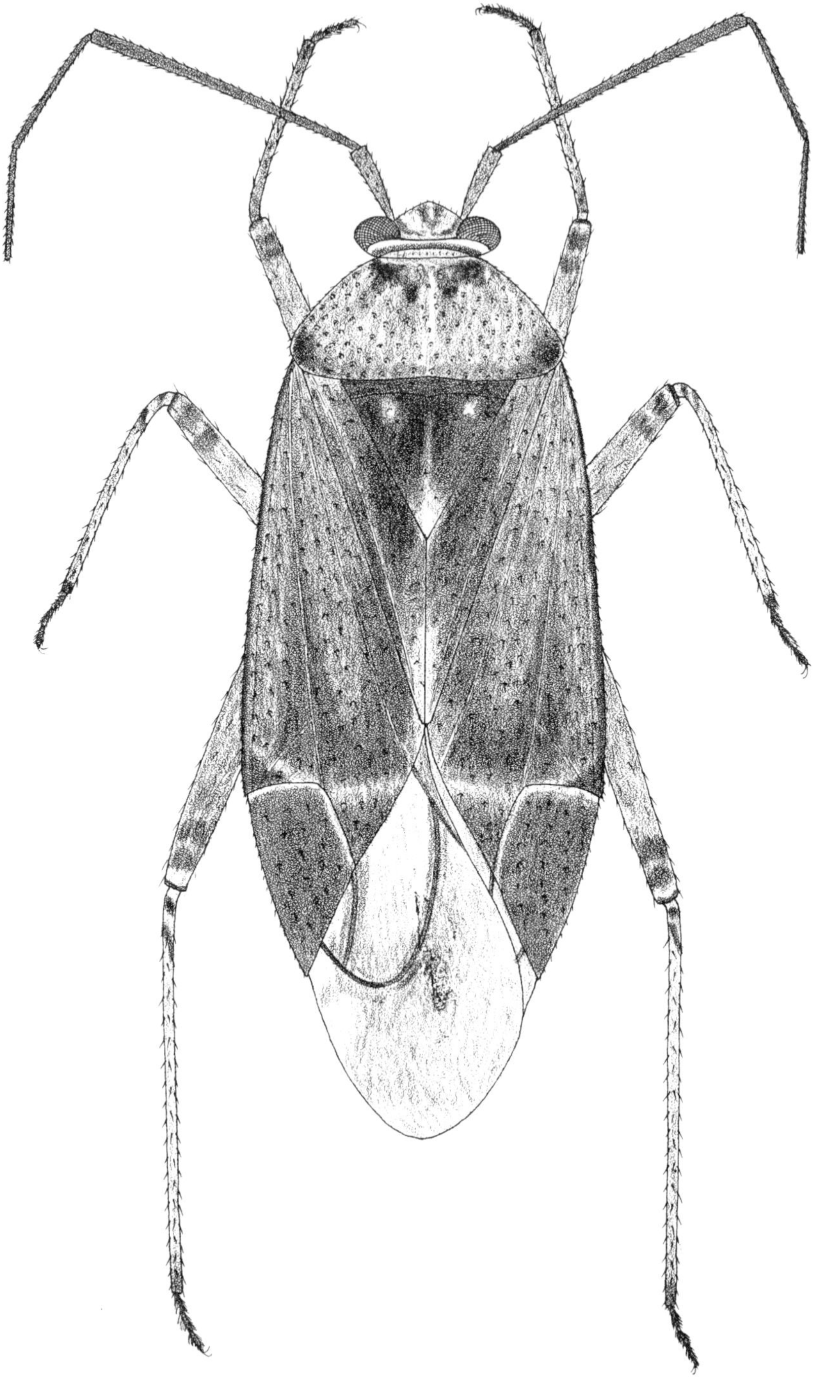

Fig. 88. *Lygus columbiensis*

Remarks. The males of this species are long and narrow, while the females are shorter and more robust.

Habitat. Collected on a great variety of plants.

Distribution. Alaska, northwestern States, Michigan, northern and Eastern Canada, British Columbia; Prairie Provinces (Map 39).

Lygus shulli Knight

Fig. 89; Map 40

Lygus shulli Knight, 1941*b*:272.

Length 5.46–6.02 mm; width 2.66–3.01 mm. Head light yellowish brown, frons smooth. Rostrum 2.38–2.80 mm long. Pronotum yellowish or brownish green, anterior angles rounded. Mesoscutum black. Hemelytra greenish yellow marked with dark brown, pubescence short, sparse (Fig. 89).

Remarks. This species is similar to *hesperus* in size, color, and general appearance, but has shorter and sparser pubescence on the hemelytra.

Habitat. Collected on a great variety of plants.

Distribution. Western States, British Columbia; Alberta and Saskatchewan (Map 40).

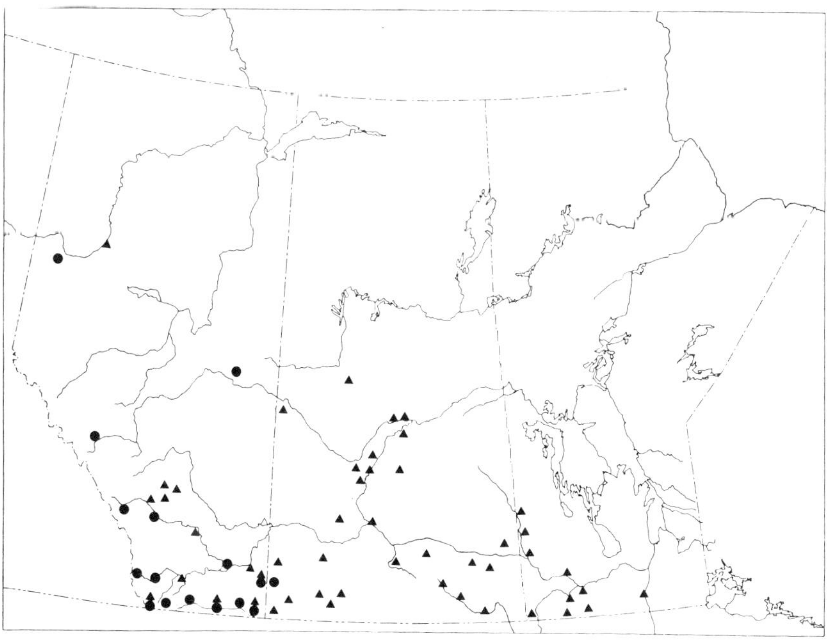

Map 40. Collection localities for *Lygus shulli* (●) and *L. solidaginis* (▲).

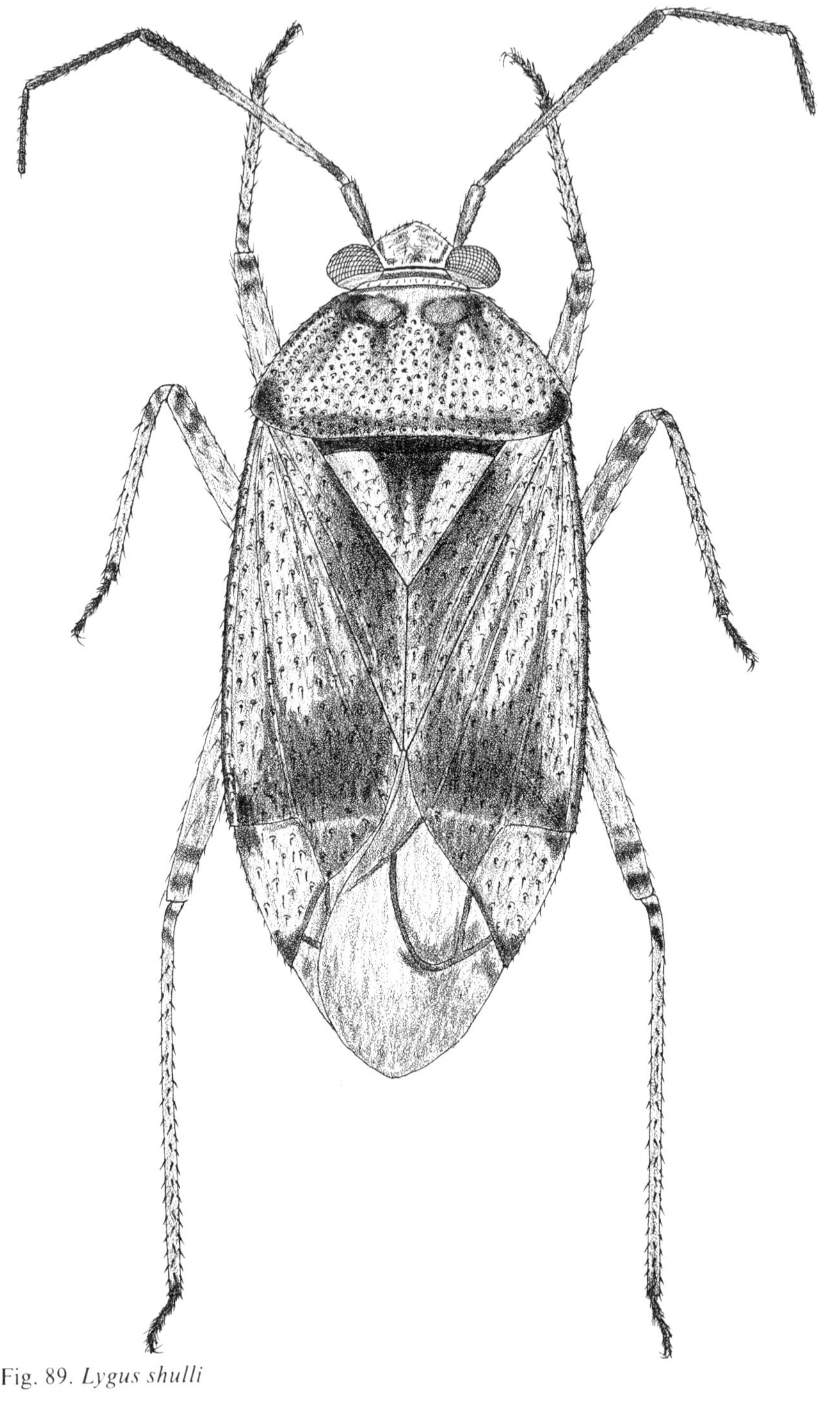

Fig. 89. *Lygus shulli*

Lygus solidaginis (Kelton)

Fig. 90; Map 40

Liocoris solidaginis Kelton, 1955*b*:489.
Lygus solidaginis: Carvalho, 1959:156.

Length 5.46–6.16 mm; width 2.38–2.94 mm. Head yellowish red, frons smooth. Rostrum 2.24–2.59 mm long. Pronotum reddish brown, anterior angles rounded. Mesoscutum black. Hemelytra reddish brown, triangular area at apex of corium brown or black, pubescence moderately long, dense (Fig. 90).

Remarks. The triangular brown or black area on the corium is distinctive.

Habitat. Collected on *Symphoricarpos occidentalis* and *Solidago* spp.

Distribution. North Central States, British Columbia; Prairie Provinces (Map 40).

Lygus rubrosignatus Knight

Fig. 91; Map 41

Lygus pratensis var. *rubrosignatus* Knight, 1923*b*:576.
Lygus rubrosignatus Knight, 1953:518.

Length 5.18–5.95 mm; width 2.38–2.94 mm. Head yellowish brown, frons smooth. Rostrum 2.24–2.66 mm long. Pronotum with reddish tinge, anterior angles prominent or rounded. Mesoscutum black. Hemelytra with reddish tinge, pubescence long, dense.

Remarks. This species is distinguished by the reddish color and by the long and dense pubescence (Fig. 91).

Habitat. Collected on *Senecio palustris* and other plants.

Distribution. Massachusetts, New York, northern Canada, Newfoundland; Prairie Provinces (Map 41).

Lygus elisus Van Duzee

Fig. 92; Map 41

Lygus pratensis var. *elisus* Van Duzee, 1914:20.
Lygus elisus Van Duzee, 1916:40.

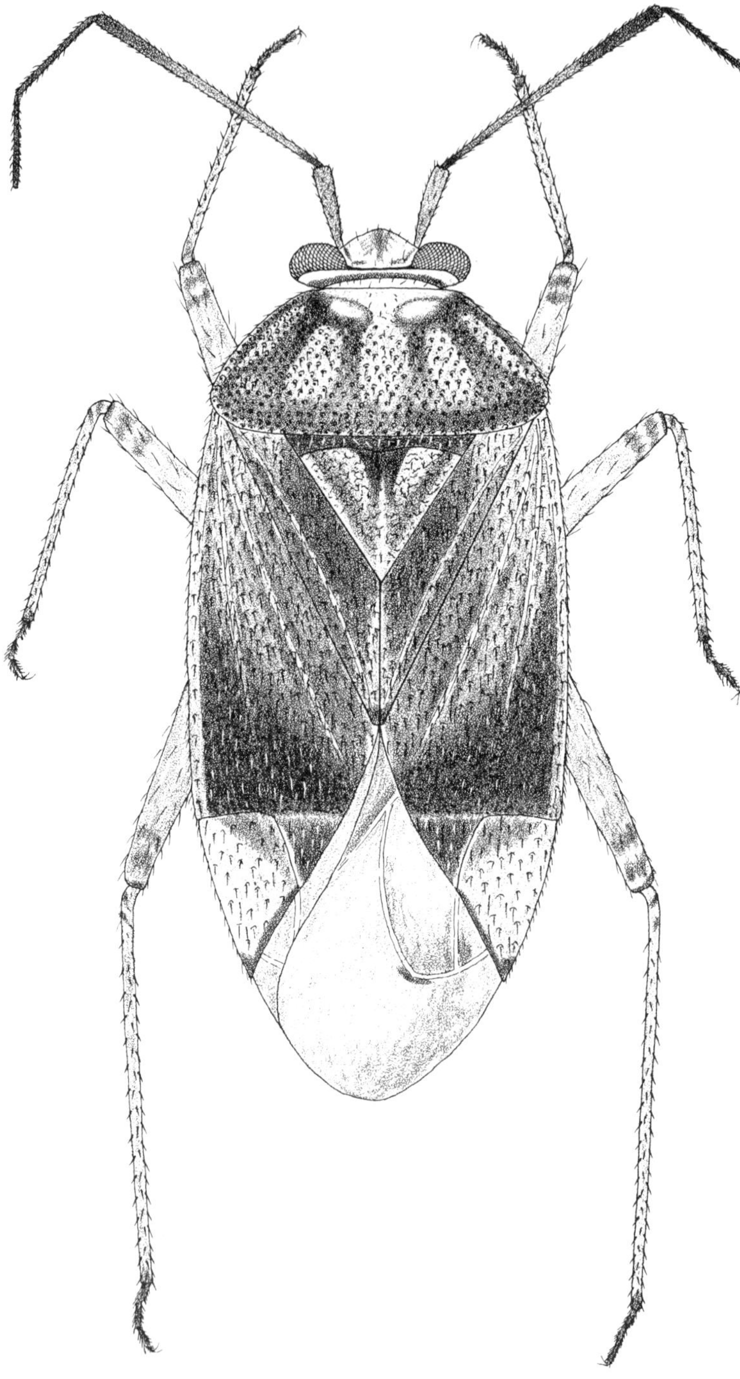

Fig. 90. *Lygus solidaginis*

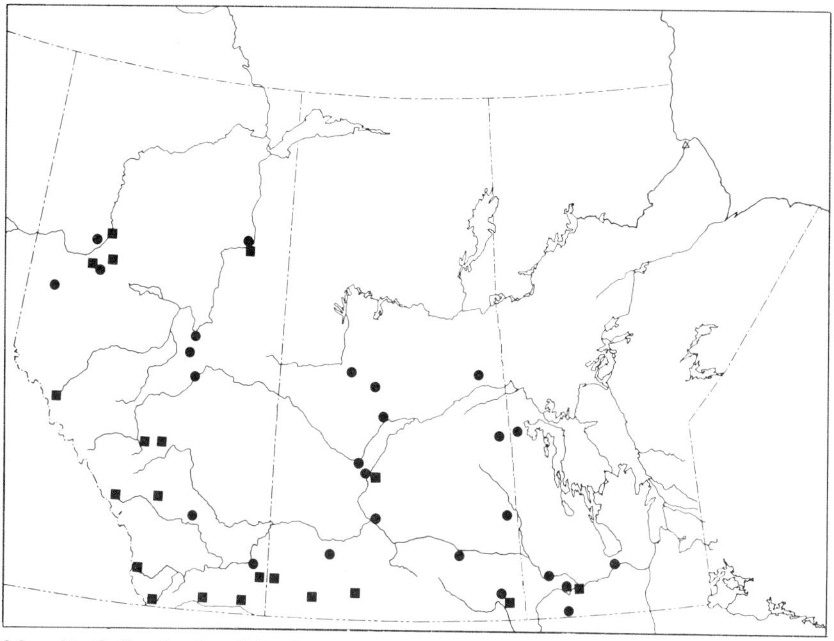

Map 41. Collection localities for *Lygus rubrosignatus* (●) and *L. elisus* (■).

Length 4.80–5.80 mm; width 2.24–3.01 mm. Head yellowish green, frons smooth. Rostrum 2.10–2.52 mm long. Pronotum green, anterior angles rounded. Mesoscutum black. Hemelytra green, pubescence moderately long, dense (Fig. 92).

Remarks. This species closely resembles *desertinus* in size and appearance, and in the length of the rostrum. However, the mesoscutum is entirely black in *elisus*, while the lateral margins are yellow in *desertinus*.

Habitat. Collected on a great variety of plants.

Distribution. Alaska, western USA, northern Canada, British Columbia; Prairie Provinces (Map 41).

Genus *Platylygus* Van Duzee

Elongate, robust, reddish brown species. Head oblique, frons striate, carina between eyes distinct. Rostrum extending beyond hind coxae. Hemelytra glabrous or pubescent. Genital segment in male with prominent tubercle on left margin.

The genus was revised by Kelton and Knight (1970) and 31 species were reported from North America. Three of these occur in the Prairie Provinces.

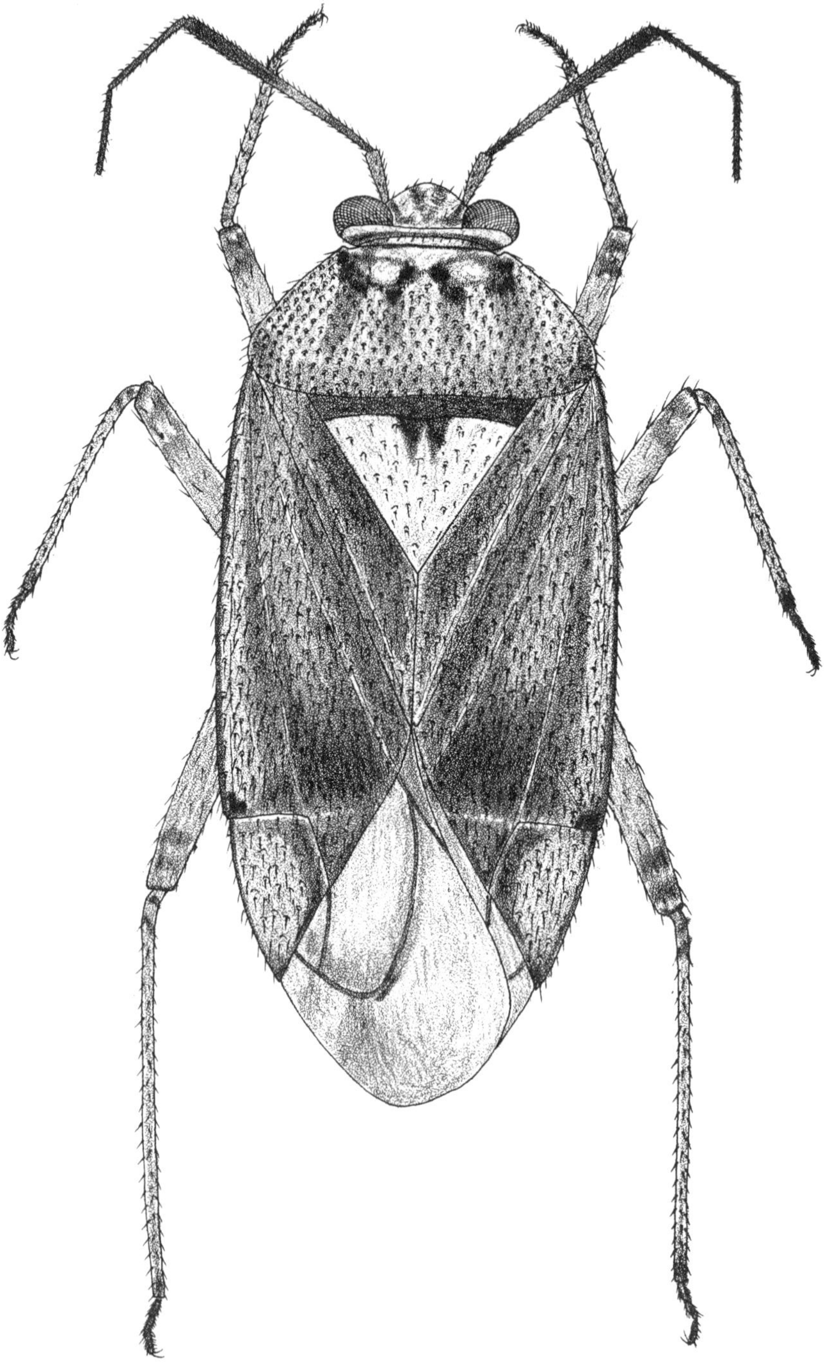

Fig. 91. *Lygus rubrosignatus*

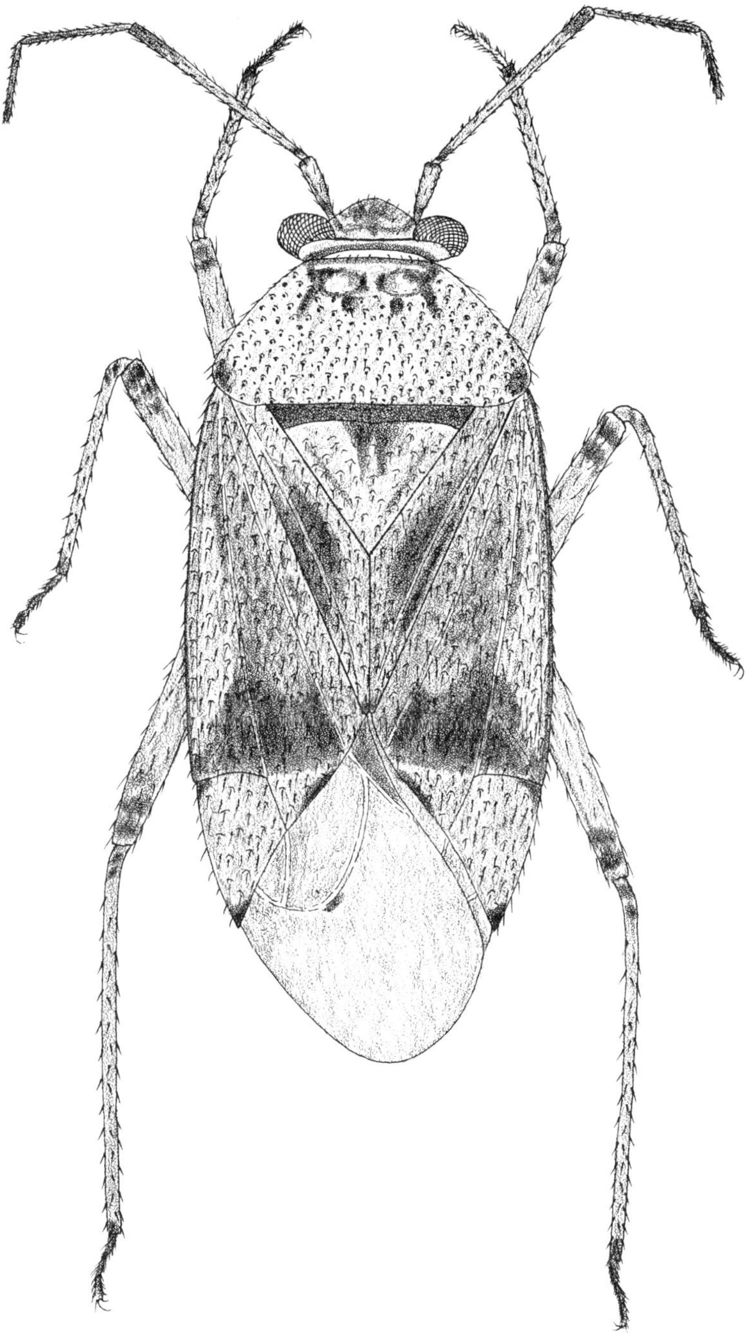

Fig. 92. *Lygus elisus*

Key to species of *Platylygus*

1. Dorsum hairy, pubescence long, dense .... ***piceicola* Kelton** (p. 138)
   Dorsum glabrous, pubescence short, sparse 2
2. Legs and antennae reddish .... ***rubripes* Knight** (p. 138)
   Legs and antennae pale or brown .... ***luridus* (Reuter)** (p. 138)

Platylygus piceicola Kelton

Map 42

Platylygus piceicola Kelton, (*in* Kelton & Knight, 1970:1436).

Length 6.44–6.65 mm; width 2.24–2.59 mm. Dorsal surface hairy, pubescence long, dense. Head, pronotum, and hemelytra reddish brown.

Remarks. This species is distinguished by the hairy dorsal surface.

Habitat. Collected on *Picea glauca*.

Distribution. Arizona, Colorado, Yukon, British Columbia; Alberta (Map 42).

Platylygus rubripes Knight

Map 42

Platylygus rubripes Knight, (*in* Kelton & Knight, 1970:1456).

Length 7.56–7.70 mm; width 2.80–2.94 mm. Dorsal surface glabrous. Second antennal segment and hind tibia red. Head, pronotum, and hemelytra yellowish brown.

Remarks. This species is distinguished by the red second antennal segment and the red hind tibia.

Habitat. Collected on *Pinus contorta*.

Distribution. Western States, British Columbia; Alberta (Map 42).

Platylygus luridus (Reuter)

Map 42

Lygidea rubecula var. *lurida* Reuter, 1909:46.
Platylygus luridus: Van Duzee, 1915:111.

Map 42. Collection localities for *Platylygus piceicola* (▲), *P. rubripes* (●), and *P. luridus* (■).

Length 6.50–7.12 mm; width 2.40–2.80 mm. Dorsal surface glabrous. Second antennal segment and hind tibia brown. Head, pronotum, and hemelytra yellowish brown to dark brown.

Remarks. This species is smaller than *rubripes* and has brown hind tibia.

Habitat. Collected on *Pinus banksiana* and *P. contorta.*

Distribution. Northeastern and western States, Yukon, Eastern Canada, British Colombia; Prairie Provinces (Map 42).

Genus *Pinalitus* Kelton

Elongate, reddish brown species. Head oblique, carina between eyes distinct. Rostrum variable in length. Pronotum finely punctate, lateral margins rounded, collar and basal margin white. Hemelytra reddish brown; pubescence long, dense.

The genus was revised by Kelton (1977). Five species are known from North America, one a European species. Two species occur in the Prairie Provinces. All are confined to the conifers.

Key to species of *Pinalitus*

1. Rostrum 2.66 mm or shorter in male, and 2.80 mm or shorter in female .... ***approximatus*** **(Stål)** (p. 140)
   Rostrum 2.80 mm or longer in male, and 2.94 mm or longer in female .... ***rostratus*** **Kelton** (p. 141)

Pinalitus approximatus (Stål)

Map 43

Deraeocoris approximatus Stål, 1858:185.
Pinalitus approximatus: Kelton, 1955*a*:282.

Length 4.90–5.25 mm; width 1.75–2.10 mm. Head brown or reddish brown. Rostrum 2.38–2.80 mm long, extending beyond hind coxae. Pronotum brown, often with lateral margins and basal angles dark brown. Scutellum dark brown, lateral margins and apex lighter brown. Hemelytra reddish brown; pubescence long, dense, slanting. Ventral surface reddish brown, legs pale, hind femur often reddish.

Remarks. The species resembles *rostratus* in color and appearance, but the rostrum is shorter.

Map 43. Collection localities for *Pinalitus approximatus* (▲) and *P. rostratus* (●).

Habitat. Collected on *Larix laricina, Picea glauca,* and *Abies balsamea.*

Distribution. Northeastern States, North Carolina, Colorado, Wyoming, Eastern Canada, British Columbia; Prairie Provinces (Map 43).

Pinalitus rostratus Kelton

Map 43

Pinalitus rostratus Kelton, 1977:1552.

Length 5.04–5.60 mm; width 1.68–2.10 mm. Head brown, clypeus and lorum often dark brown to black, frons often with reddish transverse lines. Rostrum 2.80–3.22 mm long, extending considerably beyond hind coxae. Pronotum yellow, calli and side margins brown. Scutellum mostly black, basal angles, narrow lateral margins, and apex yellow. Hemelytra light brown, diagonal line extending from tip of clavus to costal margin lighter brown; pubescence long, dense, appears silvery. Ventral surface brown; osteolar peritreme white; legs pale, hind femur often reddish.

Remarks. This species is larger than *approximatus* and has a longer rostrum.

Habitat. Collected on *Picea glauca* and *Larix laricina.*

Distribution. Colorado, New Mexico, Yukon, Eastern Canada, British Columbia; Alberta (Map 43).

Genus *Dichrooscytus* Fieber

Small to medium, flattened, reddish or green species. Head oblique, short, wide, carina between eyes distinct. Pronotum short, wide, finely punctate, lateral margins rounded. Hemelytra finely punctate; pubescence moderately long, dense.

The Canadian species were reviewed by Kelton (1972*a*). Fifty-one species are known from North America, eight occur in the Prairie Provinces. All are confined to the conifers.

Key to species of *Dichrooscytus*

1. Rostrum less than 1.60 mm 2
   Rostrum more than 1.60 mm 3
2. Tip of cuneus dark reddish brown or black .... ***ruberellus* Knight** (p. 142)
   Tip of cuneus green .... ***elegans* Heidemann** (p. 143)

3. Rostrum less than 1.90 mm .... ***alpinus* Kelton** (p. 143)
Rostrum more than 1.90 mm 4
4. Rostrum more than 2.96 mm .... ***rostratus* Kelton** (p. 144)
Rostrum less than 2.96 mm 5
5. Hemelytra bright red with trace of red on pronotum and ventral surface .... ***rubidus* Kelton** (p. 144)
Hemelytra dull reddish brown; pronotum and ventral surface green 6
6. Rostrum more than 2.60 mm .... ***suspectus* Reuter** (p. 144)
Rostrum less than 2.60 mm 7
7. Frons strongly convex; rostrum 2.10–2.24 mm long; larger species 4.83–5.39 mm .... ***convexifrons* Knight** (p. 145)
Frons gently rounded; rostrum 1.96–2.24 mm long; smaller species 4.48–4.76 mm .... ***latifrons* Knight** (p. 146)

Dichrooscytus ruberellus Knight

Map 44

Dichrooscytus ruberellus Knight, 1968*a*:199.

Length 3.50–3.90 mm; width 1.33–1.40 mm. Head and pronotum yellowish green. Rostrum 1.33–1.54 mm long. Hemelytra reddish pink;

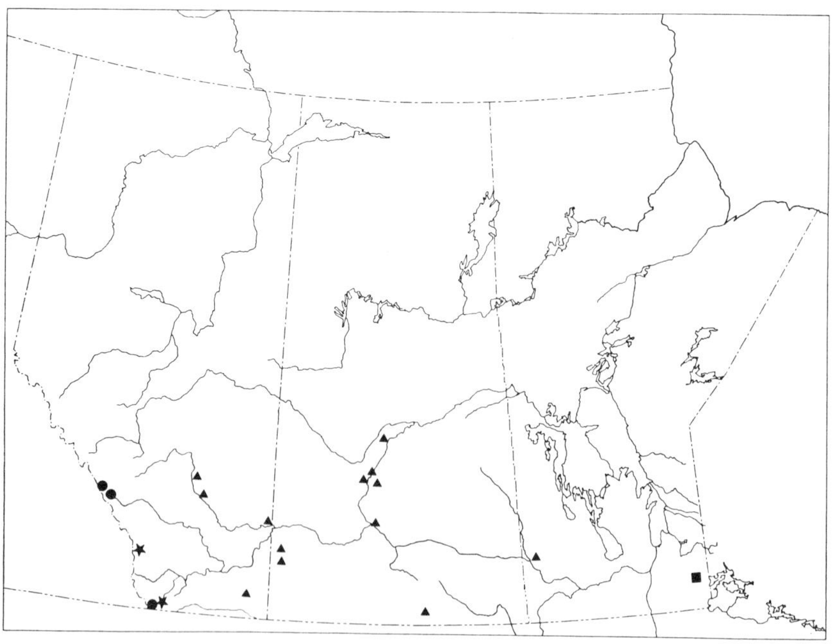

Map 44. Collection localities for *Dichrooscytus ruberellus* (▲), *D. elegans* (■), *D. alpinus* (★), and *D. rostratus* (●).

embolium pale green; tip of cuneus dark reddish brown or black; pubescence yellowish. Ventral surface yellowish green; legs pale.

Remarks. This species is distinguished by the small size, by the reddish hemelytra, by the dark reddish brown tip on the cuneus, and by the short rostrum.

Habitat. Collected on *Juniperus* spp.

Distribution. North Central States, British Columbia; Prairie Provinces (Map 44).

Dichrooscytus elegans Heidemann

Fig. 93; Map 44

Dichrooscytus elegans Heidemann, 1892:225.
Dichrooscytus tinctipennis Knight, 1927*b*:15.

Length 3.50–4.06 mm; width 1.54–1.68 mm. Head yellowish green. Rostrum 1.40–1.75 mm long. Hemelytra reddish; embolium green; cuneus reddish, tip green; pubescence yellowish. Ventral surface yellowish green; legs greenish.

Remarks. Kelton (1972*a*) referred to this species as *tinctipennis* Knight. Wheeler and Henry (1975) showed that *tinctipennis* is a junior synonym of *elegans* Heidemann.

This species resembles *ruberellus,* but lacks the dark reddish brown tip on the cuneus. The genitalia (Fig. 93) are distinctive.

Habitat. Collected on *Thuja occidentalis.*

Distribution. Eastern half of USA, Eastern Canada; Manitoba (Map 44).

Dichrooscytus alpinus Kelton

Figs. 64, 94; Map 44

Dichrooscytus alpinus Kelton, 1972*b*:1463.

Length 3.78–4.48 mm; width 1.68–1.96 mm. Head yellowish green. Rostrum 1.61–1.82 mm long. Pronotum yellowish green, often tinged with reddish brown. Hemelytra reddish brown; embolium pale; paracuneus pink; pubescence fuscous on head and pronotum, pale on hemelytra (Fig. 64). Ventral surface yellowish green; legs pale green.

Remarks. This species is distinguished by the length of the rostrum and by the male genitalia (Fig. 94).

Habitat. Collected on *Juniperus communis* and *J. horizontalis*.

Distribution. Colorado; Alberta (Map 44).

Dichrooscytus rostratus Kelton

Map 44

Dichrooscytus rostratus Kelton, 1972*a*:1041.

Length 5.04–5.60 mm; width 1.96–2.24 mm. Head yellowish green. Rostrum 2.96–3.22 mm long. Pronotum greenish brown. Hemelytra mostly red; embolium pale; paracuneus yellow; pubescence dark brown. Ventral surface yellowish green; legs light green.

Remarks. This species is distinguished by the long rostrum.

Habitat. Collected on *Pinus contorta*.

Distribution. Western States, British Columbia; Alberta (Map 44).

Dichrooscytus rubidus Kelton

Map 45

Dichrooscytus rubidus Kelton, 1972*a*:1042.

Length 4.62–5.18 mm; width 1.96–2.24 mm. Head yellowish green, speckled with red. Rostrum 2.03–2.19 mm long. Pronotum yellowish green, sprinkled with red. Hemelytra heavily sprinkled with red; embolium pale green; paracuneus white; pubescence black. Ventral surface yellowish green with trace of red; pleura and side of abdomen sprinkled with red; femora reddish, tibiae pale.

Remarks. Distinguished by the strongly reddish color.

Habitat. Collected on *Pinus contorta*.

Distribution. Washington, British Columbia; Alberta (Map 45).

Dichrooscytus suspectus Reuter

Map 45

Dichrooscytus suspectus Reuter, 1909:37.

Length 4.90–5.60 mm; width 1.89–2.24 mm. Head yellowish green. Rostrum 2.66–2.94 mm long. Pronotum yellowish green. Hemelytra reddish;

embolium and paracuneus pale; pubescence golden brown. Ventral surface yellowish green; legs green.

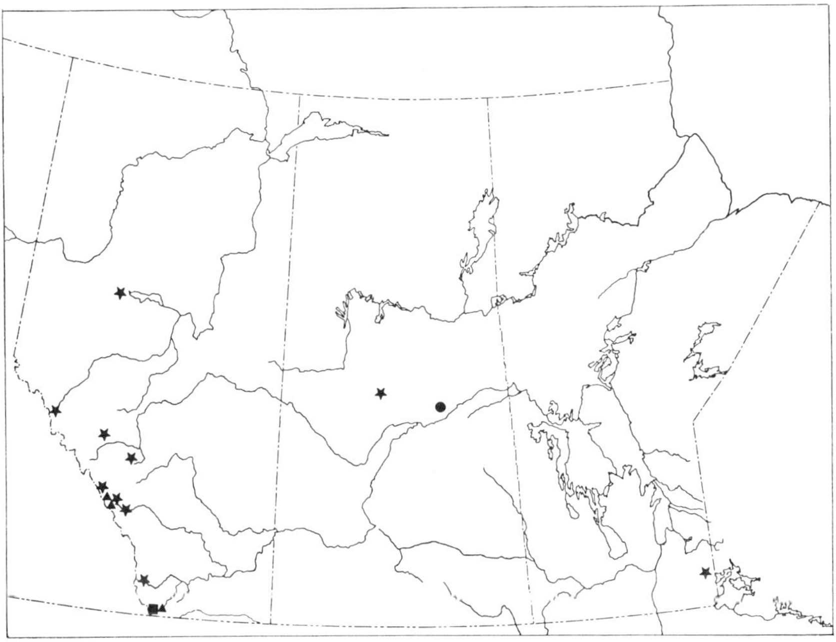

Map 45. Collection localities for *Dichrooscytus rubidus* (■), *D. suspectus* (●), *D. convexifrons* (▲), and *D. latifrons* (★).

Remarks. This species is much like *rostratus* in size and appearance, but the rostrum is shorter.

Habitat. Collected on *Pinus banksiana.*

Distribution. Northeastern States, Eastern Canada; Manitoba and Saskatchewan (Map 45).

Dichrooscytus convexifrons Knight

Map 45

Dichrooscytus convexifrons Knight, 1968*a*:198.

Length 4.83–5.39 mm; width 1.96–2.24 mm. Head yellowish green, frons strongly convex. Rostrum 2.10–2.24 mm long. Pronotum yellowish green. Hemelytra red; embolium and paracuneus pale; pubescence black. Ventral surface greenish yellow; legs green.

Remarks. This species is distinguished by the strongly convex frons and by the black pubescence on the hemelytra.

Habitat. Collected on *Pinus contorta.*

Distribution. Wyoming, British Columbia; Alberta (Map 45).

Dichrooscytus latifrons Knight

Map 45

Dichrooscytus latifrons Knight, 1968*a*:199.

Length 4.48–4.76 mm; width 1.69–1.89 mm. Head yellowish green. Rostrum 1.96–2.24 mm long. Pronotum green. Hemelytra red; embolium and paracuneus green; pubescence dark brown. Ventral surface yellowish green; legs light green.

Remarks. This species is similar to *convexifrons* in appearance, but the frons is not as convex and the rostrum is shorter.

Habitat. Collected on *Picea glauca* and *Abies lasiocarpa.*

Distribution. Arizona, Colorado, Michigan, Yukon, Eastern Canada, British Columbia; Prairie Provinces (Map 45).

Genus *Lygocoris* Reuter

Elongate-oblong, medium to large species. Head oblique, frons smooth, eyes large, conspicuous, carina between them prominent. Rostrum variable in length. Pronotum finely punctate, anterior angles rounded; calli smooth, pubescent. Hemelytra generally green, pubescence long, dense.

The genus was reviewed by Kelton (1971*c*) and information on taxonomy, distribution, and host plants may be found there.

Forty-one species are known from North America, nine occur in the Prairie Provinces.

Key to species of *Lygocoris*

1. Carina between eyes obsolete at middle (subgenus *Lygocoris*); genital claspers (Fig. 95) .... ***pabulinus* (Linnaeus)** (p. 147)
   Carina between eyes complete (subgenus *Neolygus*) 2
2. Pronotum, scutellum, and hemelytra dark brown or black (Fig. 101); genital claspers (Fig. 96) .... ***piceicola* Kelton** (p. 147)
   Pronotum, scutullum, and hemelytra with pale areas 3

3. Ventral surface mostly pale yellow or green 4
Ventral surface mostly brown or reddish 5
4. Second antennal segment yellowish green; pronotum yellowish green without dark markings (Fig. 102); genital claspers (Fig. 97) .... ***belfragii* (Reuter)** (p. 151)
Second antennal segment black at apex; pronotum yellowish green with dark rays behind calli (Fig. 103); genital claspers (Fig. 98) .... ***canadensis* (Knight)** (p. 153)
5. Tibiae with brown or black spots at bases of spines 6
Tibiae without spots at bases of spines 7
6. Abdomen with red markings; genital claspers (Fig. 99) .... ***atritylus* (Knight)** (p. 153)
Abdomen green; genital claspers (Fig. 100) .... ***contaminatus* (Fallén)** (p. 156)
7. Pronotum with dark calli or with dark rays 8
Pronotum without dark calli or with dark rays (Fig. 106); genital claspers (Fig. 101) .... ***omnivagus* (Knight)** (p. 158)
8. Pronotum with dark rays behind calli (Fig. 107); genital claspers (Fig. 102) .... ***communis* (Knight)** (p. 158)
Pronotum without dark rays behind calli (Fig. 108); genital claspers (Fig. 103) .... ***quercalbae* (Knight)** (p. 158)

Lygocoris pabulinus (Linnaeus)

Figs. 95, 104; Map 46

Cimex pabulinus Linnaeus, 1761:253.
Lygus chagnoni Stevenson, 1903:214.
Lygocoris pabulinus: Carvalho et al., 1961:283.

Length 5.18–6.16 mm; width 1.82–2.38 mm. Head yellowish green, often yellowish, carina between eyes obsolete at middle. Rostrum 2.10–2.17 mm long. Pronotum and hemelytra yellowish green or green.

Remarks. This species is distinguished by the overall greenish color and slender form (Fig. 104), by the partly obsolete carina, and by the genital claspers (Fig. 95).

Habitat. Collected on a great variety of herbaceous plants.

Distribution. Widespread in North America; Prairie Provinces (Map 46).

Lygocoris piceicola Kelton

Figs. 96, 105; Map 46

Lygocoris piceicola Kelton, 1971*b*:1107.

Length 5.25–5.74 mm; width 2.17–2.45 mm. Head yellowish brown, often brown to black. Second antennal segment yellowish brown. Pronotum

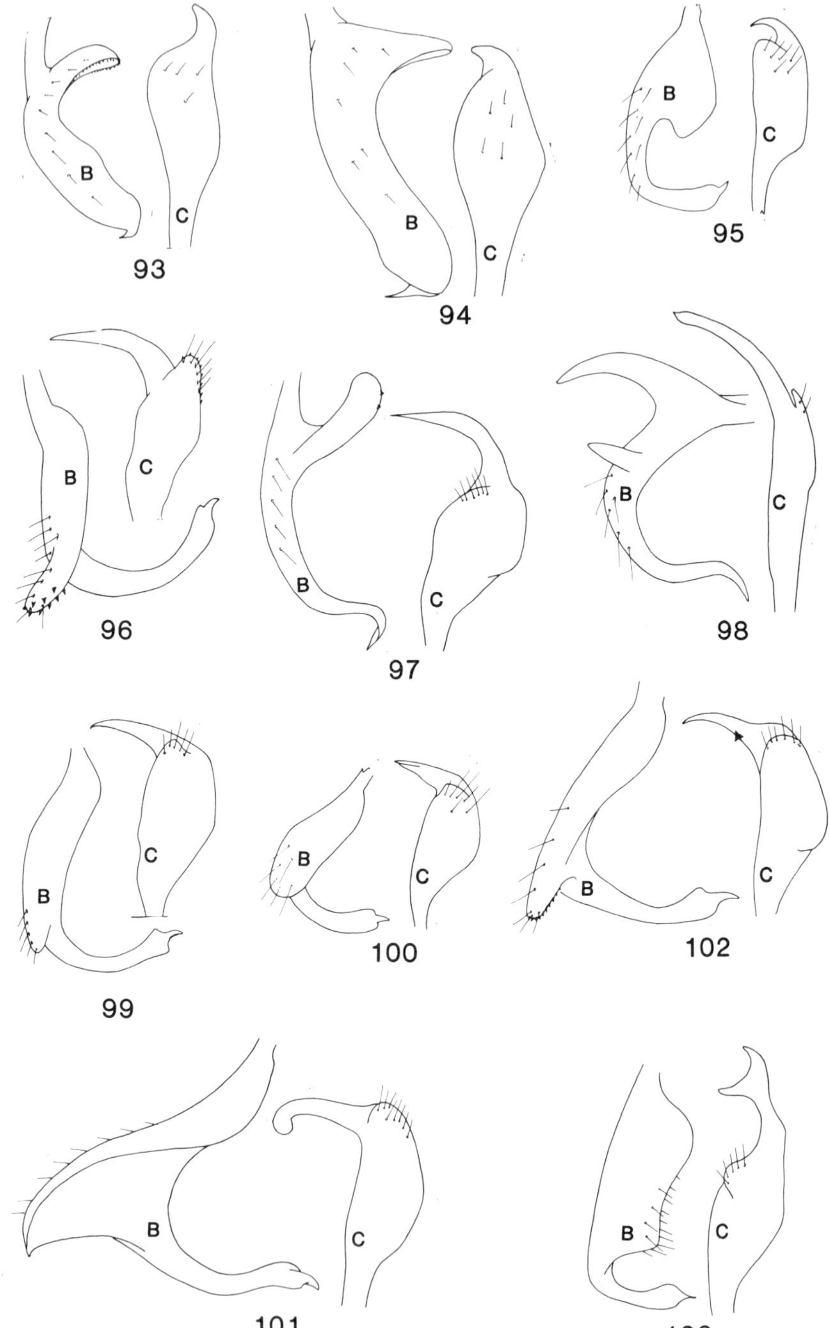

Figs. 93–103. Male claspers of *Dichrooscytus* and *Lygocoris* spp. 93, *Dichrooscytus elegans*; 94, *D. alpinus*; 95, *Lygocoris pabulinus*; 96, *L. piceicola*; 97, *L. belfragii*; 98, *L. canadensis*; 99, *L. atritylus*; 100, *L. contaminatus*; 101, *L. omnivagus*; 102, *L. communis*; 103, *L. quercalbae*

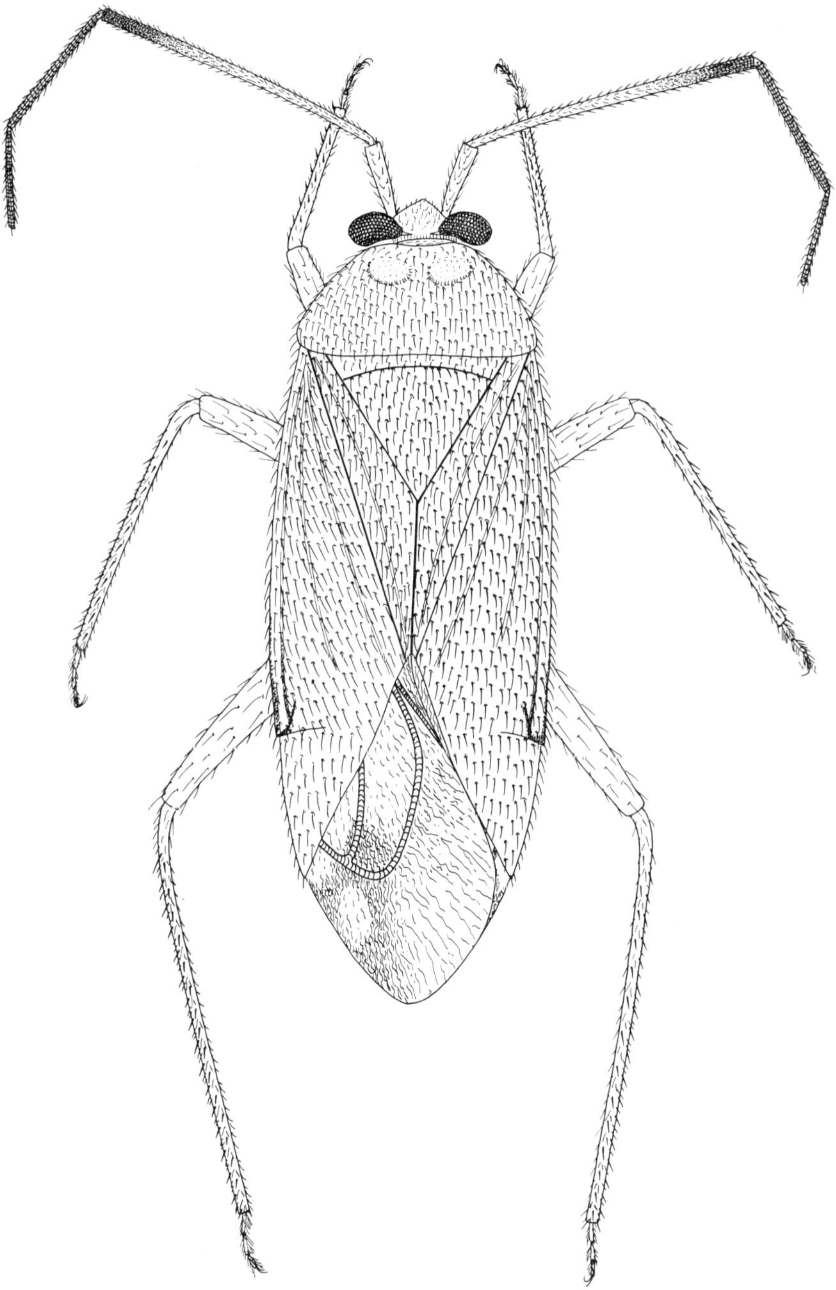

Fig. 104. *Lygocoris pabulinus*

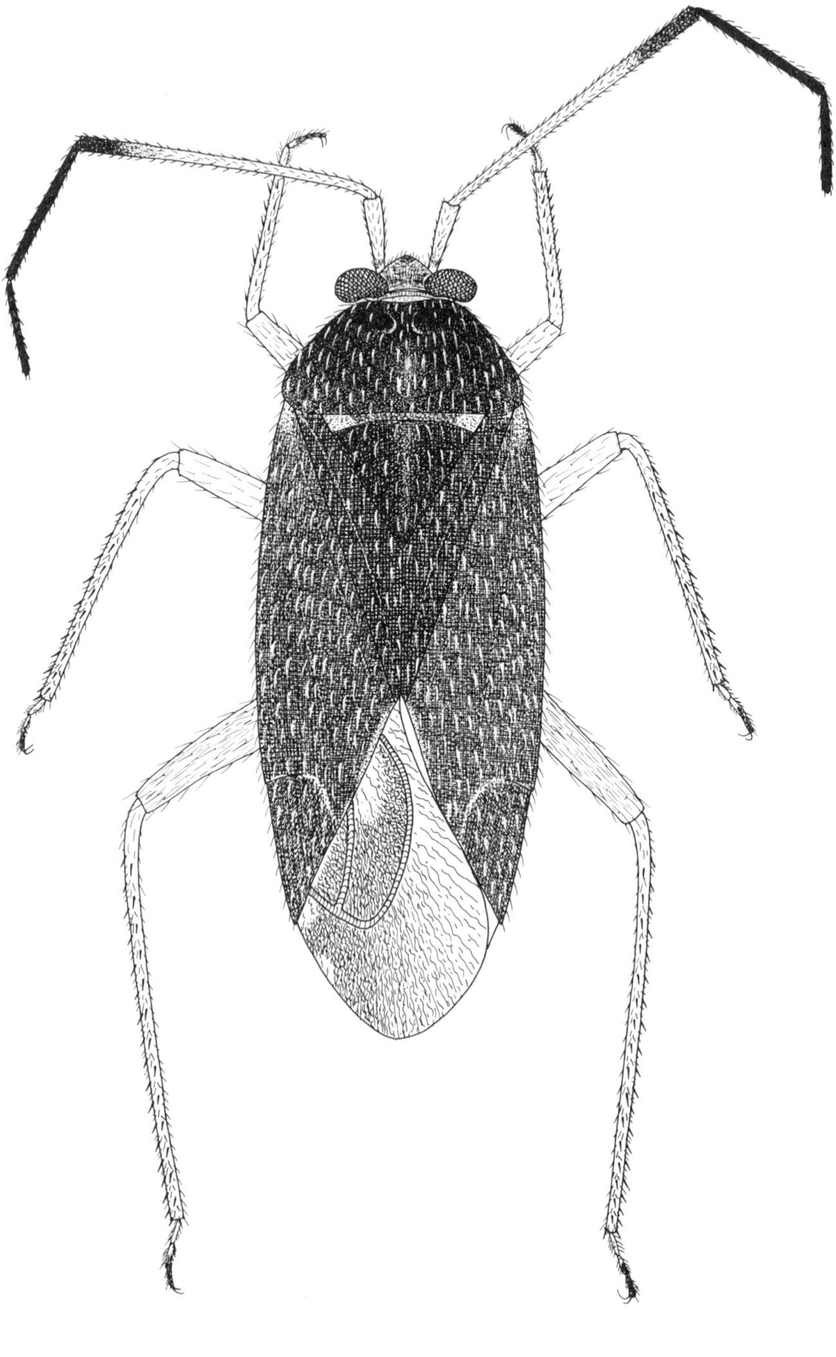

Fig. 105. *Lygocoris piceicola*

dark brown, in teneral specimens light brown with lateral margins and calli black. Scutellum and hemelytra dark brown to black.

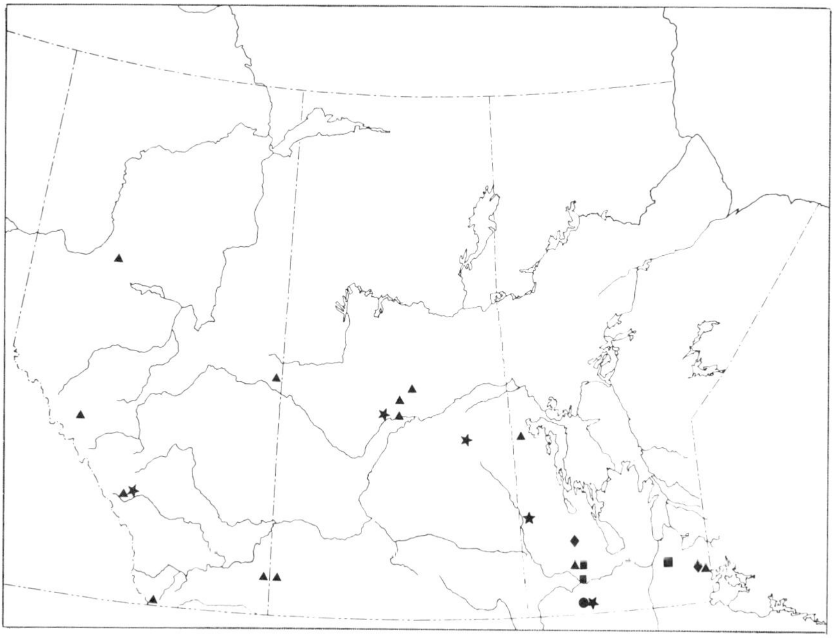

Map 46. Collection localities for *Lygocoris pabulinus* (▲), *L. piceicola* (◆), *L. belfragii* (●), *L. canadensis* (■), and *L. atritylus* (★).

Remarks. This species is distinguished by the overall dark color (Fig. 105), and by the genital claspers (Fig. 96).

Habitat. Collected on *Picea glauca*.

Distribution. Eastern Canada; now known to occur in Manitoba (Map 46).

Lygocoris belfragii (Reuter)

Figs. 97, 106; Map 46

Lygus belfragii Reuter, 1875*c*:71.
Lygocoris belfragii: Carvalho, 1959:141.

Length 5.40–6.02 mm; width 2.24–2.52 mm. Head greenish yellow or yellowish brown. Second antennal segment yellowish green. Pronotum yel-

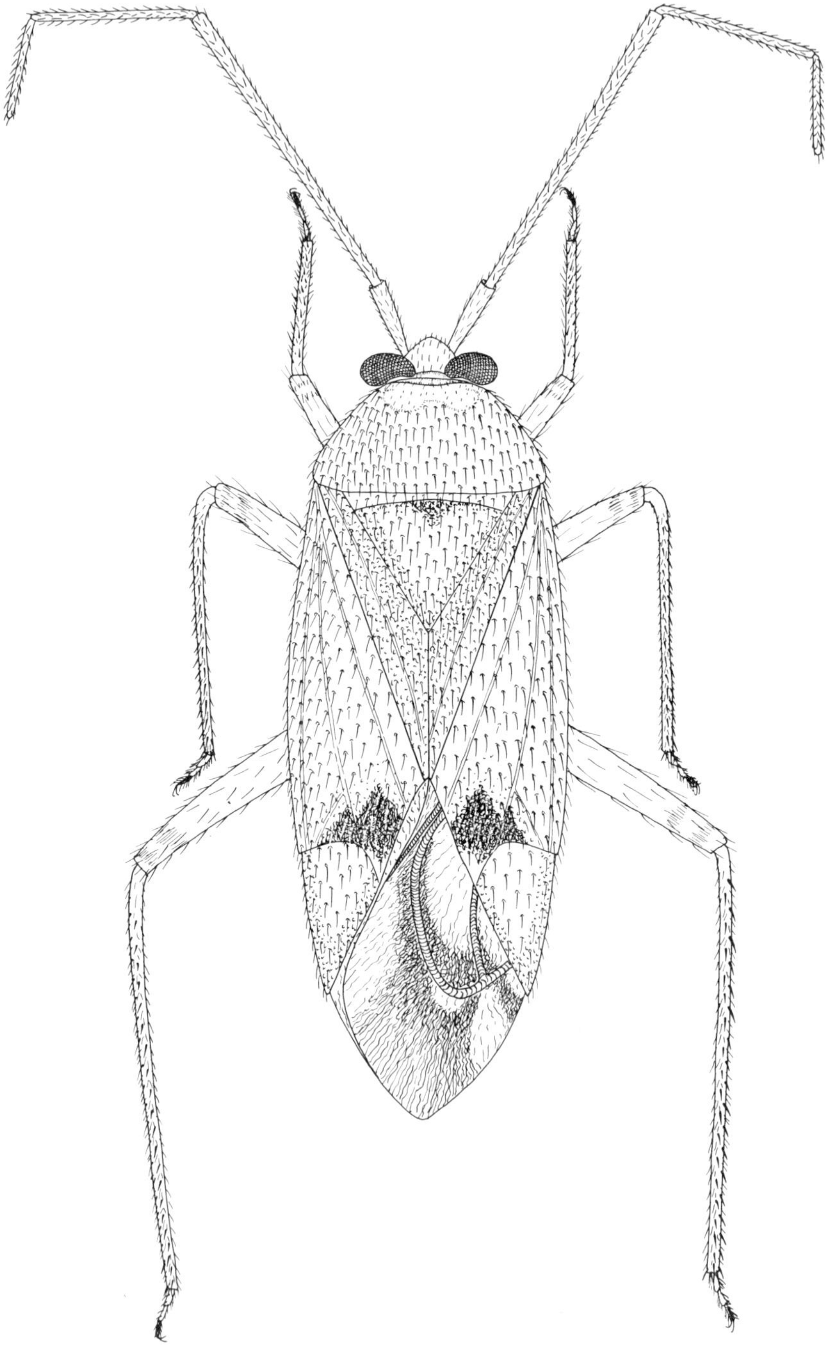

Fig. 106. *Lygocoris belfragii*

lowish green. Hemelytra yellowish green, triangular spot at apex of corium brown.

Remarks. This species is distinguished by the yellowish green color, by the triangular spot at the apex of corium, by the yellowish green second antennal segment (Fig. 106), and by the genital claspers (Fig. 97).

Habitat. Collected on *Corylus americana.*

Distribution. Eastern half of USA, Eastern Canada; Manitoba (Map 46).

Lygocoris canadensis (Knight)

Figs. 98, 107; Map 46

Lygus (Neolygus) canadensis Knight, 1917*b*:634.
Lygocoris canadensis: Carvalho, 1959:141.

Length 5.25–5.60 mm; width 2.10–2.45 mm. Head yellowish green. Second antennal segment black at apex. Pronotum yellowish green, rays behind calli brown or black. Hemelytra pale green, triangular spot at apex of corium brown.

Remarks. This species is distinguished by the black apex of the second antennal segment (Fig. 107), and by the genital claspers (Fig. 98).

Habitat. Collected on *Corylus americana*

Distribution. Northeastern States, Eastern Canada; Manitoba (Map 46).

Lygocoris atritylus (Knight)

Figs. 99, 108; Map 46

Lygus (Neolygus) atritylus Knight, 1917*b*:606.
Lygocoris atritylus: Carvalho, 1959:140.

Length 4.69–5.46 mm; width 1.96–2.52 mm. Head light yellowish brown marked with reddish tinge, apical half of clypeus black. Pronotum yellowish green, often sprinkled with red. Hemelytra yellowish brown, corium with reddish brown bar along radial vein. Abdomen with red markings, legs yellowish green, tibiae with dark spots at bases of spines.

Remarks. This species is distinguished by the reddish markings on the head and pronotum, by the black tip on the clypeus, by the dark spots at the bases of spines on the tibiae (Fig. 108), and by the genital claspers (Fig. 99).

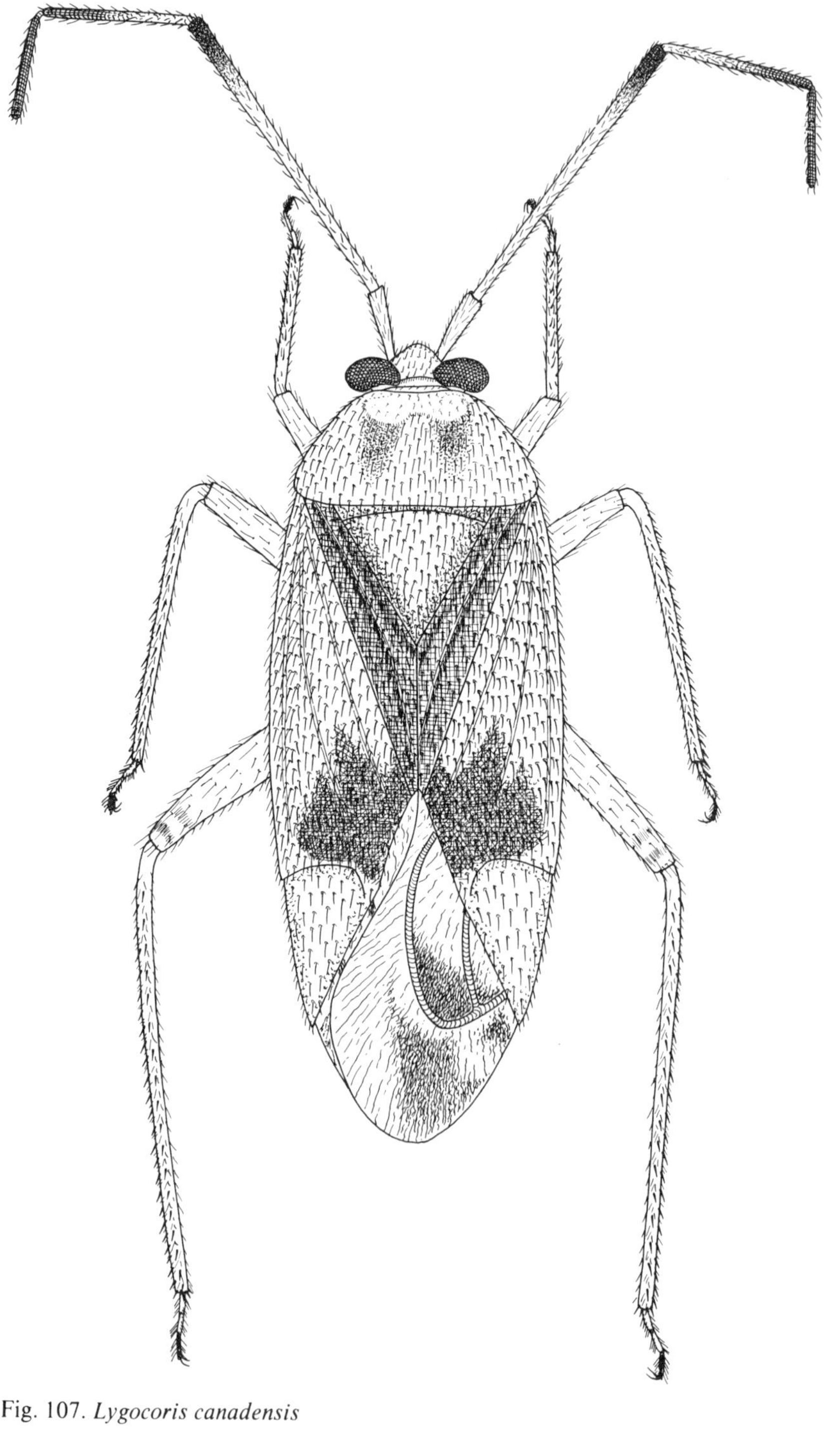

Fig. 107. *Lygocoris canadensis*

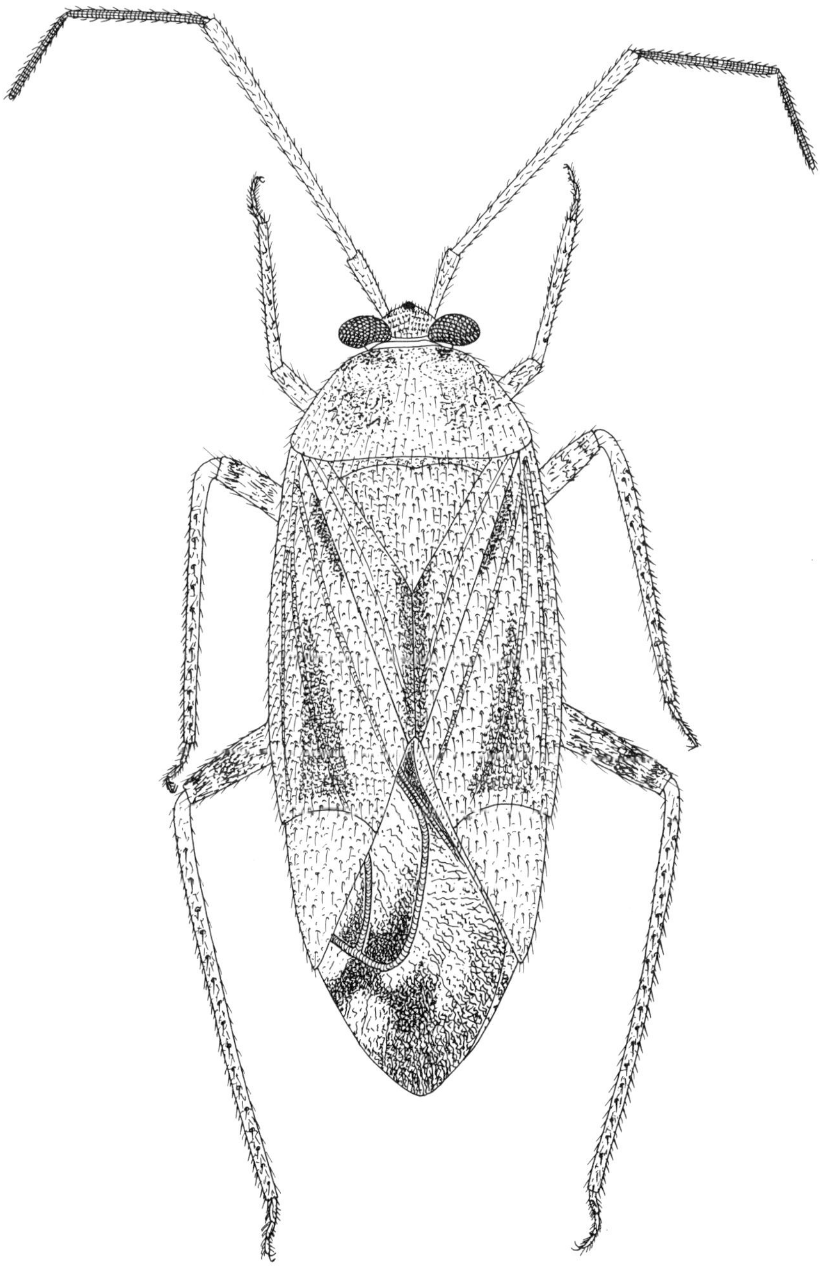

Fig. 108. *Lygocoris atritylus*

Habitat. Collected on *Salix* spp.

Distribution. Northeastern States, Eastern Canada, British Columbia; Prairie Provinces (Map 46).

Lygocoris contaminatus (Fallén)

Figs. 100, 109; Map 47

Lygaeus contaminatus Fallén, 1807:76.
Lygus confusus Knight, 1917*b*:606.
Lygocoris contaminatus: Southwood & Leston, 1959:280.

Length 5.25–5.95 mm; width 2.10–2.24 mm. Head yellowish green, clypeus often brownish. Pronotum yellowish green. Hemelytra yellowish green, inner corium with fuscous patch at apex. Legs greenish yellow, spines on tibiae brownish with distinct brown spots.

Remarks. This species is distinguished by the fuscous patch at the apex of corium, by the brown spots at the bases of spines on the tibiae (Fig. 109), and by the genital claspers (Fig. 100).

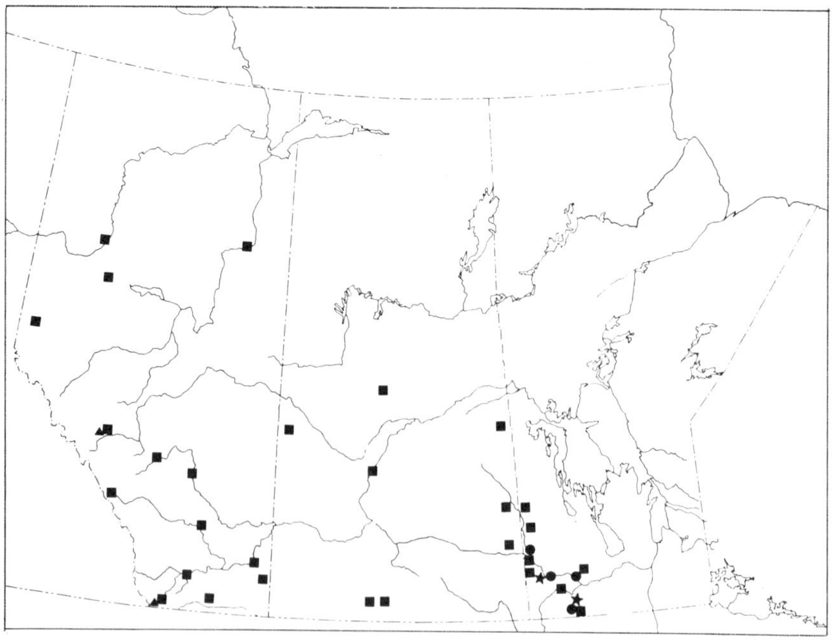

Map 47. Collection localities for *Lygocoris contaminatus* (▲), *L. omnivagus* (●), *L. communis* (■), and *L. quercalbae* (★).

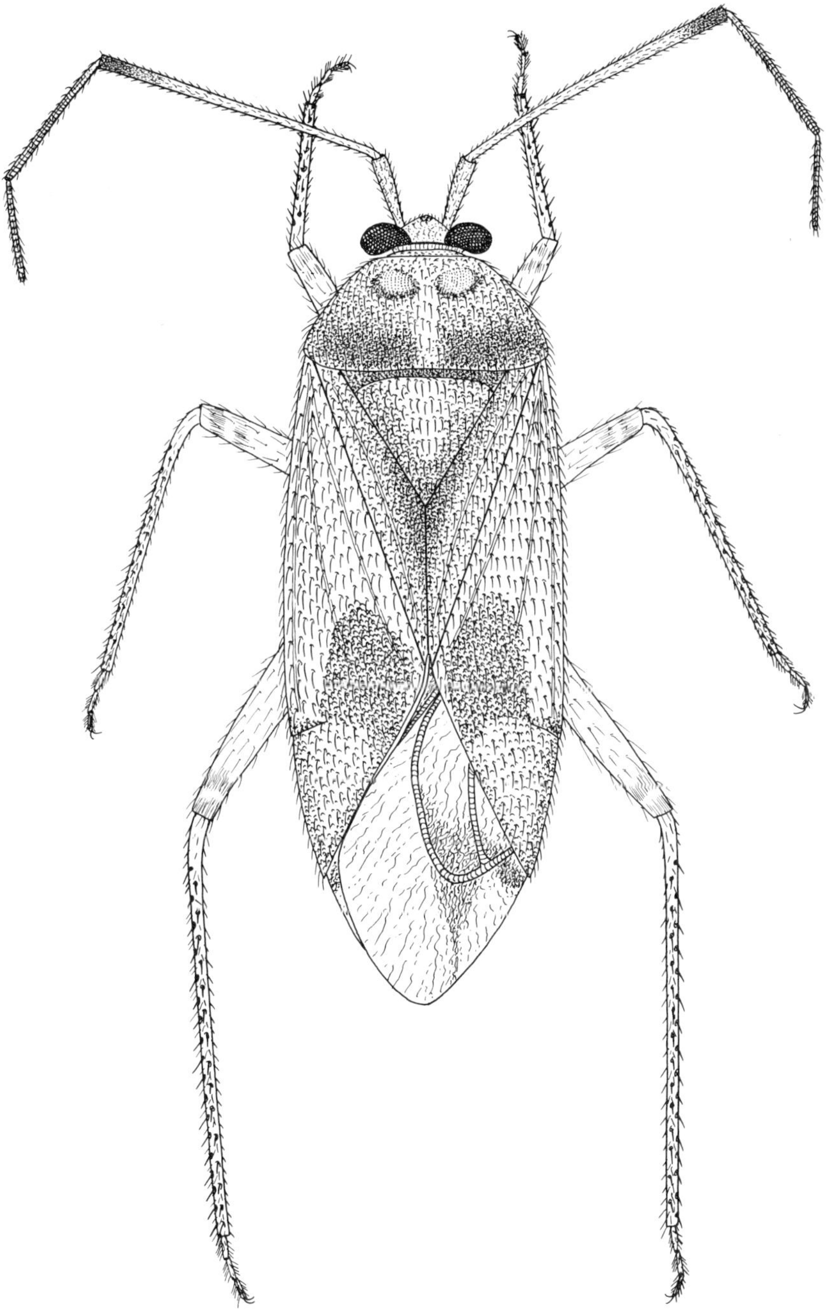

Fig. 109. *Lygocoris contaminatus*

Habitat. Collected on *Alnus tenuifolia*.

Distribution. Alaska, northeastern States, Yukon, Eastern Canada, British Columbia; Alberta (Map 47).

Lygocoris omnivagus (Knight)

Figs. 101, 110; Map 47

Lygus (Neolygus) omnivagus Knight, 1917*a*:627.
Lygocoris omnivagus: Carvalho, 1959:144.

Length 4.90–5.60 mm; width 2.10–2.45 mm. Head yellowish brown. Pronotum yellowish brown, usually darker behind callus. Hemelytra strongly marked with brown especially on clavus and on apical half of corium.

Remarks. This species is distinguished by the color pattern on the hemelytra (Fig. 110), and by the genital claspers (Fig. 101).

Habitat. Collected on a great variety of trees and shrubs.

Distribution. Eastern States, Eastern Canada; Manitoba (Map 47).

Lygocoris communis (Knight)

Figs. 102, 111; Map 47

Lygus communis Knight, 1916: 346.
Lygocoris communis: Carvalho, 1959:141.

Length 5.11–5.95 mm; width 2.10–2.66 mm. Head brown, frons usually marked with transverse reddish lines. Pronotum yellowish green, rays reddish brown. Hemelytra dark reddish brown.

Remarks. This species is distinguished by the reddish lines on the frons, by the rays on the pronotum (Fig. 111), and by the genital claspers (Fig. 102).

Habitat. Collected on a great variety of trees and shrubs.

Distribution. Alaska, transcontinental in North America; Prairie Provinces (Map 47).

Lygocoris quercalbae (Knight)

Figs. 103, 112; Map 47

Lygus (Neolygus) quercalbae Knight, 1917*a*:624.
Lygocoris quercalbae: Carvalho, 1959:145.

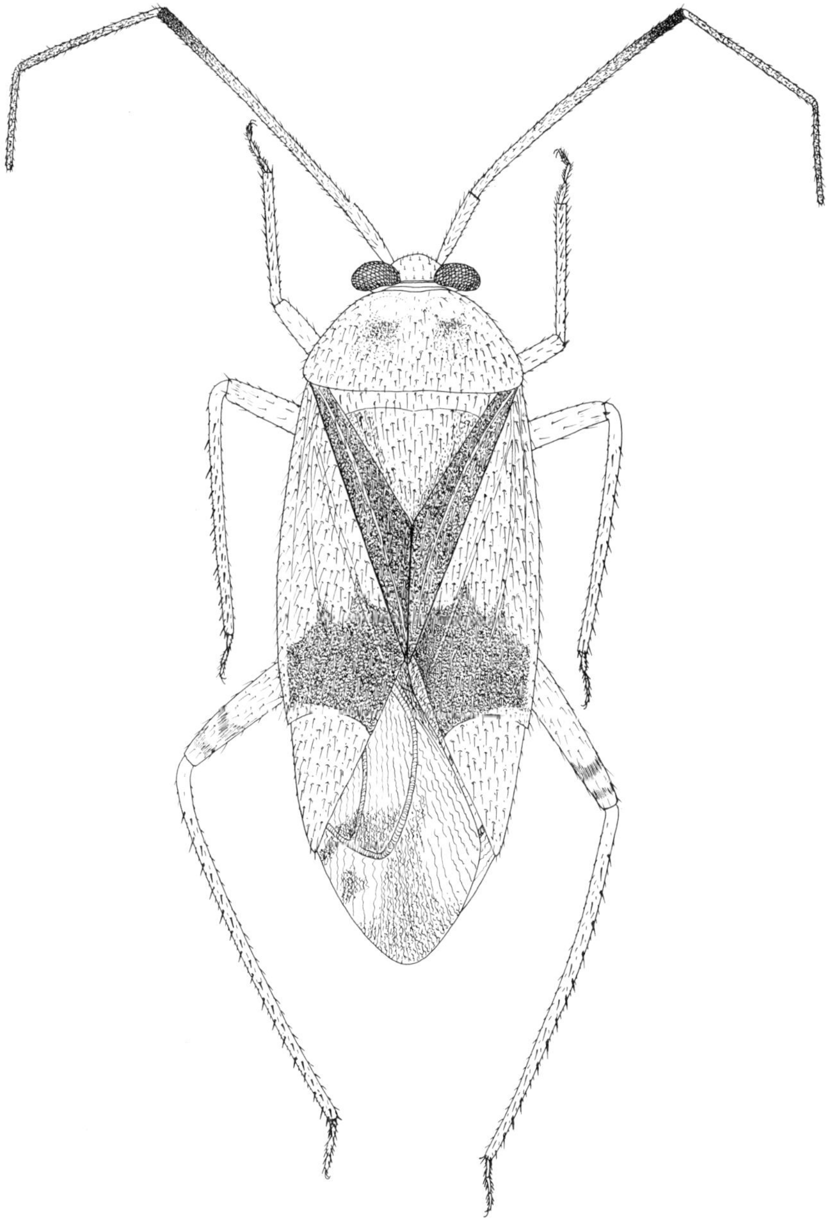

Fig. 110. *Lygocoris omnivagus*

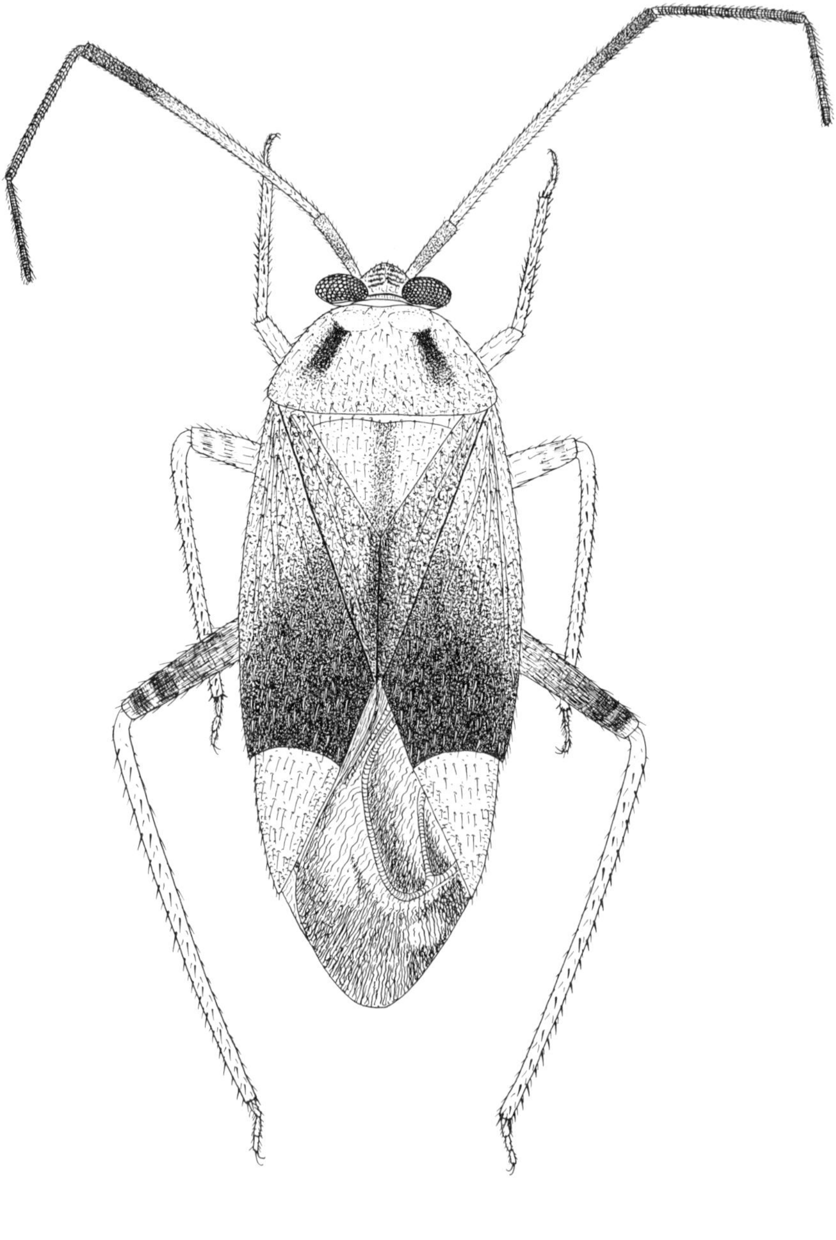

Fig. 111. *Lygocoris communis*

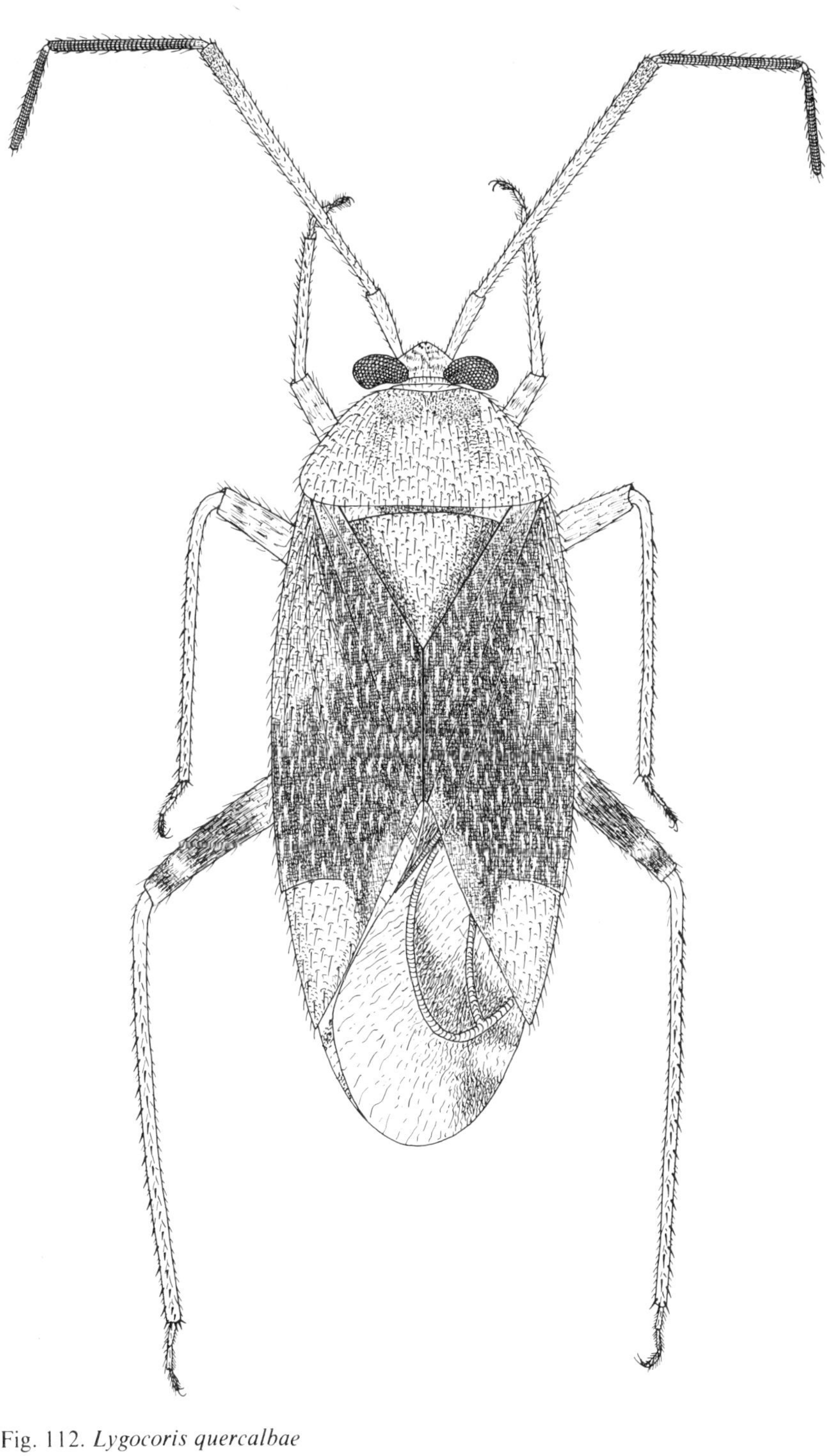

Fig. 112. *Lygocoris quercalbae*

Length 4.76–5.74 mm; width 2.10–2.59 mm. Head light yellowish brown marked with reddish tinge. Pronotum yellowish brown marked with red; callus often brown. Hemelytra yellowish brown, apical half of corium dark brown tinged with red.

Remarks. This species is distinguished by the reddish markings on the head and pronotum, by the brown calli (Fig. 112), and by the genital claspers (Fig. 103).

Habitat. Collected on *Quercus macrocarpa.*

Distribution. Northeastern States, Eastern Canada; Manitoba (Map 47).

Genus *Metriorrhynchomiris* Kirkaldy

Glabrous, black, or black and red species. Head oblique, carina between eyes absent. Second antennal segment incrassate. Pronotum and hemelytra shiny, almost smooth and glabrous. Ventral surface black; femora black, tibiae palish.

Three species are now included in this genus, but *dislocatus* alone has 13 varietal names. Two of these varieties, *goniphorus* and *nigritus* occur in the Prairie Provinces.

Metriorrhynchomiris dislocatus (Say)

Fig. 113; Map 48

Capsus dislocatus Say, 1832:21.
Horcias dislocatus Reuter, 1909:41.
Metriorrhynchomiris dislocatus: Carvalho & Jurberg, 1974:50.

Length 5.88–6.44 mm; width 2.87–3.36 mm. Head black, frons pale. Pronotum black, often red along margins. Scutellum black. Hemelytra black, often red along wide costal margin including cuneus (Fig. 100).

Remarks. This species was formerly placed in *Horcias* Distant. It is distinguished by the glabrous and shiny appearance, and by the black or black and red color.

Habitat. Collected on a great variety of herbaceous plants.

Distribution. Northeastern and North Central States, Texas, Ontario, Quebec; now known to occur in the Prairie Provinces (Map 48).

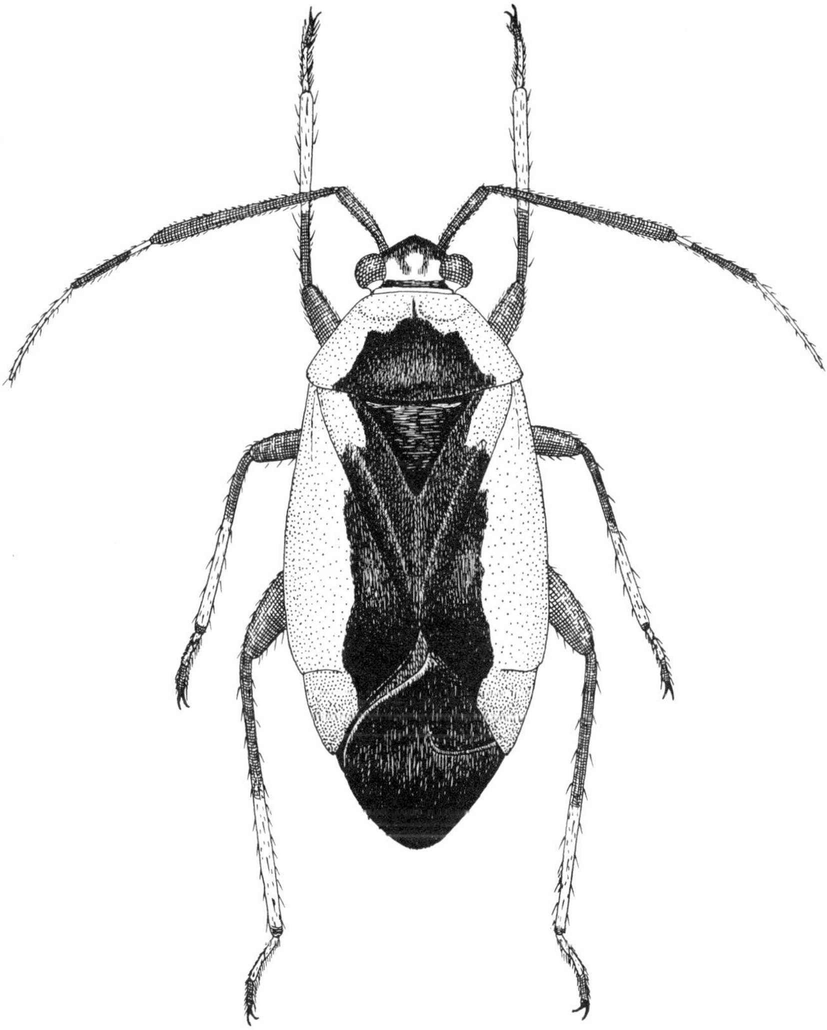

Fig. 113. *Metriorrhynchomiris dislocatus*

Genus *Stenotus* Jakovlev

Elongate, green and black, pubescent species. Head oblique, frons weakly striate, eyes large, carina between them absent. Rostrum extending beyond hind coxae. Pronotum finely punctate, lateral margins rounded. Hemelytra finely punctate, densely pubescent. Legs with short, black hairs.

One species occurs in North America. It was probably introduced from Europe.

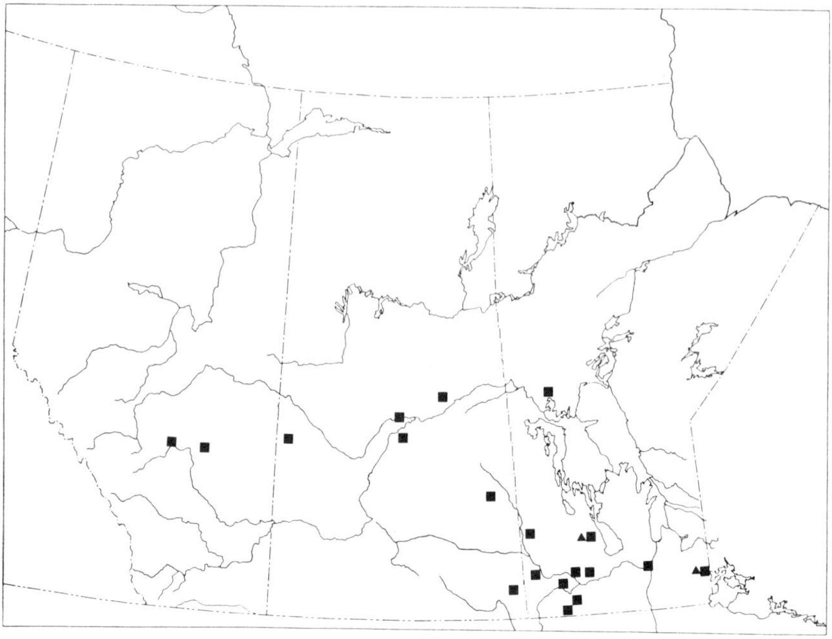

Map 48. Collection localities for *Metriorrhynchomiris dislocatus* (■) and *Stenotus binotatus* (▲).

Stenotus binotatus (Fabricius)

Map 48

Lygaeus binotatus Fabricius, 1794:172.
Stenotus binotatus: Reuter, 1888:636.

Length 5.95–6.44 mm; width 1.82–2.10 mm. Head mostly green, clypeus and often part of frons black; first antennal segment pale with short, black hairs. Pronotum, scutellum, and hemelytra green or yellowish; males with black stripe extending from each callus to base of cuneus, not as pronounced in female. Ventral surface green, legs pale, first segment of hind tarsus longer than third.

Remarks. This species is distinguished by the two black stripes.

Habitat. Collected on grasses.

Distribution. Transcontinental in USA, Ontario, Quebec, British Columbia; now known to occur in Manitoba (Map 48).

Genus *Phytocoris* Fallén

Elongate, parallel species. Head oblique, short; lora inflated; eyes large and prominent, carina between them absent. Pronotum impunctate, side margins rounded, calli inconspicuous. Hemelytra variable in color, pubescence generally of two types, appressed, sericeous pubescence, mixed with erect, longer hairs. Legs long, slender.

The genus is very large and the species are often difficult to identify. In many cases only the males may be identified with certainty by using the genital structures.

Approximately 220 species are known from North America, 20 occur in the Prairie Provinces.

Key to species of *Phytocoris*

1. Species yellow; genital claspers (Fig. 115) .... ***laevis*** **(Uhler)** (p. 166)
   Species not yellow 2
2. First antennal segment thickened (Fig. 114), as thick as front femur 3
   First antennal segment slender, not as thick as front femur 5
3. Apex of second antennal segment black; genital claspers (Fig. 116) .... ***lasiomerus*** **Reuter** (p. 167)
   Second antennal segment yellow 4
4. Pubescence on first antennal segment black; head and first antennal segment yellow with red markings; genital claspers (Fig. 117) .... ***pallidicornis*** **Reuter** (p. 169)
   Pubescence on first antennal segment pale; head and first antennal segment grayish with brown markings; genital claspers (Fig. 118) .... ***validus*** **Reuter** (p. 169)
5. First antennal segment much shorter than head width 6
   First antennal segment as long as or longer than head width 9
6. Apex of corium with clump of pale spots; pronotum with distinct brown submargin; scutellum uniformly pale; genital claspers (Fig. 119) .... ***stellatus*** **Van Duzee** (p. 171)
   Apex of corium without clump of pale spots; pronotum without distinct brown submargin; scutellum mostly reddish brown 7
7. Brown species; rostrum 2.80 mm or longer; genital claspers (Fig. 120) .... ***listi*** **Knight** (p. 172)
   Reddish species; rostrum 2.10 mm or shorter 8
8. Scutellum uniformly reddish brown; vein on wing membrane uniformly reddish; genital claspers (Fig. 121) .... ***pinicola*** **Knight** (p. 173)
   Scutellum with apical one third pale; vein on wing membrane pale at apex; genital claspers (Fig. 122) .... ***conspersipes*** **Reuter** (p. 173)
9. Second antennal segment without pale bands 10
   Second antennal segment with pale bands 14
10. First antennal segment and pronotum mostly black; hind tibia black, with two pale rings; genital claspers (Fig. 123) .... ***alpinus*** **Kelton** (p. 173)
    First antennal segment and pronotum mostly pale 11
11. Species reddish brown; genital claspers (Fig. 124) .. ***michiganae*** **Knight** (p. 174)
    Species black or brown 12

12. Tubercle at base of left clasper broadly rounded, (Fig. 125) .... ***albertae* Knight** (p. 175)
Tubercle at base of left clasper erect 13
13. Tubercle at base of right clasper rounded, inconspicuous (Fig. 126) .... ***brimleyi* Knight** (p. 175)
Tubercle at base of right clasper erect, prominent (Fig. 127) .... ***dreisbachi* Knight** (p. 177)
14. Second antennal segment with pale band at base and middle 15
Second antennal segment with pale band only at base 16
15. Tubercle at base of left clasper long, slender (Fig. 128) .... ***conspurcatus* Knight** (p. 178)
Tubercle at base of left clasper absent (Fig. 129) .... ***borealis* Knight** (p. 179)
16. Left clasper with long and slender basal projection (Fig. 130) .... ***erectus* Van Duzee** (p. 179)
Left clasper with small basal projection 17
17. Right clasper slender, basal projection absent (Fig. 131) .... ***brooksi* Kelton** (p. 180)
Right clasper with prominent basal projection 18
18. Right clasper with rounded basal projection (Fig. 132) .. ***salicis* Knight** (p. 183)
Right clasper with pointed basal projection 19
19. Left clasper with triangular basal projection (Fig. 133) .... ***neglectus* Knight** (p. 183)
Left clasper with rounded basal projection (Fig. 134) .... ***onustus* Van Duzee** (p. 184)

Phytocoris laevis (Uhler)

Fig. 115; Map 49

Callodemus laevis Uhler, 1895:33.
Phytocoris laevis: Reuter, 1909:14.

Length 7.00–8.40 mm; width 2.24–2.45 mm. Head yellow. Rostrum 2.52–2.59 mm long. First antennal segment slender, almost as long as width of pronotum at base. Pronotum and scutellum yellow, spot above each callus and two spots at apex of scutellum brown. Hemelytra yellow; inner clavus, apical corium, and tip of cuneus marked with orange; pubescence short, golden. Legs yellow, femora often spotted with brown.

Remarks. This is the only yellow species in the Prairie Provinces. The females are smaller than the males. The frons is elevated, giving the head its blunt appearance. The jugum bears a short tubercle. Male genital segment without tubercles; genital claspers (Fig. 115).

Habitat. Collected by sweeping prairie vegetation, and at light.

Distribution. Colorado; now known to occur in Saskatchewan (Map 49).

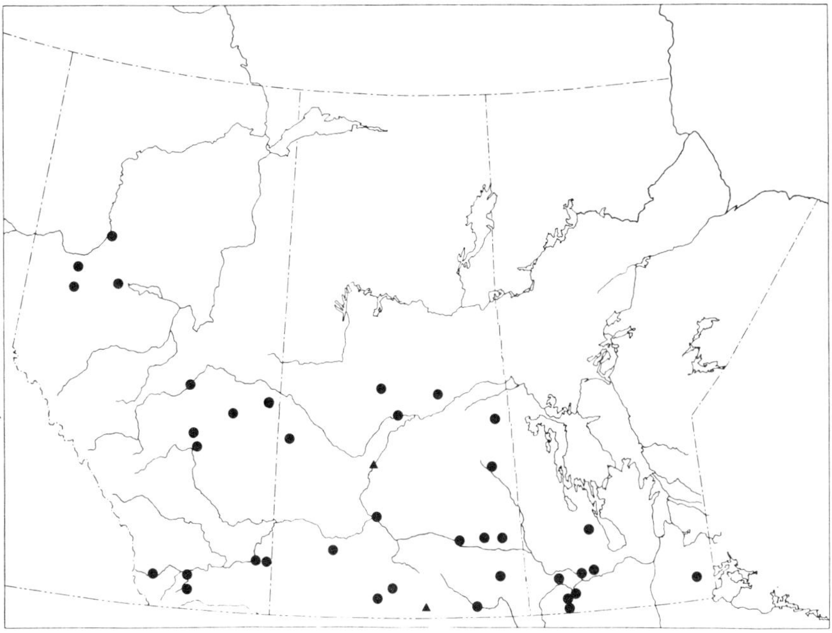

Map 49. Collection localities for *Phytocoris laevis* (▲) and *P. lasiomerus* (●).

Phytocoris lasiomerus Reuter

Figs. 114, 116; Map 49

Phytocoris lasiomerus Reuter, 1909:34.
Phytocoris scrupeus Provancher, 1887:108.

Length 7.35–8.05 mm; width 2.10–2.24 mm. Head yellowish brown marked with red. Rostrum 3.22–3.50 mm long. First antennal segment thickened with long, black bristles. Second antennal segment pale, apical one fourth black. Pronotum, scutellum, and hemelytra pale, often with reddish tinge; pubescence silvery, recumbent, mixed with erect hairs. Legs pale, femora often marked with reddish tinge.

Remarks. This species is distinguished by the thickened first antennal segment and by the black apex of the second segment (Fig. 114). The genital claspers (Fig. 116) are distinctive.

Habitat. Collected on herbaceous plants; probably predaceous.

Distribution. Transcontinental in northern USA, Ontario, Quebec; now known to occur in the Prairie Provinces (Map 49).

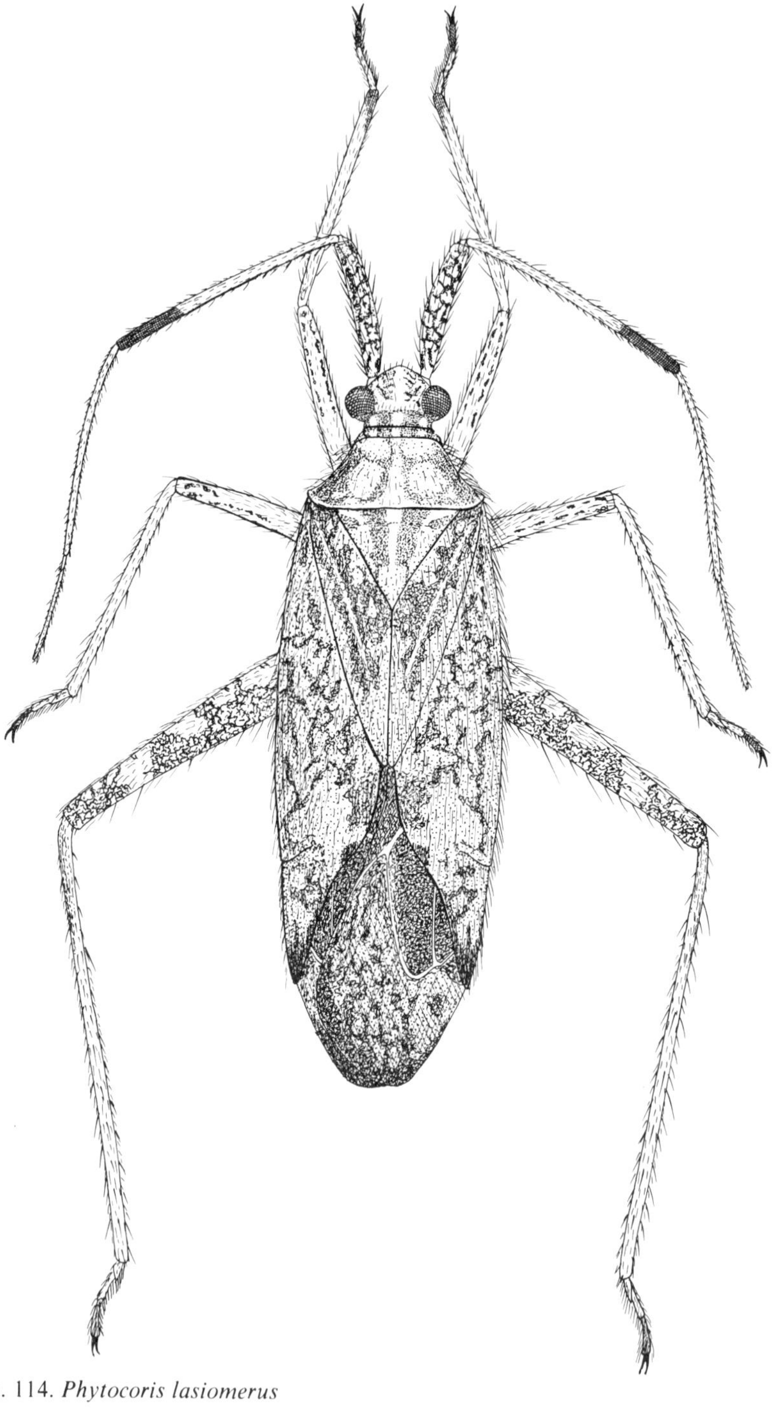

Fig. 114. *Phytocoris lasiomerus*

Phytocoris pallidicornis Reuter

Fig. 117; Map 50

Phytocoris pallidicornis Reuter, 1875*c*:69.

Length 6.72–7.00 mm; width 1.96–2.24 mm. Head yellow marked with red. Rostrum 2.94–3.15 mm long. First antennal segment yellow, thickened, marked with red, pubescence black, long, dense; second segment pale yellow. Pronotum and scutellum pale, often marked with reddish brown. Hemelytra pale marked with reddish brown; pubescence silvery, recumbent. Legs pale, femora marked with reddish brown.

Remarks. This species resembles *lasiomerus* and the two are found in similar habitats. It is easily separated from *lasiomerus* by the unicolored second antennal segment and by the absence of a tubercle at the base of the left clasper. The genital claspers (Fig. 117) are distinctive.

Habitat. Collected on herbaceous plants; probably predaceous.

Distribution. Transcontinental in northern USA, Quebec, British Columbia; Alberta, now known to occur in Saskatchewan and Manitoba (Map 50).

Phytocoris validus Reuter

Fig. 118; Map 50

Phytocoris validus Reuter, 1909:31.

Length 6.30–6.65 mm; width 1.68–1.96 mm. Head grayish marked with brown. Rostrum 2.66–2.80 mm long. First antennal segment thickened, grayish marked with brown, pubescence pale, long and erect only on inner surface; second segment pale. Pronotum and scutellum gray marked with brown. Hemelytra gray, lightly marked with brown; pubescence silvery, appressed. Legs gray, femora marked with brown.

Remarks. Females are brachypterous and the hemelytra do not extend beyond the abdomen. The tubercle above the base of the left clasper is prominent, and the genital claspers (Fig. 118) are distinctive.

Habitat. Collected on herbaceous plants; probably predaceous.

Distribution. Colorado; now known to occur in the Prairie Provinces (Map 50).

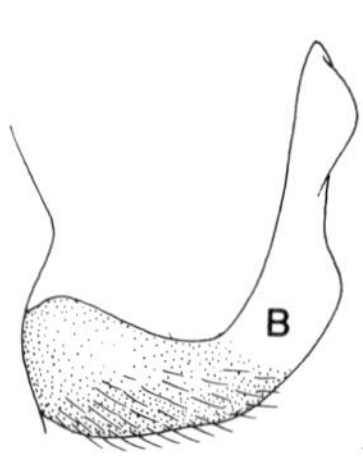

115

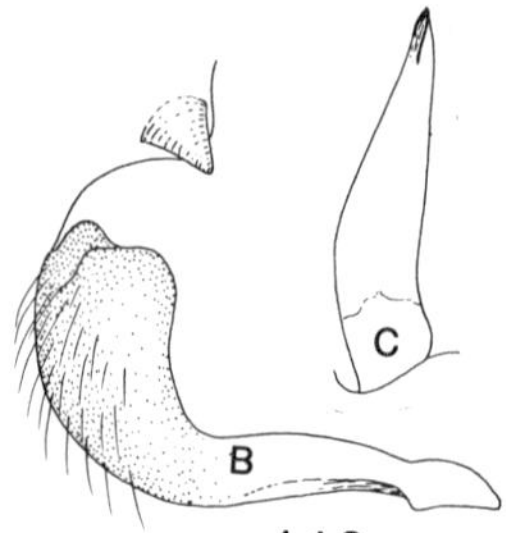

116

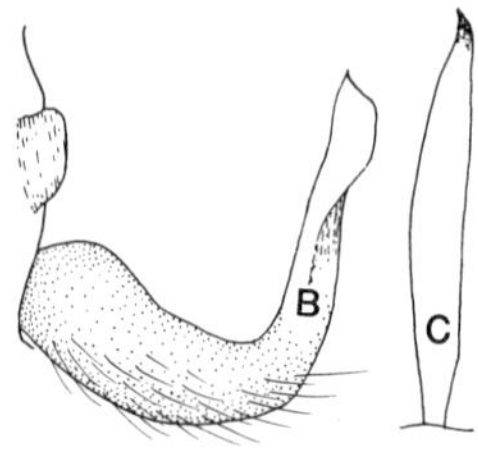

117

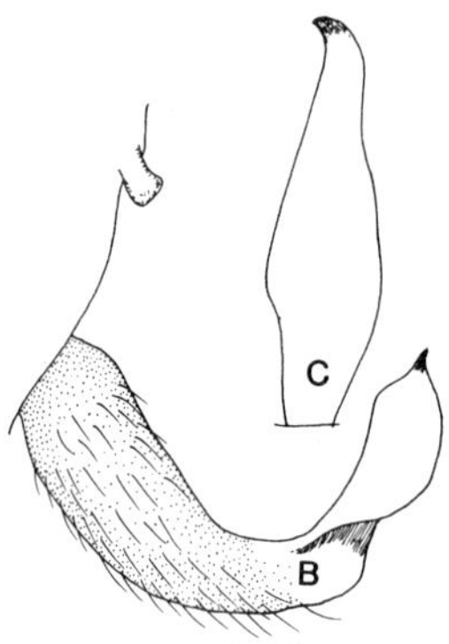

118

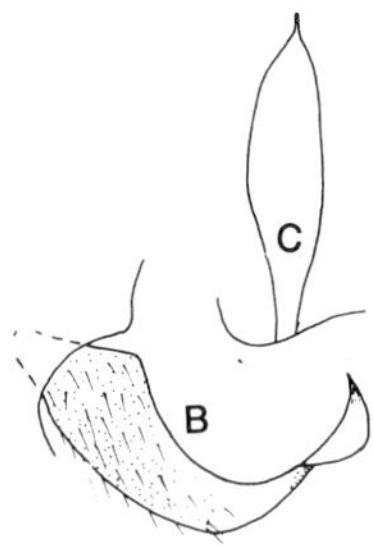

119

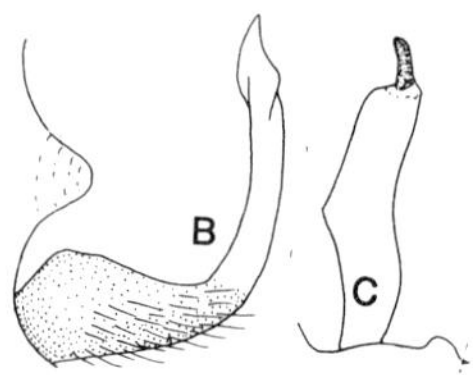

120

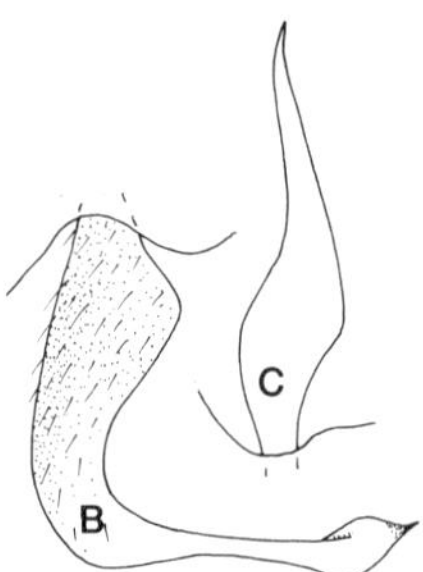

121

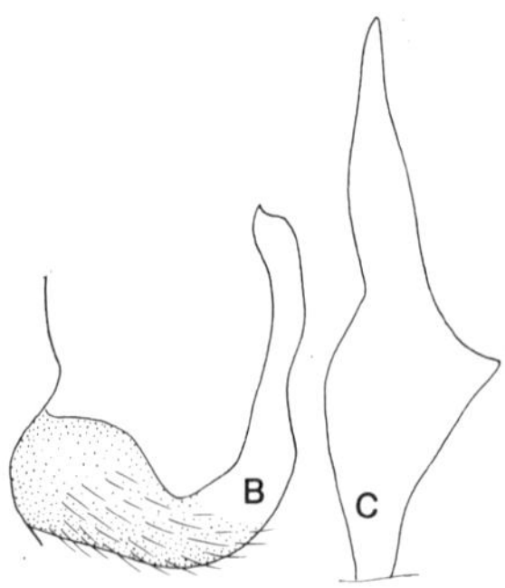

122

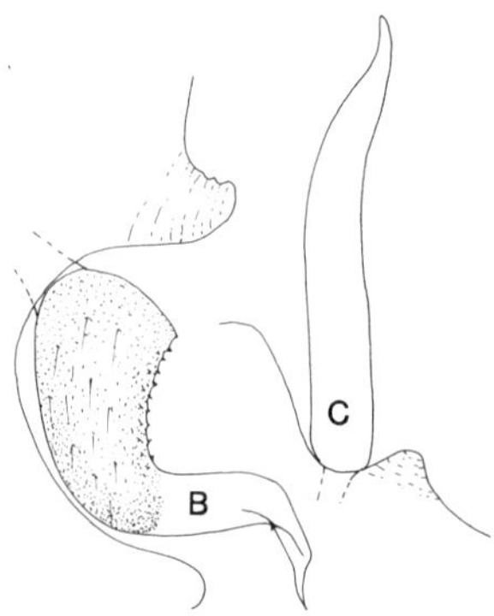

123

Figs. 115–123. Male claspers of *Phytocoris* spp. 115, *laevis*; 116, *lasiomerus*; 117, *pallidicornis*; 118, *validus*; 119, *stellatus*; 120, *listi*; 121, *pinicola*; 122, *conspersipes*; 123, *alpinus*.

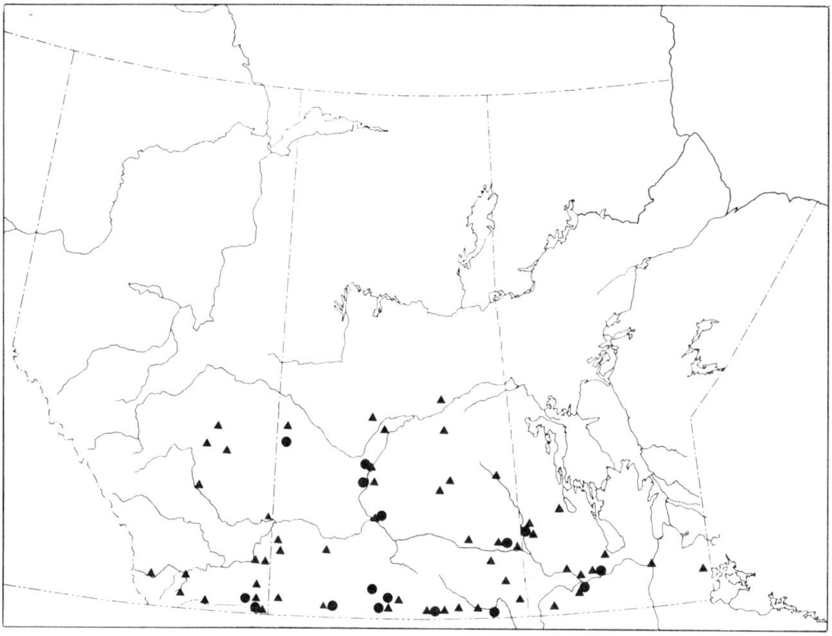

Map 50. Collection localities for *Phytocoris pallidicornis* (▲) and *P. validus* (●).

Phytocoris stellatus Van Duzee

Fig. 119; Map 51

Phytocoris stellatus Van Duzee, 1920:350.

Length 5.60–6.30 mm; width 1.96–2.10 mm. Head marked with reddish brown. Rostrum 2.10–2.24 mm long. First antennal segment shorter than head width, pubescence long and erect on inner surface; second segment brown. Pronotum marked with reddish brown, submargin brown. Scutellum pale, often marked with two brown spots near apex. Hemelytra reddish, often with two pale areas on corium, one near middle and one near apex; pubescence silvery, mixed with erect, black hairs. Legs marked with red.

Remarks. This species is distinguished by the short and slender first antennal segment and by the genital claspers (Fig. 119).

Habitat. Collected on *Pinus contorta*; predaceous on aphids.

Distribution. California, British Columbia; now known to occur in Alberta (Map 51).

Map 51. Collection localities for *Phytocoris stellatus* (●), *P. listi* (★), and *P. pinicola* (■).

Phytocoris listi Knight

Fig. 120; Map 51

Phytocoris listi Knight, 1928*a*:30.

Length 5.18–5.46 mm; width 1.54–1.70 mm. Head grayish marked with brown. Rostrum 2.80–2.94 mm long. First antennal segment shorter than head width. Pronotum grayish brown. Scutellum dark brown, basal angles and apex pale. Hemelytra gray marked with brown; pubescence silvery, mixed with erect, black hairs. Tibiae pale banded with black.

Remarks. This species is distinguished by the short first antennal segment, by the grayish brown color, by the long rostrum, and by the genital claspers (Fig. 120).

Habitat. Unknown.

Distribution. Colorado, South Dakota; now known to occur in Saskatchewan (Map 51).

Phytocoris pinicola Knight

Fig. 121; Map 51

Phytocoris pinicola Knight, 1920:59.

Length 4.34–4.90 mm; width 1.68–1.75 mm. Head marked with red. Rostrum 1.96–2.10 mm long. First antennal segment shorter than head width, pubescence long and erect on inner surface; second segment brown. Pronotum and scutellum reddish brown. Hemelytra with reddish tinge; pubescence silvery, recumbent, intermixed with slanting, brown hairs. Legs tinged with red.

Remarks. This species and *stellatus* are similar in appearance, but may be separated from each other by the genital claspers (Fig. 121).

Habitat. Collected on *Pinus banksiana*; predaceous on aphids.

Distribution. Northeastern States, Missouri, Mississippi; now known to occur in Manitoba (Map 51).

Phytocoris conspersipes Reuter

Fig. 122; Map 52

Phytocoris conspersipes Reuter, 1909:22.

Length 4.76–4.90 mm; width 1.68–1.82 mm. Head marked with red. Rostrum 1.82–1.96 mm long. First antennal segment short, pubescence on inner surface erect; second segment black. Pronotum with reddish tinge. Scutellum marked with red, apex pale. Hemelytra marked with red; pubescence silvery, mixed with erect, black hairs. Legs marked with red.

Remarks. This species and *pinicola* are similar in color and appearance, but are separated by the genital claspers (Fig. 122).

Habitat. Collected on *Pinus banksiana* and *P. contorta.*

Distribution. Eastern States, Ontario; now known to occur in the Prairie Provinces (Map 52).

Phytocoris alpinus Kelton

Figs. 123, 135; Map 52

Phytocoris alpinus Kelton, 1979*a*:689.

Length 8.68–8.96 mm; width 2.61–2.80 mm. Head mostly black, vertex partly pale; frons faintly striate. Rostrum 3.50–3.64 mm long. Antennae black, first segment with few pale spots. Pronotum black, median line between calli often pale. Scutellum black, basal angles, median line, and apex often pale. Hemelytra mostly black, corium with several pale patches, cuneus pale at base (Fig. 135); pubescence silvery, appressed, intermixed with slanting, black hairs. Coxae pale; femora pale irrorate with black; tibiae black, hind tibia with two pale rings.

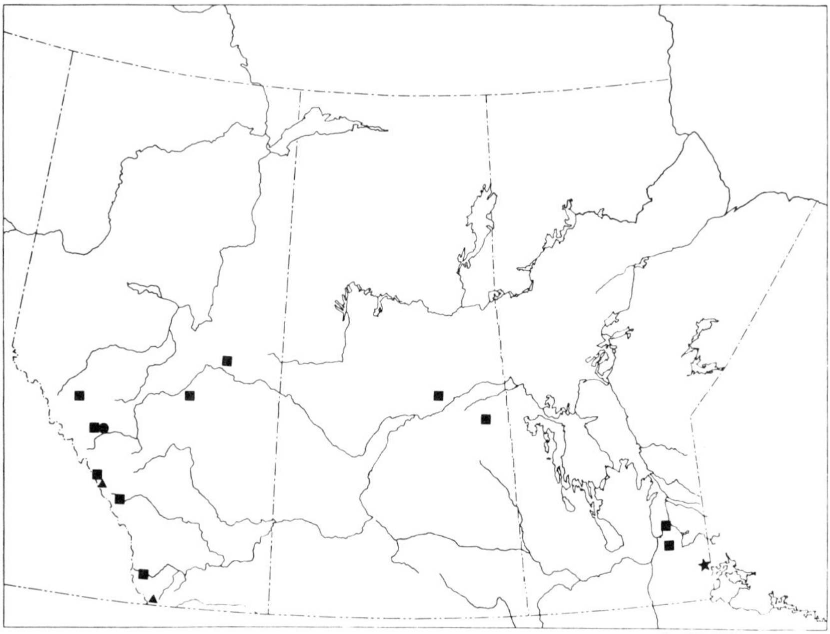

Map 52. Collection localities for *Phytocoris conspersipes* (■), *P. alpinus* (▲), *P. michiganae* (★), and *P. albertae* (●).

Remarks. This is the darkest species encountered in the *junceus* Knight group. It is distinguished by the black antennae and pronotum, by the biannulate hind tibia, and by the prominent tubercles on the genital segment (Fig. 123).

Habitat. Collected on *Pinus contorta*.

Distribution. Alberta (Map 52).

Phytocoris michiganae Knight

Fig. 124; Map 52

Phytocoris michiganae Knight, 1974:128.

Length 7.35–8.05 mm; width 2.24–2.38 mm. Head light yellowish brown, clypeus and juga marked with red. Rostrum 3.22–3.50 mm long.

First antennal segment mostly pale, lightly marked with brown, ventral surface with longitudinal brown line; second segment light brown, apex dark brown; terminal segments black. Pronotum light reddish brown. Scutellum reddish brown, basal angles, median line, and apex white. Hemelytra light reddish brown; corium with three white spots, one near base, one at center, and one near apex; cuneus pale on basal half, reddish brown on apical half; pubescence silvery, mixed with longer black hairs. Legs pale, femora irrorate with reddish brown; tibiae pallid.

Remarks. This is the most reddish brown species of the group. The tubercles at the bases of claspers are prominent (Fig. 124).

Habitat. Collected on *Pinus banksiana.*

Distribution. Michigan; now known to occur in Manitoba (Map 52).

Phytocoris albertae Knight

Fig. 125; Map 52

Phytocoris albertae Knight, 1974:131.

Length 8.68–9.10 mm; width 2.66–2.80 mm. Head mostly black, vertex with few pale areas. Rostrum 3.64–3.78 mm long. Antennae mostly black, first segment with several pale spots. Pronotum pale at middle, lateral margins and basal submargin black. Scutellum black, basal angles and median line often pale. Hemelytra fuscous white with irregular patches of black; cuneus tinged with red on basal half; pubescence silvery, appressed, intermixed with slanting, black hairs. Coxae pale, femora irrorate with brown, hind femur often with subapical pale band; tibiae fuscous to black, hind tibia often with two pale rings.

Remarks. This species closely resembles *alpinus*, but is not as dark, and the tubercle at the base of left clasper is different (Fig. 125). It was originally included in the series with *junceus*, but Knight (1974) showed that the Nordegg specimens were distinct species.

Habitat. Collected on *Picea glauca.*

Distribution. Alberta (Map 52).

Phytocoris brimleyi Knight

Fig. 126; Map 53

Phytocoris brimleyi Knight, 1974:130.

Length 7.42–7.84 mm; width 2.52–2.66 mm. Head pale yellow, frons, clypeus, lora, and juga red or black. Rostrum 3.36–3.64 mm long. First

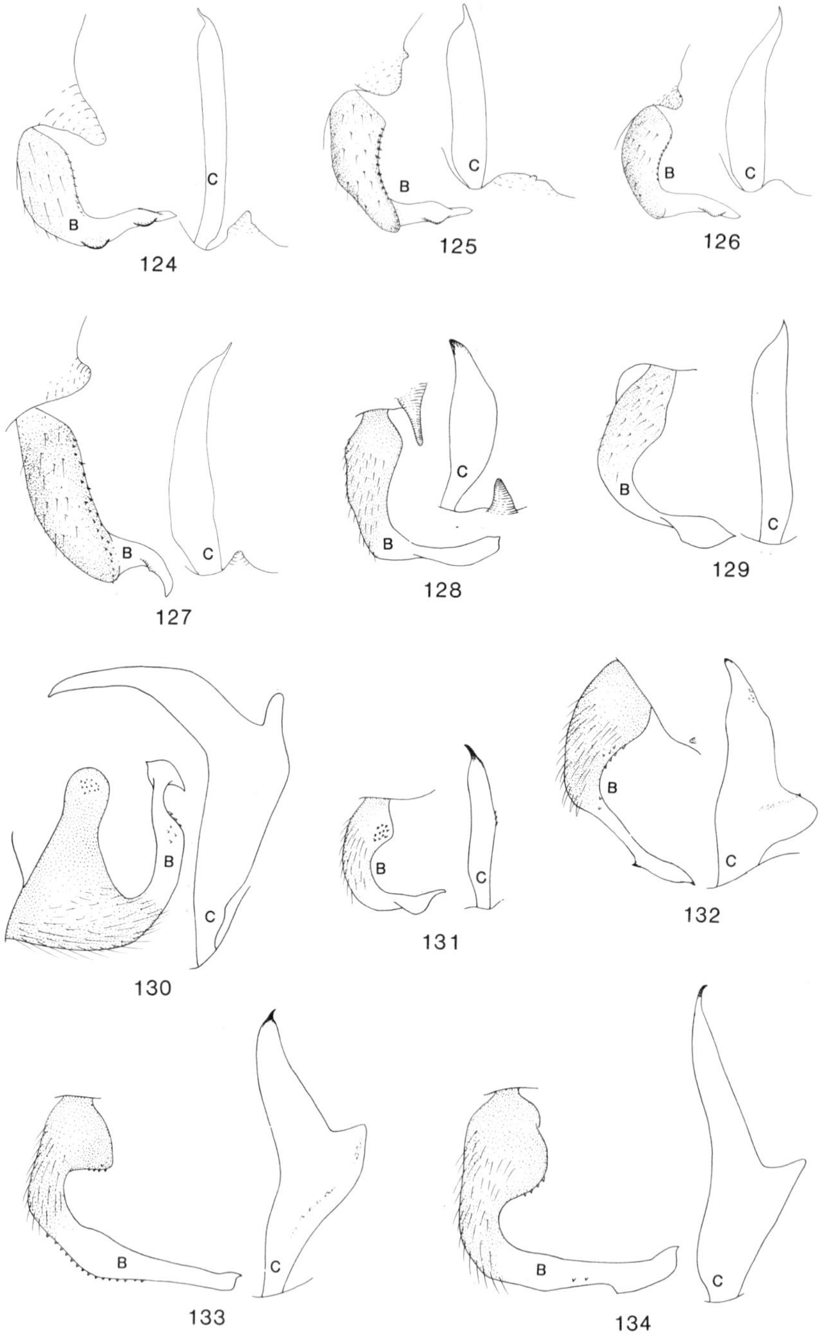

Figs. 124–134. Male claspers of *Phytocoris* spp. 124, *michiganae*; 125, *albertae*; 126, *brimleyi*; 127, *dreisbachi*; 128, *conspurcatus*; 129, *borealis*; 130, *erectus*; 131, *brooksi*; 132, *salicis*; 133, *neglectus*; 134, *onustus*.

antennal segment pale, ventral surface with broken longitudinal brown stripe; second segment mostly pale, tip fuscous; terminal segments black. Pronotum light yellowish brown, lateral margins and subbasal margin brown. Scutellum brown, lateral angles, median line, and apex pale. Hemelytra mostly brown with irregular patches of white, cuneus tinged with red. Legs pallid, femora irrorate with reddish brown.

Remarks. This species resembles *michiganae*, but is more brownish, and the left tubercle is short and rounded (Fig. 126).

Habitat. Collected on *Picea glauca*.

Distribution. Ontario; now known to occur in Manitoba and Alberta (Map 53).

Phytocoris dreisbachi Knight

Fig. 127; Map 53

Phytocoris dreisbachi Knight, 1974:125.

Length 7.42–7.56 mm; width 2.38–2.59 mm. Head mostly black, area between eyes yellowish. Rostrum 3.43–3.64 mm long. First antennal seg-

Map 53. Collection localities for *Phytocoris brimleyi* (▲) and *P. dreisbachi* (●).

ment pallid, lightly marked with brown, ventral surface with longitudinal brown line; second and terminal segments brown to black. Pronotum pallid, lateral margins and subbasal margin dark brown. Scutellum dark brown, lateral margins and apex white. Hemelytra dark brown, with pale mottling; corium with three irregular white spots, one near base, one at center, and one near apex; cuneus pallid on basal half; pubescence silvery, mixed with longer erect hairs. Legs pallid; femora reticulate with brown.

Remarks. This species closely resembles *brimleyi*, but differs in having a broader tubercle at the base of the left clasper (Fig. 127). This tubercle is similar to that of *alpinus*, but the latter has uniformly black pronotum.

Habitat. Collected on *Pinus banksiana* in Manitoba, and on *P. contorta* in Alberta.

Distribution. Michigan; now known to occur in the Prairie Provinces (Map 53).

Phytocoris conspurcatus Knight

Fig. 128; Map 54

Phytocoris conspurcatus Knight, 1920:61.

Length 5.74–5.95 mm; width 2.17–2.31 mm. Head gray marked with brown and reddish brown. Rostrum 2.66–2.80 mm long. First antennal

Map 54. Collection localities for *Phytocoris conspurcatus* (■), *P. borealis* (●), and *P. erectus* (▲).

segment with long, erect, white hairs; second segment with white band at base and middle. Pronotum gray, subbasal margin often black. Hemelytra gray marked with brown; pubescence sericeous, white and black, mixed with slanting, black hairs. Legs marked with brown.

Remarks. The banded second antennal segment, and the long and slender tubercle at the base of left clasper (Fig. 128) readily distinguish this species.

Habitat. Collected on *Salix* spp.; predaceous on aphids.

Distribution. Northeastern and North Central States, Mississippi, Ontario, British Columbia; now known to occur in the Prairie Provinces (Map 54).

Phytocoris borealis Knight

Fig. 129; Map 54

Phytocoris borealis Knight, 1926*c*:158.

Length 6.44–6.80 mm; width 2.24–2.38 mm. Head marked with black, lower half of face white. Rostrum 2.80–2.90 mm long. First antennal segment with erect, white hairs; second segment with white band at base and middle. Pronotum marked with black. Hemelytra mostly marked with black; pubescence white, matted, intermixed with black, slanting hairs. Legs marked with black.

Remarks. This species resembles *conspurcatus*, but lacks the black, sericeous pubescence on the hemelytra, and the tubercle near the base of left clasper is absent (Fig. 129).

Habitat. Unknown.

Distribution. New York, Ontario; now known to occur in Manitoba (Map 54).

Phytocoris erectus Van Duzee

Fig. 130; Map 54

Phytocoris erectus Van Duzee, 1920:345.

Length 5.46–5.88 mm; width 1.96–2.24 mm. Head marked with brown and black. Rostrum 2.52–2.66 mm long. First antennal segment with several long, pale hairs; second antennal segment with pale band at base. Pronotum light brown, subbasal margin dark brown. Hemelytra brown, apex of corium with diagonal dark brown bar; pubescence silvery, appressed, intermixed with slanting, darker hairs. Legs marked with brown.

Remarks. This species lacks the pale, middle band on the second antennal segment, and the tubercle at the base of the left clasper is absent. The left clasper (Fig. 130) has a long, basal projection.

Habitat. Collected on *Salix* spp.; predaceous on aphids.

Distribution. Northeastern and Central States, Ontario, Quebec; now known to occur in Saskatchewan (Map 54).

Phytocoris brooksi Kelton

Figs. 131, 136; Map 55

Phytocoris brooksi Kelton, 1979*a*:689.

Length 5.46–5.67 mm; width 1.89–2.10 mm. Head marked with red and brown. Rostrum 2.52–2.66 mm long. First antennal segment with several long, white bristles; second segment brown or black with pale band at base. Pronotum brown, calli often marked with red. Hemelytra marked with brown or black; pubescence silvery, intermixed with slanting, black hairs. Ventral surface white, pleura and side of abdomen black. Legs spotted with brown.

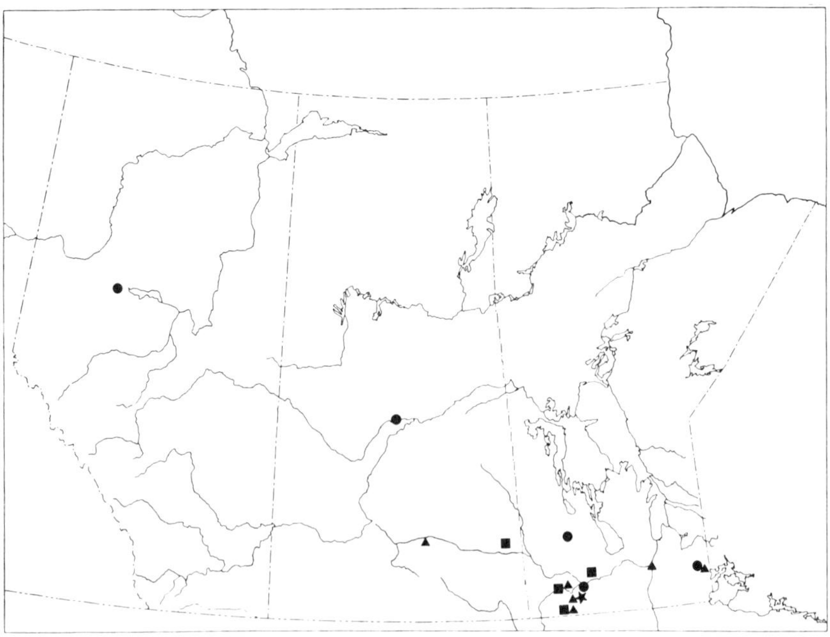

Map 55. Collection localities for *Phytocoris brooksi* (■), *P. salicis* (▲), *P. neglectus* (●), and *P. onustus* (★).

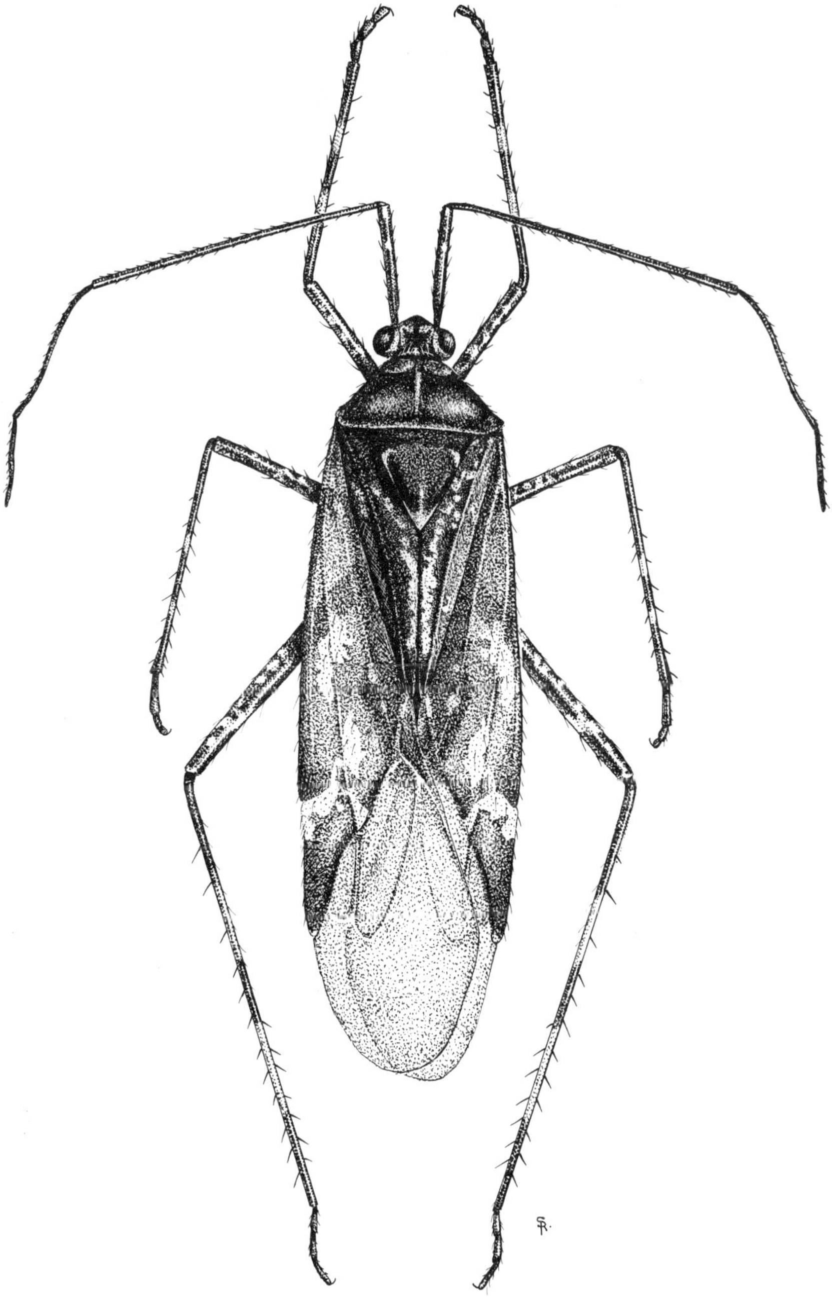

Fig. 135. *Phytocoris alpinus*

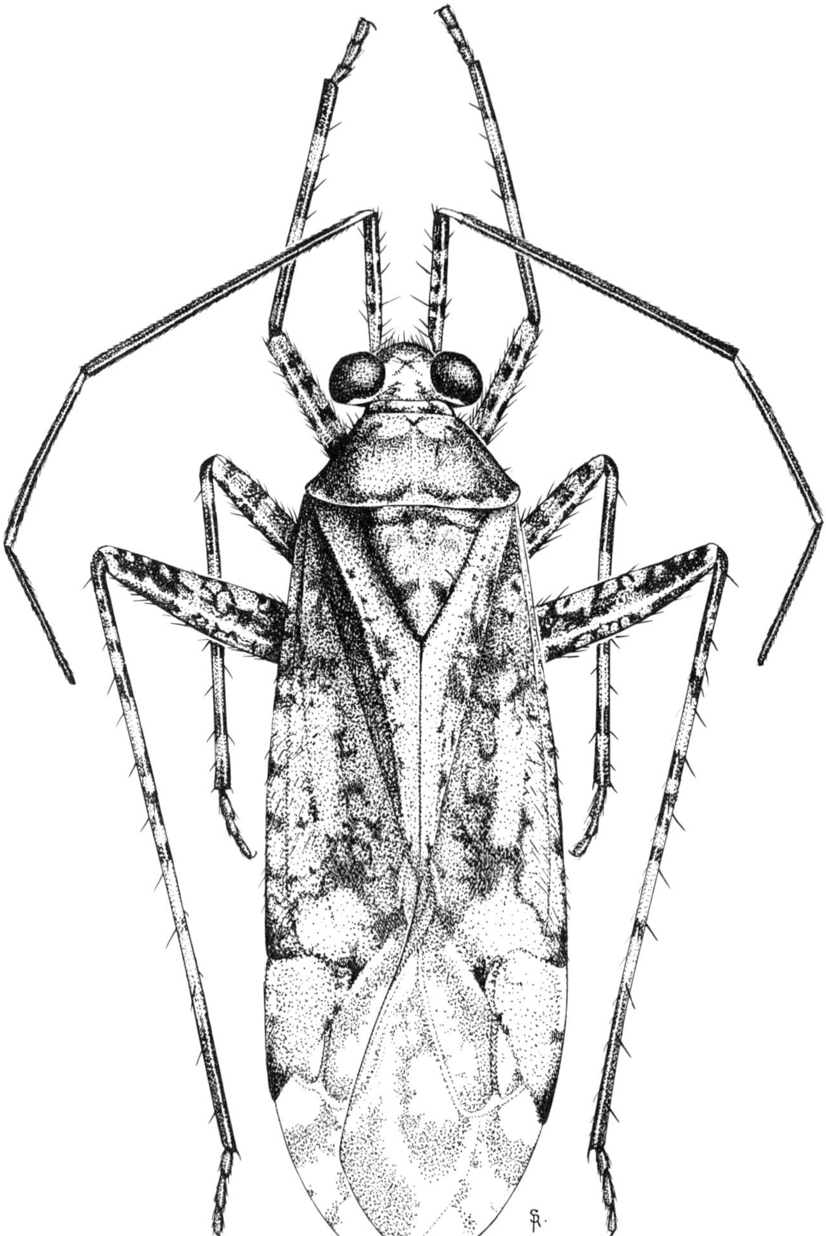

Fig. 136. *Phytocoris brooksi*

Remarks. This species is distinguished by the pale band at the base of second antennal segment (Fig. 136), and by the slender, right clasper (Fig. 131).

Habitat. Collected on *Quercus macrocarpa*; predaceous on aphids.

Distribution. Saskatchewan and Manitoba (Map 55).

Phytocoris salicis Knight

Fig. 132; Map 55

Phytocoris salices Knight, 1920:56.

Length 5.60–6.16 mm; width 2.10–2.38 mm. Head marked with brown. Rostrum 2.59–2.80 mm long. First antennal segment spotted with brown, bristles on inner surface white, long; second segment with pale band at base. Pronotum marked with brown. Hemelytra brown; pubescence silvery, intermixed with slanting, golden hairs. Legs marked with brown, coxae pale.

Remarks. This species is distinguished by the pale band at the base of second antennal segment, by the tiny tubercle near the base of left clasper, and by the broad and rounded basal projection on the right clasper (Fig. 132).

Habitat. Collected on *Salix* spp.; predaceous on aphids.

Distribution. Eastern and Central States, Ontario; now known to occur in Manitoba and Saskatchewan (Map 55).

Phytocoris neglectus Knight

Fig. 133; Map 55

Phytocoris neglectus Knight, 1920:54.

Length 6.16–6.58 mm; width 2.10–2.38 mm. Head brown marked with darker brown. Rostrum 2.80–2.94 mm long. First antennal segment spotted with black, bristles on inner surface white, long; second segment with pale band at base. Pronotum brown, subbasal margin darker brown with clumps of silvery pubescence. Hemelytra brown; pubescence silvery, appressed, mostly in small clumps, intermixed with slanting, darker hairs. Legs marked with brown, coxae usually pale.

Remarks. This species is distinguished by the pale band at the base of second antennal segment, and by the broad and triangular basal projection on the left clasper (Fig. 133).

Habitat. Collected on *Picea glauca* and *Abies balsamea*; predaceous on aphids.

Distribution. Eastern half of USA, Ontario; now known to occur in the Prairie Provinces (Map 55).

Phytocoris onustus Van Duzee

Fig. 134; Map 55

Phytocoris onustus Van Duzee, 1920:344.

Length 7.00–7.28 mm; width 2.38–2.52 mm. Head marked with black. Rostrum 2.94–3.08 mm long. First antennal segment spotted with brown, bristles on inner surface white, long; second segment with pale band at base. Pronotum brown, subbasal margin black, interrupted by pale spaces. Hemelytra brown, outer margin of apex darker brown, inner apex pale; pubescence silvery, appressed, intermixed with slanting, black hairs. Legs marked with brown, coxae pale.

Remarks. This species is distinguished by the pale band at the base of second antennal segment, and by the sharp, triangular projection on the right clasper (Fig. 134).

Habitat. Unknown.

Distribution. Northeastern States, Georgia, Illinois, Ontario; now known to occur in Manitoba (Map 55).

Genus *Adelphocoris* Reuter

Elongate, robust, pubescent species. Head oblique, frons smooth, carina between eyes absent, vertex with shallow longitudinal groove. Rostrum extending to hind coxae. Pronotum finely punctate, rugose, lateral margins rounded. Hemelytra punctate, densely pubescent. Dorsolateral margin of genital segment with short, sharp tubercle.

Three species occur in North America, and all are found in the Prairie Provinces; one of them an introduced European species.

Key to species of *Adelphocoris*

1. Dorsal surface greenish yellow, pronotum usually with pair of black spots, scutellum with two brown lines .... ***lineolatus*** **(Goeze)** (p. 185)
   Dorsal surface brown or reddish 2
2. Hemelytra brown, costal margin yellow .... ***rapidus*** **(Say)** (p. 186)
   Hemelytra brown, costal margin and cuneus red, or hemelytra yellow or red .... ***superbus*** **(Uhler)** (p. 187)

Adelphocoris lineolatus (Goeze)

Map 56

Cimex lineolatus Goeze, 1778:267.
Adelphocoris lineolatus: Reuter, 1896:222.

Length 7.00–9.10 mm; width 2.52–3.08 mm. Head and pronotum yellowish green, pronotum often with two black spots behind calli. Scutellum yellowish green with two brown lines. Hemelytra greenish yellow, clavus and triangular area on apical half of corium often fuscous. Ventral surface yellowish green; legs green; femora spotted with black.

Remarks. This species was introduced from Europe. It is distinguished by the greenish yellow color and by the black spotting on the femora.

Habitat. Collected on *Medicago sativa*, *Melilotus* spp., *Trifolium* spp., and on other herbaceous plants.

Distribution. North Central States, Eastern Canada; Prairie Provinces (Map 56).

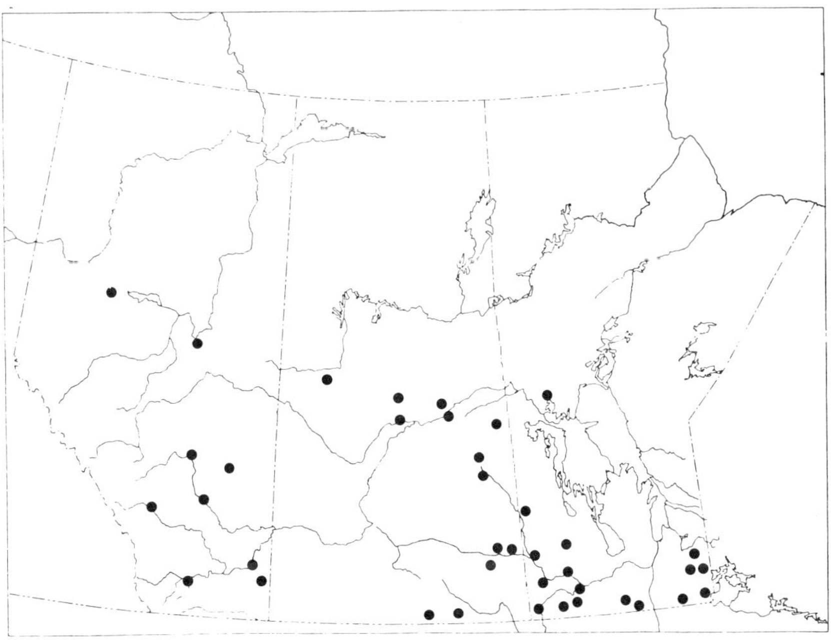

Map 56. Collection localities for *Adelphocoris lineolatus*.

Adelphocoris rapidus (Say)

Map 57

Capsus rapidus Say, 1832:20.
Adelphocoris rapidus: Van Duzee, 1908:113.

Length 6.65–8.05 mm; width 2.66–3.08 mm. Head reddish brown, clypeus darker brown. Pronotum yellowish brown, basal half often with two large, brown spots. Hemelytra brown, costal margin yellow. Ventral surface reddish brown; legs lighter brown.

Remarks. The color of this species varies considerably. The brown spots on the pronotum may be absent. The hemelytra may be diffuse brown with no demarkation from the yellow costal margin and the darker clavus and corium.

Habitat. Collected on many herbaceous plants.

Distribution. Widespread in USA, Eastern Canada; Prairie Provinces (Map 57).

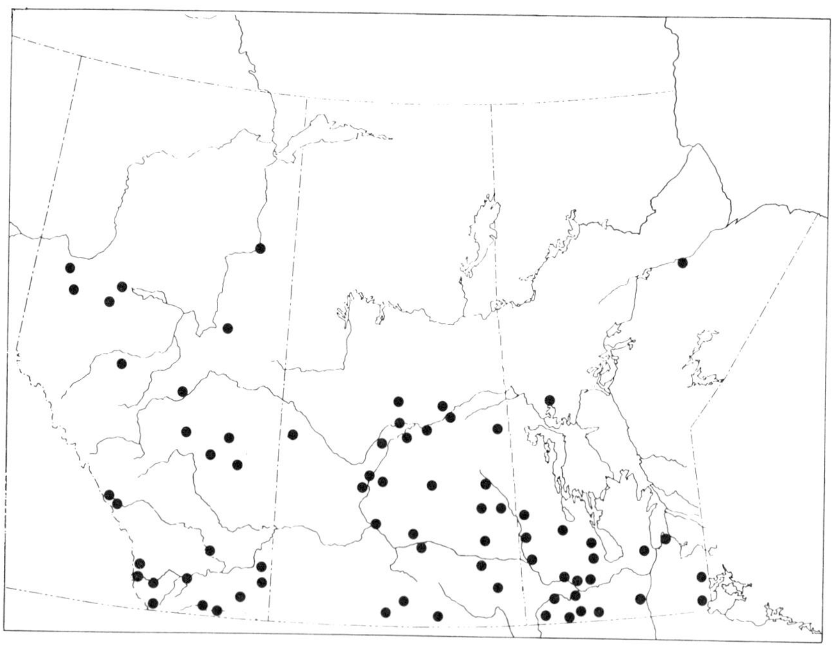

Map 57. Collection localities for *Adelphocoris rapidus*.

Adelphocoris superbus (Uhler)

Map 58

Calocoris superbus Uhler, 1875:838.
Adelphocoris superbus: Tucker, 1907:60.

Length 6.65–8.26 mm; width 2.59–3.08 mm. Head red, clypeus darker red. Pronotum red, seldom with two brown spots as in *rapidus*. Hemelytra red, median portion and scutellum often darker. Ventral surface red; femora red, tibiae lighter red.

Remarks. Although this species and *rapidus* are structurally similar, they are strikingly different in color.

Habitat. Collected on *Glycyrrhiza lepidota*, *Monarda fistulosa*, *Lupinus* spp., and *Medicago sativa*.

Distribution. North Central and western States; Saskatchewan and Alberta (Map 58).

Genus *Calocoris* Fieber

Elongate, reddish brown, pubescent species. Head oblique, frons smooth, carina between eyes absent, vertex smooth. Rostrum extending to

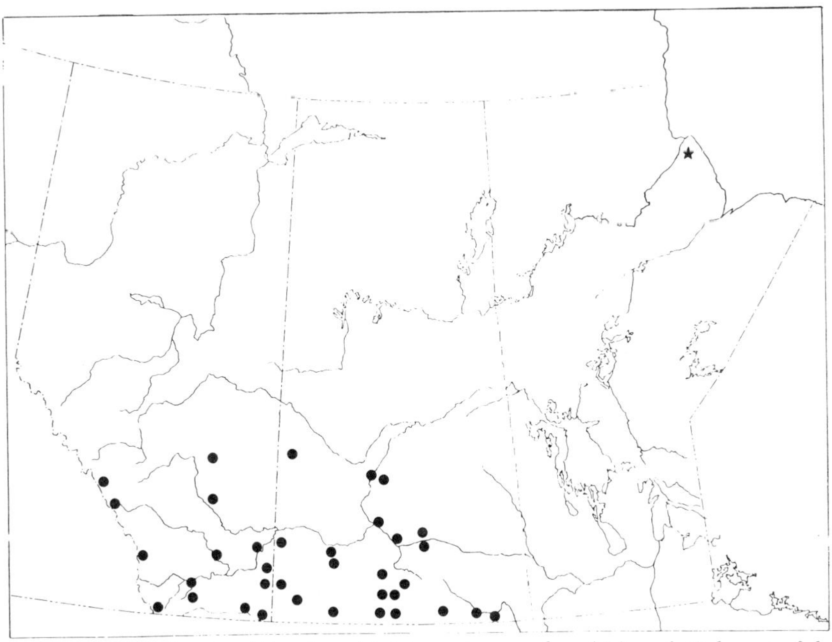

Map 58. Collection localities for *Adelphocoris superbus* (●) and *Calocoris fulvomaculatus* (★).

hind coxae. Pronotum smooth, lateral margins rounded. Hemelytra very finely punctate, densely pubescent. Ventral surface mostly black, legs pale, spotted with black.

Seven species are known from North America, two of these are Holarctic in distribution. *Calocoris norvegicus* occurs in British Columbia and Eastern Canada. *C. fulvomaculatus* occurs in northern Canada and Alaska.

Calocoris fulvomaculatus (De Geer)

Map 58

Cimex fulvomaculatus De Geer, 1773:294.
Calocoris fulvomaculatus: Fieber, 1861:253.

Length 6.16–7.00 mm; width 2.24–2.66 mm. Head mostly black or reddish brown, two spots on vertex between eyes often pale. Second antennal segment long, slender. Pronotum reddish brown to black. Hemelytra reddish brown to black, cuneus pale or reddish. Genital segment with small tubercle on left dorsal margin.

Remarks. This species is distinguished by the reddish brown color, by the long second antennal segment, by the smooth pronotum, and by the spotted legs.

Habitat. Unknown.

Distribution. Alaska, Yukon, British Columbia; now known to occur in Manitoba (Map 58).

Subfamily Orthotylinae Van Duzee

In North America this is the second largest subfamily with 55 genera and approximately 500 species. Most of the species are phytophagous.

The following are the subfamily characteristics: 1) large, free parempodia converging at apices; 2) small and depressed pronotal collar; 3) male genitalia basically with flexible ductus seminis, with or without spicula.

In the Prairie Provinces the subfamily is represented by 3 tribes, 24 genera, and 87 species.

Key to tribes of Orthotylinae

1. Compact, black species with silvery, sericeous pubescence; head dorsoventrally elongated; hind femur saltatorial (Fig. 137) .... **Halticini** (p. 189)

Elongate or oval species, if black, without sericeous pubescence; head not as above; hind femur not saltatorial 2
2. Hemelytra without transverse bands of silvery, sericeous pubescence .... **Orthotylini** (p. 194)
Hemelytra with transverse bands of silvery, sericeous pubescence (Fig. 138) .... **Pilophorini** (p. 274)

Tribe Halticini

In the Prairie Provinces the tribe is represented by two genera and six species.

Key to genera of Halticini

1. Eyes pedunculate (Fig. 139); dull black species; second antennal segment shorter than width of pronotum at base .... ***Labops*** **Burmeister** (p. 189)
Eyes not pedunculate; shiny species; second antennal segment longer than width of pronotum at base .... ***Halticus*** **Hahn** (p. 194)

Genus *Labops* Burmeister

Black, compact, pubescent species. Head vertical, eyes protruding outwards beyond anterior angles of pronotum, carina between them distinct. Pronotum subrectangular, lateral margins rounded. Hemelytra pruinose, dimorphic, in brachypterous forms wing membrane absent. Ventral surface with sericeous pubescence.

The genus was revised by Slater (1954) recording nine species from North America; five occur in the Prairie Provinces.

Key to species of *Labops*

1. Pubescence short, sparse; broad, pale costal margin interrupted by black at middle of corium .... ***verae*** **Knight** (p. 190)
Pubescence long, dense; narrow, pale costal margin not interrupted 2
2. Hind tibia black or brownish black 3
Hind tibia partly reddish brown or yellow 4
3. Jugum greatly inflated in males, less so in females; jugum with dense hair .... ***hesperius*** **Uhler** (p. 191)
Jugum not inflated; jugum with sparse, short hair .... ***brooksi*** **Slater** (p. 191)
4. Clypeus with prominent clump of white hair; frons black; hind tibia not pilose .... ***tumidifrons*** **Knight** (p. 192)
Clypeus without prominent hair; frons with pale median line; hind tibia strongly pilose .... ***hirtus*** **Knight** (p. 192)

Labops verae Knight

Map 59

Labops verae Knight, 1929*d*:214.

Length 3.36–3.78 mm; width 1.61–1.82 mm. Head black, apical half of clypeus, jugum, margin along eye, and carina next to eye often pale or reddish. Rostrum 1.40–1.54 mm long. Antennae black, second antennal segment reddish. Pronotum and scutellum black. Hemelytra black, basal half and apical margin of corium pale; pubescence short. Ventral surface black, legs mostly black, tibiae reddish. Brachypterous.

Remarks. This species is distinguished by the almost glabrous appearance, by the reddish second antennal segment and tibiae, and by the color pattern on the hemelytra.

Habitat. Unknown.

Distribution. Washington, Northwest Territories, British Columbia; Alberta, now known to occur in Manitoba (Map 59).

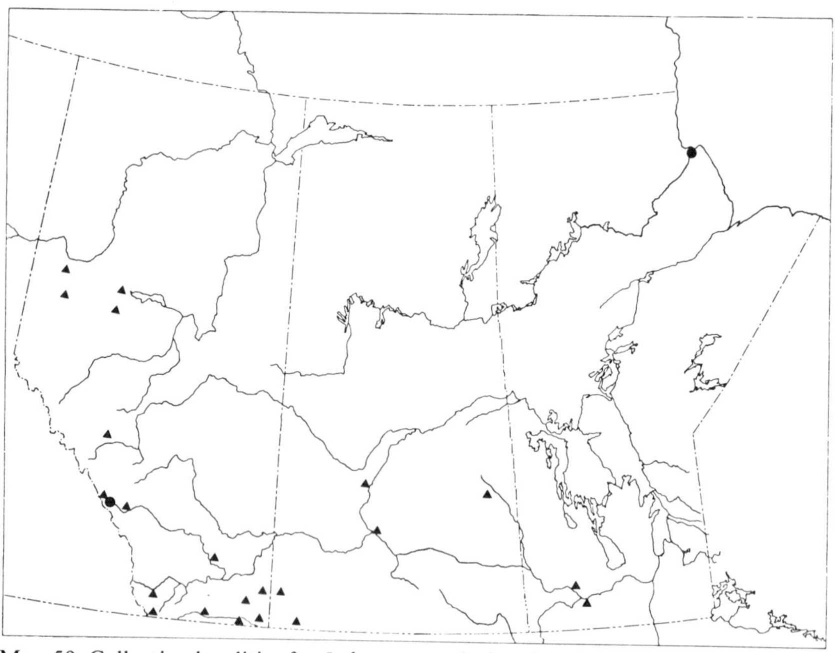

Map 59. Collection localities for *Labops verae* (●) and *L. hesperius* (▲).

Labops hesperius Uhler

Map 59

Labops hesperius Uhler, 1872:416.

Length 3.92–4.20 mm; width 1.68–2.24 mm. Head black, yellow areas similar to those of *brooksi*, jugum greatly inflated in males, less so in females. Rostrum 1.54–1.68 mm long. Antennae black. Pronotum and scutellum black. Hemelytra similar to *brooksi* in color and pubescence. Ventral surface and legs mostly black, hind tibia mostly pale. Macropterous and brachypterous forms.

Remarks. This species is distinguished by the greatly inflated jugum.

Habitat. Collected on *Agropyron cristatum*, range grasses, and *Rosa arkansana*.

Distribution. Transcontinental in USA, Northwest Territories, Ontario, Quebec, British Columbia; Alberta, Manitoba, now known to occur in Saskatchewan (Map 59).

Labops brooksi Slater

Map 60

Labops brooksi Slater, 1954:64.

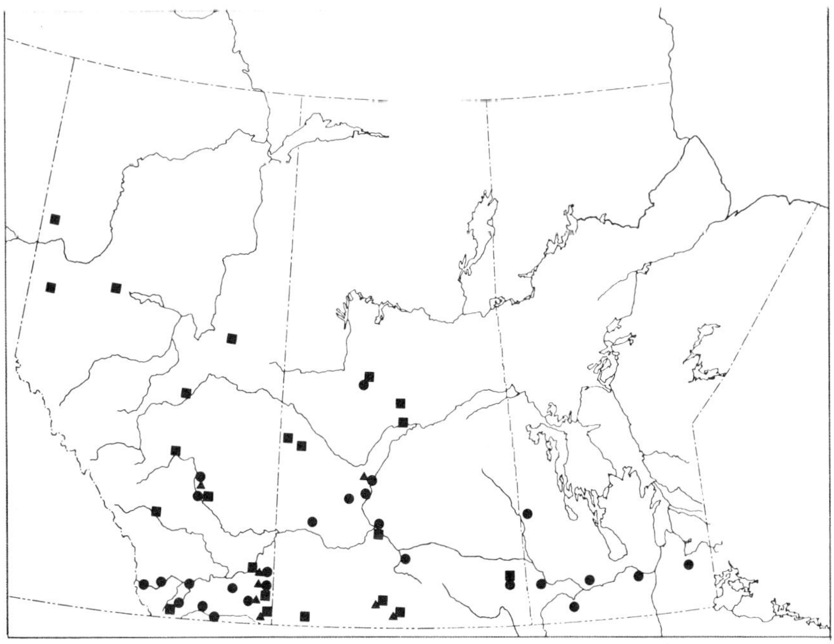

Map 60. Collection localities for *Labops brooksi* (●), *L. tumidifrons* (▲), and *L. hirtus* (■).

Length 3.91–4.83 mm; width 1.75–2.13 mm. Head black, tip of clypeus, jugum, short median line on frons, base of lorum and area under eye, and spot next to eye on top yellow. Rostrum 1.40–1.47 mm long. Antennae black. Pronotum and scutellum black. Hemelytra black, costal margin pale; pruinose; pubescence long, slanting. Ventral surface and legs mostly black. Males macropterous, females brachypterous.

Remarks. This species is distinguished by the flattened jugum and the black hind tibia.

Habitat. Collected on grasses.

Distribution. British Columbia; Prairie Provinces (Map 60).

Labops tumidifrons Knight

Map 60

Labops tumidifrons Knight, 1922:259.

Length 3.08–4.20 mm; width 1.19–1.68 mm. Head black, jugum, area beneath eye, and spot next to eye on top yellow. Rostrum 0.98–1.12 mm long. Antennae black. Pronotum and scutellum black. Hemelytra black, costal margin pale; pruinose; pubescence long, semierect. Ventral surface black, tibiae often pale. Males macropterous, females macropterous and brachypterous.

Remarks. This is a small species with no pale area on the frons. The clypeus has long, dense pubescence.

Habitat. Collected on range grasses.

Distribution. South Dakota, British Columbia; Alberta, now known to occur in Saskatchewan (Map 60).

Labops hirtus Knight

Fig. 139; Map 60

Labops hirtus Knight, 1922:258.

Length 3.50–5.60 mm; width 1.82–2.38 mm. Head black, shiny, jugum, base of lorum, spot under eye, longitudinal median line on frons, and crescent mark next to eye on top yellow. Scales absent. Rostrum 1.40–1.54 mm long. Antennae black. Pronotum and scutellum black. Hemelytra black, costal margin yellow; pruinose; pubescence pale, erect, long, dense. Ventral surface black, legs mostly pale, tibiae pilose. Brachypterous and macropterous forms.

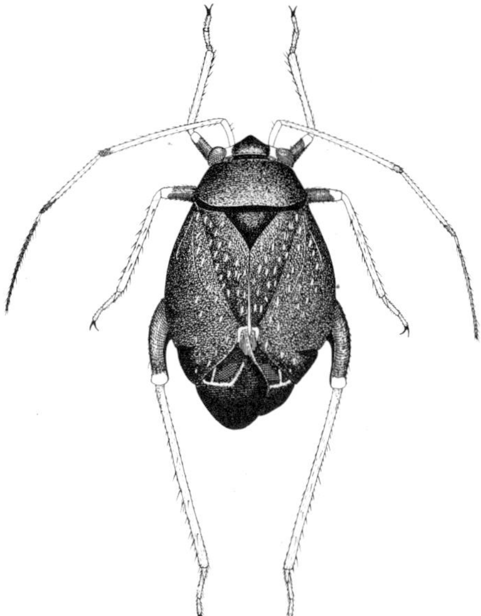

137

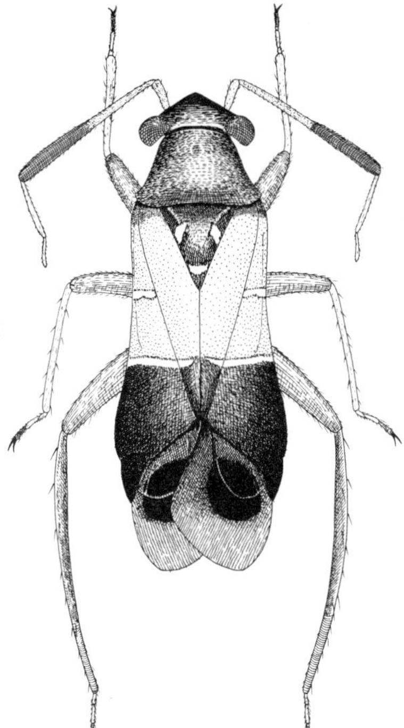

138

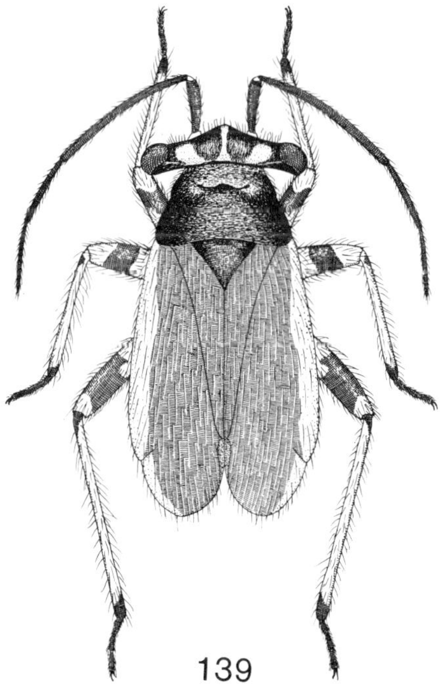

139

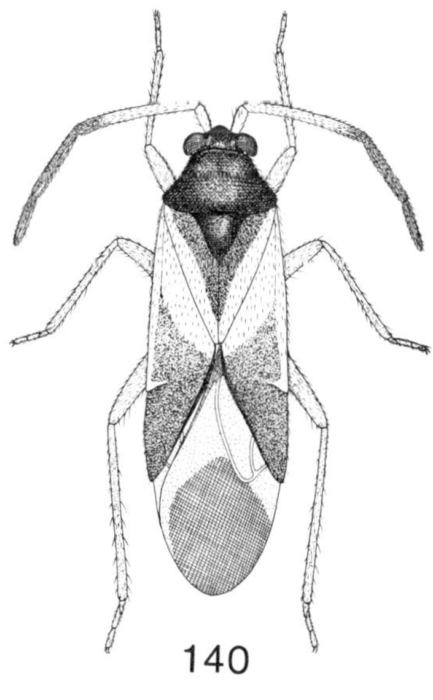

140

Figs. 137–140. Adult Miridae. 137, *Halticus intermedius*; 138, *Pilophorus piceicola*; 139, *Labops hirtus*; 140, *Ceratocapsus geminatus*.

Remarks. This species is readily distinguished by the pale median line on the frons, by the long, dense pubescence on the hemelytra, and by the pilose tibiae (Fig. 139).

Habitat. Collected on range grasses.

Distribution. Northeastern and northwestern States, Ontario, Quebec, British Columbia; Alberta, now known to occur in Saskatchewan (Map 60).

Genus *Halticus* Hahn

Short, oval, black, shiny species. Head vertical, dorsoventrally elongated, carina sharp, eyes overlap front margin of pronotum. Second antennal segment longer than basal width of pronotum. Pronotum trapeziform, smooth, lateral margins sharply angled, basal margin convex. Hemelytra with patches of sericeous pubescence. Femora black, hind femur saltatorial, tibiae pale.

Three species are known from North America, one occurs in the Prairie Provinces.

Halticus intermedius Uhler

Fig. 137; Map 61

Halticus intermedius Uhler, 1904:360.

Length 2.94–3.15 mm; width 1.82–2.10 mm. Head reddish, clypeus and frons black. Rostrum 0.98–1.12 mm long. Antennae pale, terminal segments often brown. Pronotum and scutellum black. Hemelytra black, shiny, widest in the middle; pubescence short, black, intermixed with patches of sericeous hairs (Fig. 137). Ventral surface black, tibiae pale. Macropterous.

Remarks. This species is readily distinguished by the black color, by the vertical head, by the clumps of sericeous pubescence, and by the saltatorial hind femora.

Habitat. Collected on *Anemone canadensis.*

Distribution. Eastern and Central States, Ontario; now known to occur in Manitoba and Saskatchewan (Map 61).

Tribe Orthotylini

In the Prairie Provinces the tribe is represented by 20 genera and 24 species.

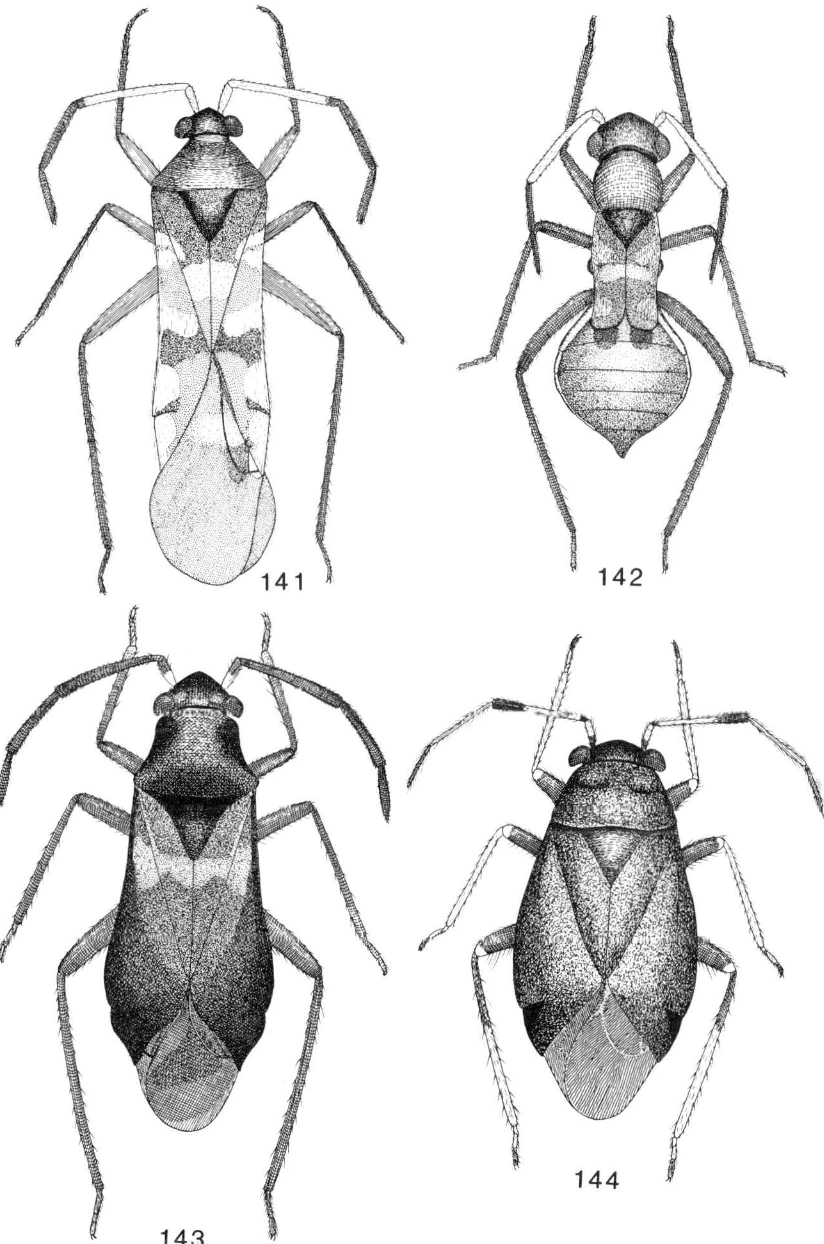

Figs. 141–144. Adult Miridae. 141, *Sericophanes heidemanni*, male; 142, *S. heidemanni*, female; 143, *Schaffneria davisi*; 144, *Slaterocoris stygicus*.

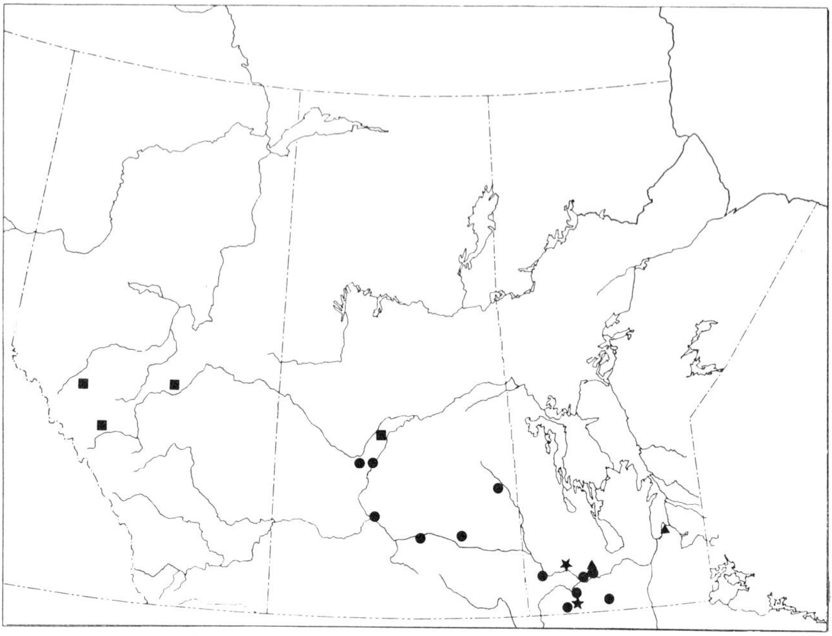

Map 61. Collection localities for *Halticus intermedius* (●), *Ceratocapus pilosulus* (▲), *C. digitalus* (★), and *C. drakei* (■).

Key to genera of Orthotylini

1. Antennal segments thickened, all nearly of equal thickness (Fig. 140) 2
   Second antennal segment incrassate, third segment thinner than apex of second (Fig. 144) 4
2. Hemelytra without distinct pollinose, silvery bands (Fig. 140) .... ***Ceratocapsus* Reuter** (p. 197)
   Hemelytra with distinct pollinose, silvery markings (Figs. 141, 143); females apterous (Fig. 142) 3
3. Cuneus and corium not polished; clavus with white spot at middle (Fig. 141); females apterous (Fig. 142) .... ***Sericophanes* Reuter** (p. 211)
   Cuneus and apical half of corium polished; clavus without white spot (Fig. 143) .... ***Schaffneria* Knight** (p. 212)
4. Compact, black species; costal margins of hemelytra convex (Fig. 144) .... ***Slaterocoris* Wagner** (p. 214)
   Not compact, black species; hemelytra straight 5
5. Hemelytra with black scales, intermixed with regular pubescence (Figs. 145, 146) 6
   Hemelytra without black scales 8
6. Pronotum behind calli and base of scutellum with patches of black scales (Fig. 145) .... ***Ilnacora* Reuter** (p. 217)
   Pronotum and scutellum without patches of black scales (Fig. 146) 7
7. First antennal segment much longer than vertex between eyes; species more than 5.0 mm in length; second antennal segment longer than width of pronotum at base .... ***Ilnacorella* Reuter** (p. 221)

First antennal segment shorter than width of vertex between eyes; species less than 4.9 mm in length; second antennal segment equal to or shorter than width of pronotum at base .... ***Brooksetta*** **Kelton** (p. 221)

8. Lateral margins of pronotum carinate 9

   Lateral margins of pronotum angulate or rounded 10

9. Genal suture extending from antennal socket directly to eye (Fig. 49); carina between eyes with stiff, black bristles (Fig. 50) .... ***Hadronema*** **Uhler** (p. 225)

   Genal suture extending from antennal socket below eye (Fig. 51); carina between eyes without stiff bristles .... ***Lopidea*** **Uhler** (p. 231)

10. Head subvertical, vertex wide; genal suture extending from antennal socket below eye (Fig. 52) .... ***Labopidea*** **Uhler** (p. 237)

    Head oblique or subhorizontal, if subvertical, genal suture extending from antennal socket directly to eye 11

11. Pronotum and hemelytra with sericeous pubescence, intermixed with simple hairs 12

    Pronotum and hemelytra with simple hairs 16

12. Hemelytra with black and silvery, sericeous pubescence, intermixed with simple, black hairs .... ***Noctuocoris*** **Knight** (p. 246)

    Hemelytra with silvery, sericeous pubescence only, intermixed with simple hairs 13

13. Head horizontal; vertex wide; genal suture extending from antennal socket below eye (Fig. 51) .... ***Labopidicola*** **Kelton** (p. 244)

    Head oblique; vertex narrow; genal suture extending from antennal socket directly to eye (Fig. 49) 14

14. Carina between eyes absent .... ***Parthenicus*** **Reuter** (p. 247)

    Carina between eyes present 15

15. Large robust species .... ***Hesperocapsus*** **Knight** (p. 249)

    Slender, narrow species (Fig. 148) .... ***Melanotrichus*** **Reuter** (p. 251)

16. First antennal segment with black longitudinal lines (Fig. 149) .... ***Reuteria*** **Puton** (p. 259)

    First antennal segment without black longitudinal lines 17

17. Head nearly vertical; (Figs. 150–153) 18

    Head oblique 20

18. Hemelytra pale green .... ***Fieberocapsus*** **Carvalho & Southwood** (p. 259)

    Hemelytra with some black 19

19. Cuneus as long as width of head, or longer; females often brachypterous (Figs. 150–153) .... ***Mecomma*** **Fieber** (p. 261)

    Cuneus much shorter than width of head, only as long as width of vertex plus eye; females macropterous (Fig. 154) .... ***Cyrtorhinus*** **Fieber** (p. 262)

20. Small, delicate species, eyes near middle of head (Fig. 155) .... ***Diaphnocoris*** **Kelton** (p. 263)

    Large, robust species, eyes near anterior angles of pronotum (Fig. 156) .... ***Orthotylus*** **Fieber** (p. 265)

Genus *Ceratocapsus* Reuter

Elongate or oval species. Head oblique, basal margin overlaps apical portion of pronotum; eyes prominent, carina between them distinct. Antennae stout, of nearly equal thickness throughout. Pronotum trapeziform, smooth or punctate, lateral margins rounded, calli indistinct. Hemelytra smooth or punctate, glabrous or pubescent, variable in color. Males macropterous, females often brachypterous.

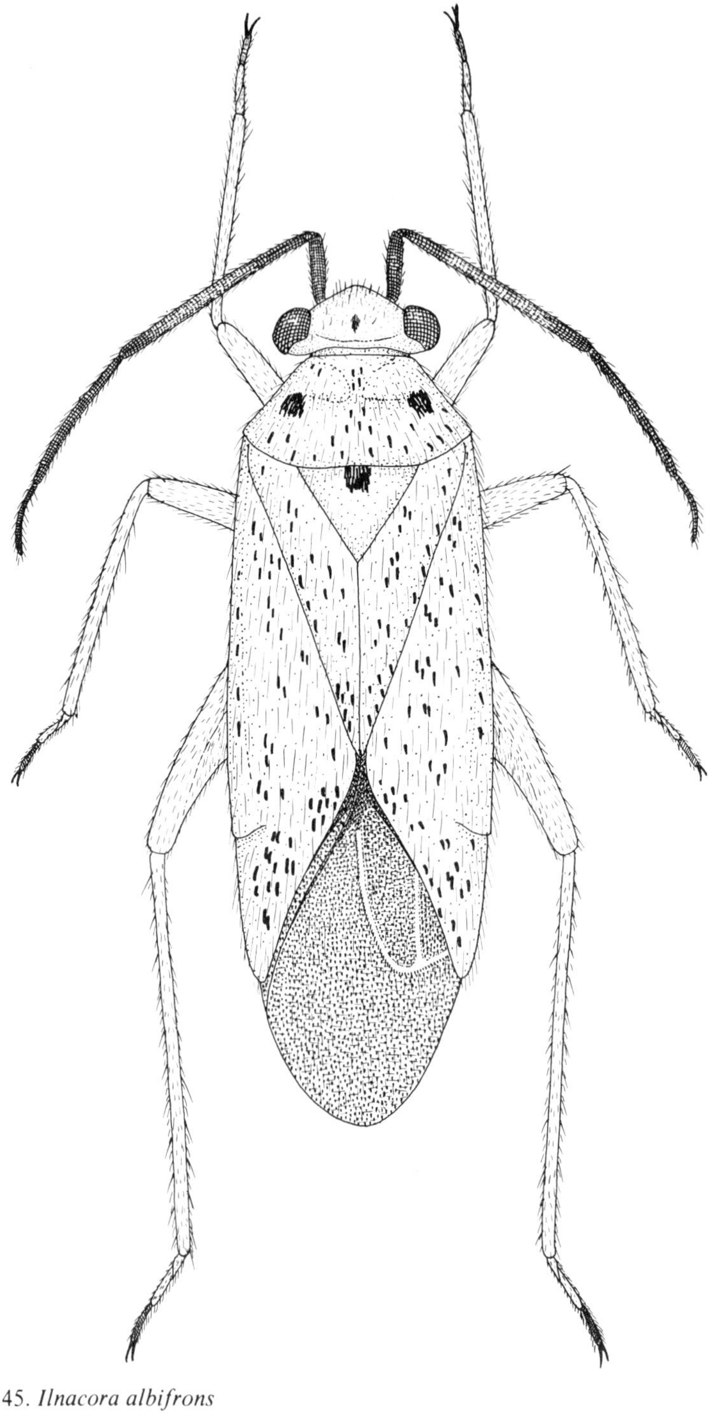

Fig. 145. *Ilnacora albifrons*

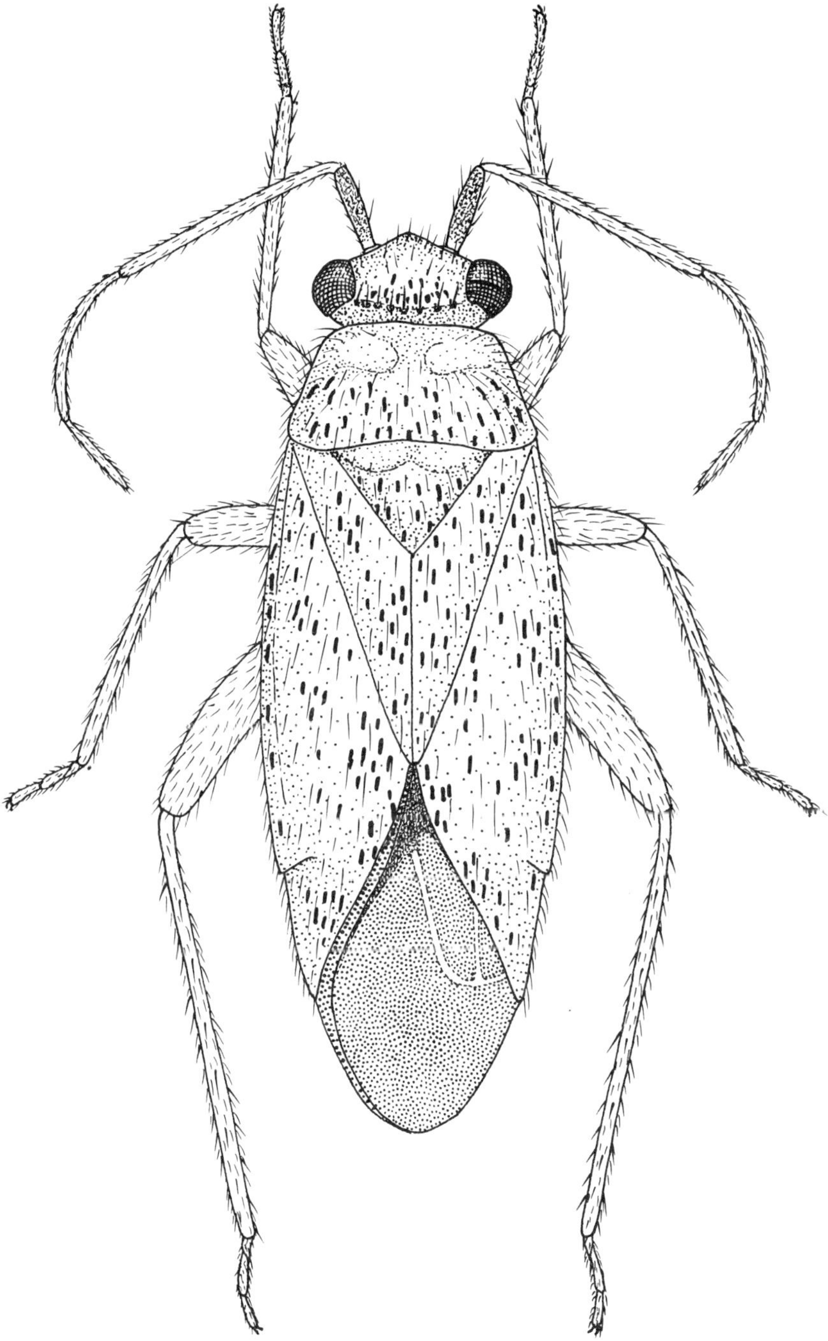

Fig. 146. *Brooksetta inconspicua*

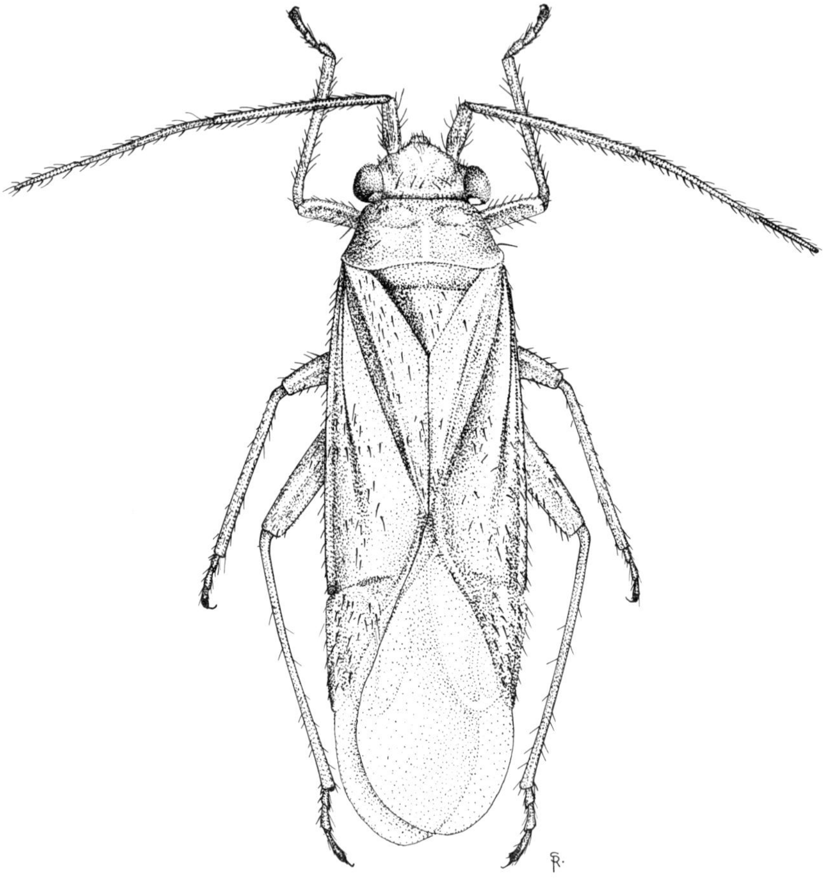

Fig. 147. *Labopidicola idahoensis*

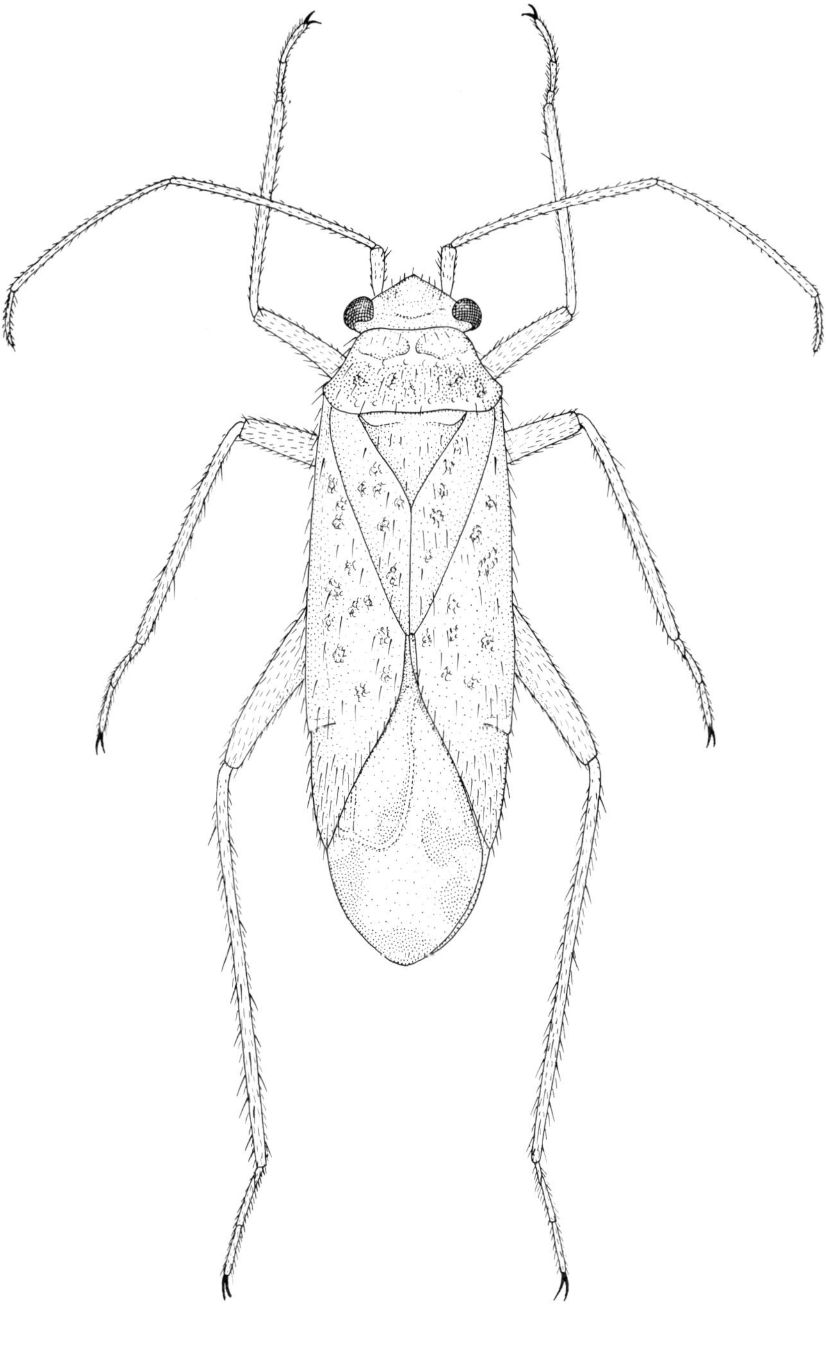

Fig. 148. *Melanotrichus coagulatus*

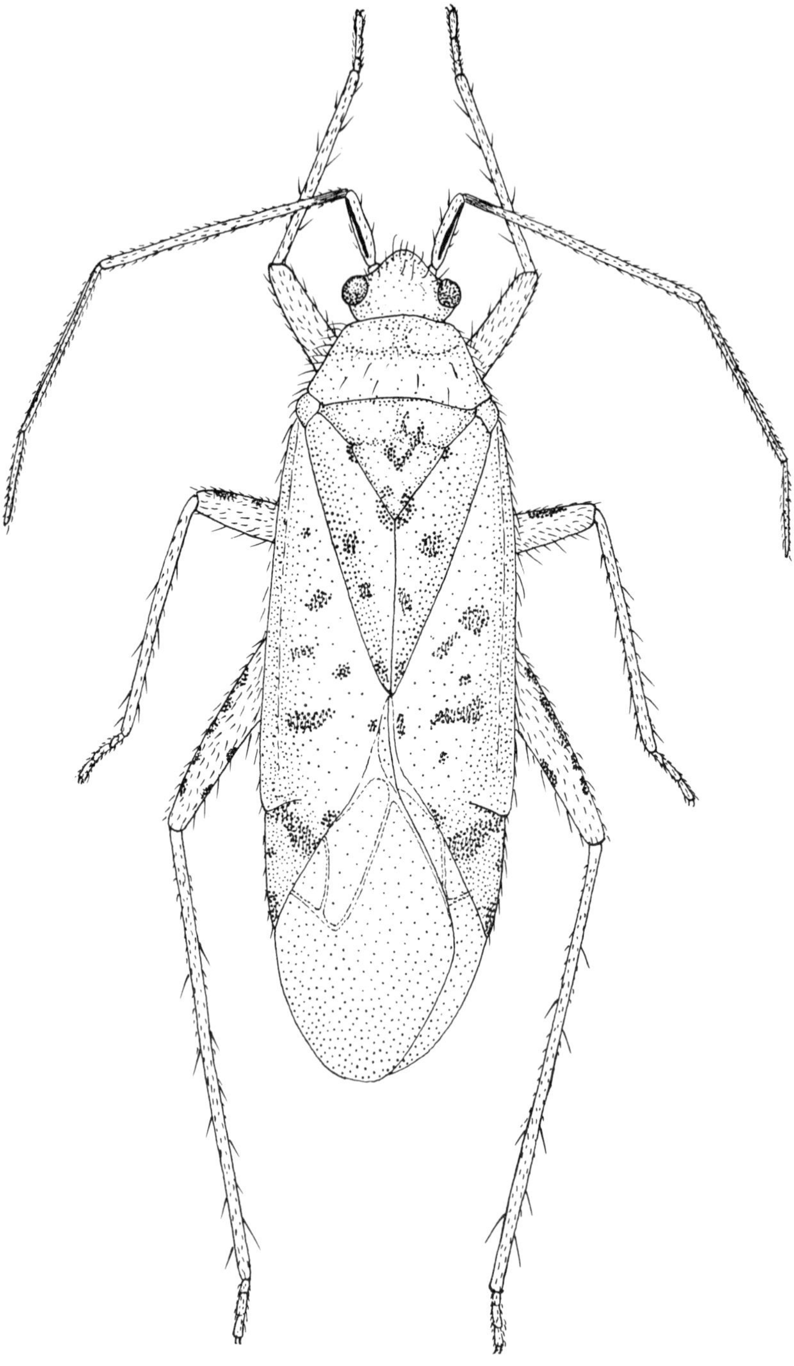

Fig. 149. *Reuteria querci*

Approximately 59 species are known from North America, six occur in the Prairie Provinces.

Key to species of *Ceratocapsus*

1. Hemelytra with wide, pale, transverse band just beyond tip of scutellum; genital claspers (Fig. 157) .... ***pilosulus* Knight** (p. 203)
   Hemelytra without pale, transverse band 2
2. Pronotum and hemelytra distinctly punctate, each puncture black 3
   Pronotum and hemelytra not distinctly punctate 4
3. Short, robust species; cuneus colored as corium; pubescence long, dense, intermixed with shorter reclining hairs; genital claspers (Fig. 158) .... ***digitulus* Knight** (p. 205)
   Elongate, slender species; hemelytra with long, sparse, pilose hairs, intermixed with short, sericeous hairs; head, cuneus, and legs reddish; genital claspers (Fig. 159) .... ***drakei* Knight** (p. 205)
4. Large, robust species; head light to dark brown, pronotum and scutellum dark brown; hemelytra light to dark brown; almost glabrous with few, long, pilose hairs; genital claspers (Fig. 160) .... ***modestus* (Uhler)** (p. 205)
   Slender species; hemelytra with short, appressed pubescence 5
5. Femur and tibia reddish; second antennal segment 1.12–1.26 mm long; pubescence on hemelytra short; genital claspers (Fig. 161) .... ***geminatus* Knight** (p. 209)
   Femur and tibia yellow or brown; second antennal segment 1.40–1.54 mm long; pubescence on hemelytra long; genital claspers (Fig. 162) .... ***nigrocephalus* Knight** (p. 209)

Ceratocapsus pilosulus Knight

Fig. 157; Map 61

Ceratocapsus pilosus Knight, 1923*b*:526.
Ceratocapsus pilosulus Knight, 1930*b*:198.

Length 3.22–3.50 mm; width 1.33–1.61 mm. Head, pronotum, and scutellum brown. Eyes hairy. Terminal antennal segments reddish. Hemelytra with wide, pale, transverse band. Pubescence on dorsum pilose, intermixed with appressed, silvery hairs. Legs reddish.

Remarks. This is the only species of the genus in the Prairie Provinces with banded hemelytra. The genital claspers (Fig. 157) are distinctive.

Habitat. Collected on *Quercus macrocarpa*.

Distribution. Northeastern and Central States, Ontario; now known to occur in Manitoba (Map 61).

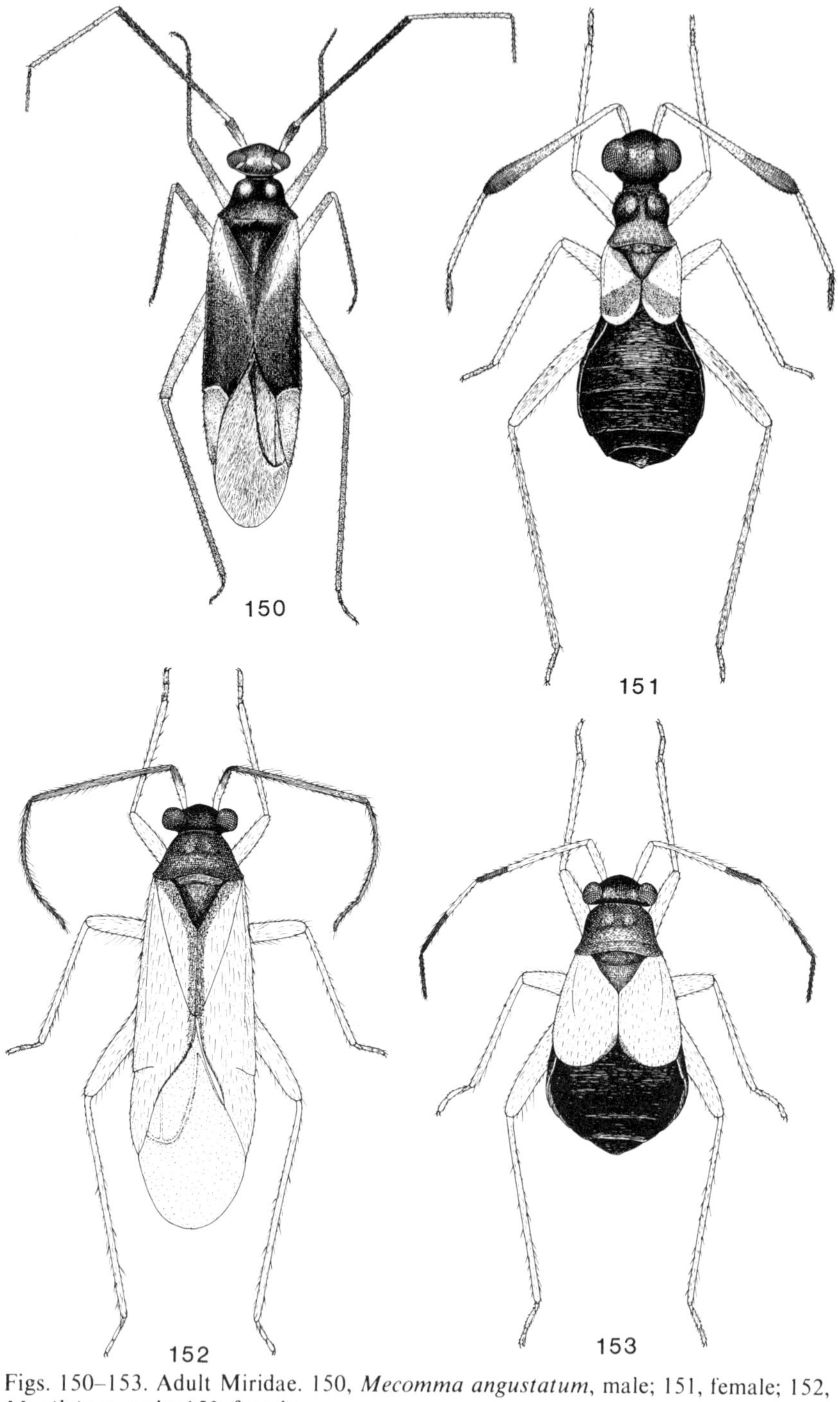

Figs. 150–153. Adult Miridae. 150, *Mecomma angustatum*, male; 151, female; 152, *M. gilvipes*, male; 153, female.

Ceratocapsus digitulus Knight

Fig. 158; Map 61

Ceratocapsus digitulus Knight, 1923*b*:533.

Length 3.50–3.85 mm; width 1.61–1.82 mm. Head, pronotum, scutellum, and hemelytra brown. Eyes hairy. First and second antennal segments pale, first segment with red bar near base. Pubescence on dorsum dense, slanting, intermixed with appressed, sericeous hairs. Hemelytra and pronotum punctate. Legs pale.

Remarks. This is the only robust and brown, densely pubescent species in the Prairie Provinces. The pronotum and hemelytra are punctate. The genital claspers (Fig. 158) are distinctive.

Habitat. Collected on *Salix interior*.

Distribution. Eastern and Central States, Ontario; now known to occur in Manitoba (Map 61).

Ceratocapsus drakei Knight

Fig. 159; Map 61

Ceratocapsus drakei Knight, 1923*b*:533.

Length 4.55–5.04 mm; width 1.54–1.68 mm. Head usually red. First antennal segment and basal half of second pale, terminal segments brown. Pronotum, scutellum, and hemelytra brown, cuneus reddish. Pronotum and hemelytra punctate. Pubescence on dorsum pilose, intermixed with sericeous, appressed hairs. Legs reddish.

Remarks. This slender species has punctate pronotum and hemelytra. The genital claspers (Fig. 159) are distinctive.

Habitat. Unknown.

Distribution. New York; Alberta, now known to occur in Saskatchewan (Map 61).

Ceratocapsus modestus (Uhler)

Fig. 160; Map 62

Melinna modesta Uhler, 1887:69.
Ceratocapsus modestus: Smith, 1910:161.

Length 4.34–4.55 mm; width 1.68–1.89 mm. Head light to dark brown. Antennae brown. Pronotum and scutellum dark brown. Hemelytra light to dark brown. Pronotum and hemelytra impunctate, sparsely pilose, other pubescence short. Legs brown.

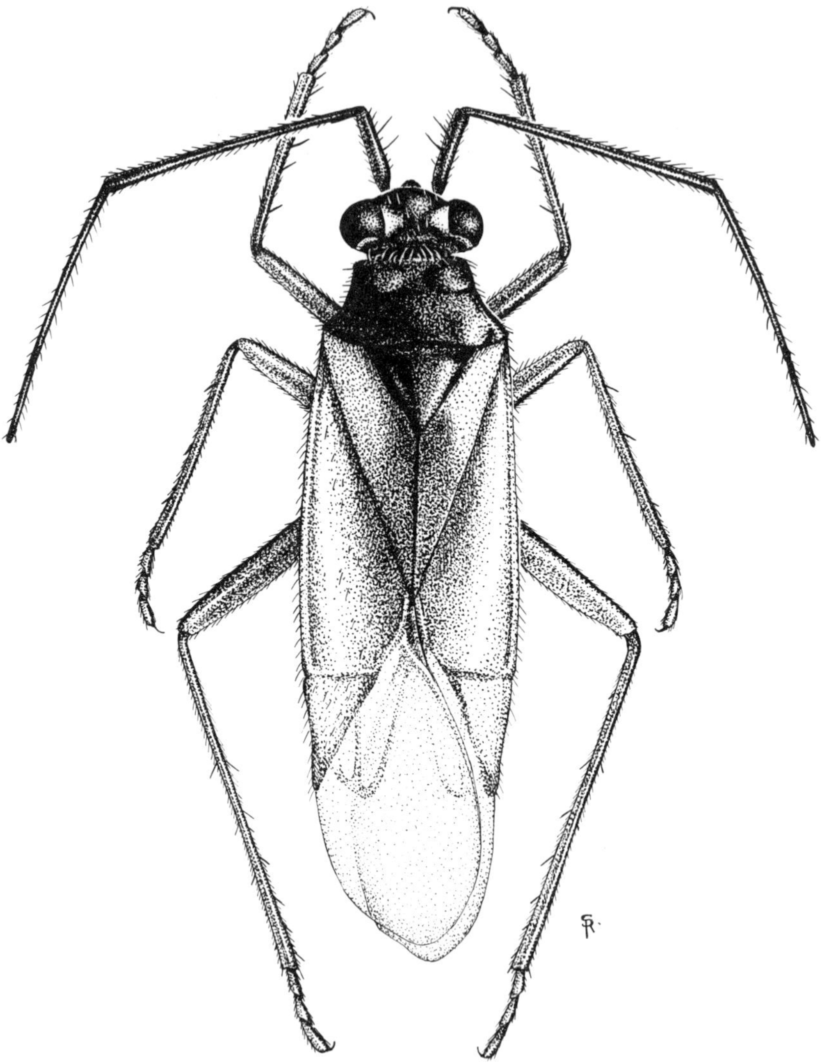

Fig. 154. *Cyrtorhinus caricis*

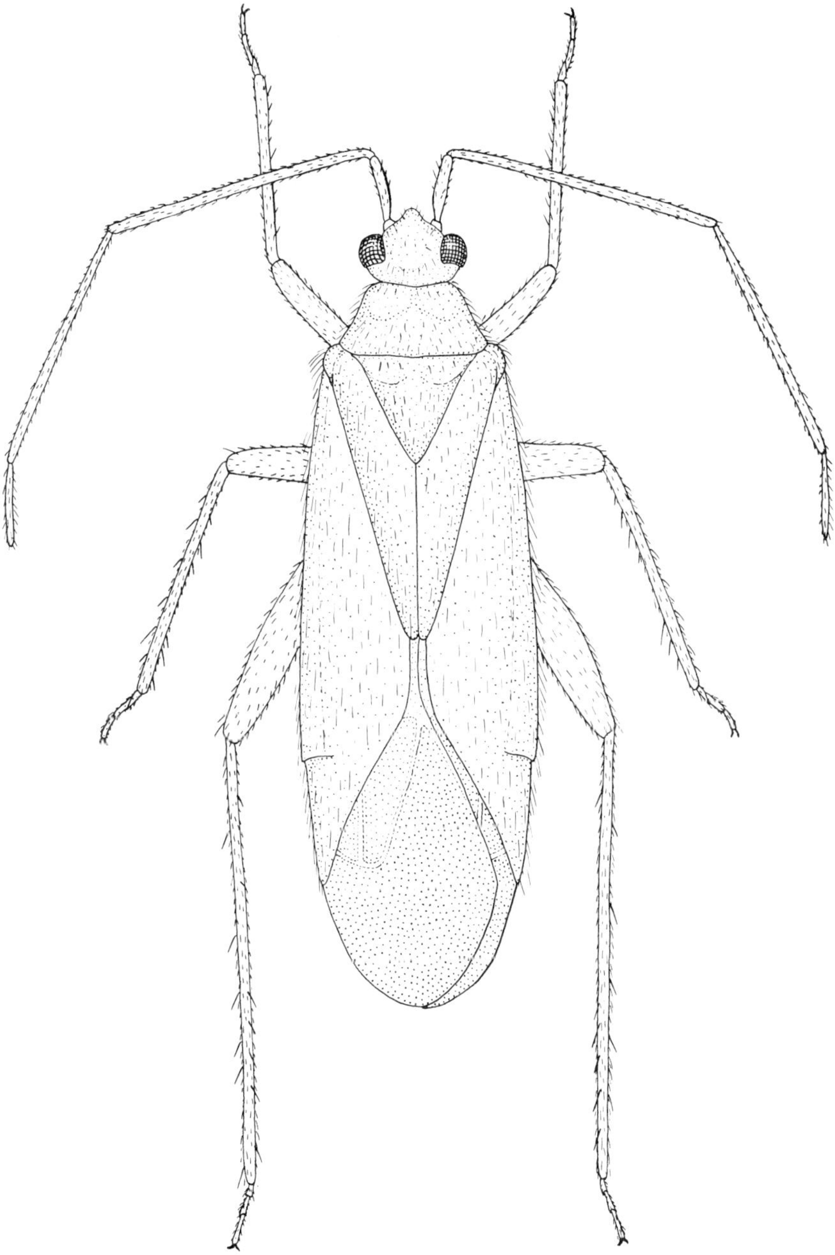

Fig. 155. *Diaphnocoris provancheri*

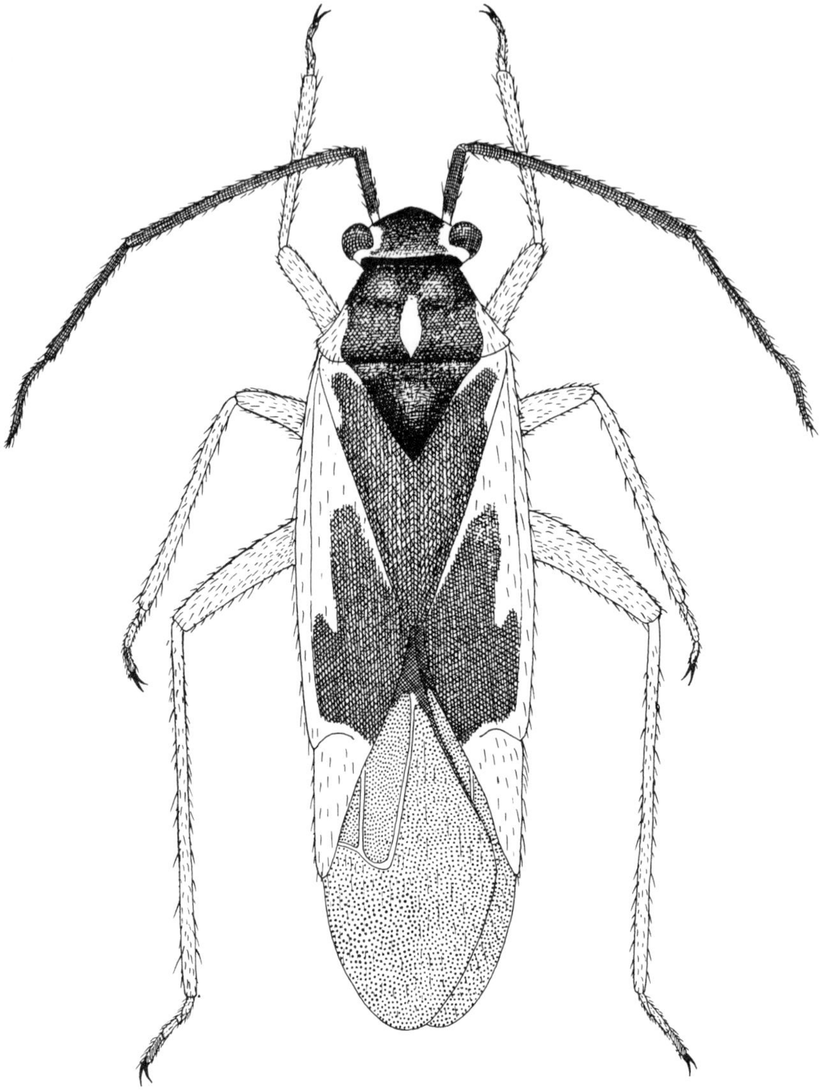

Fig. 156. *Orthotylus dorsalis*

Remarks. This is the largest brown species in the Prairie Provinces. The impunctate pronotum and sparingly pilose hemelytra also help to distinguish the species. The genital claspers (Fig. 160) are distinctive.

Habitat. Collected on *Quercus macrocarpa*.

Distribution. Eastern and Central States, Ontario, Quebec; now known to occur in Manitoba and Saskatchewan (Map 62).

Ceratocapsus geminatus Knight

Figs. 140, 161; Map 62

Ceratocapsus geminatus Knight, 1930*b*:192.

Length 3.78–3.92 mm; width 1.19–1.26 mm. Head, pronotum, and scutellum dark brown. First antennal segment and base of second pale, remainder brown. Hemelytra brown, pale along claval suture (Fig. 140). Pronotum and hemelytra impunctate, pubescence short, appressed. Legs reddish.

Remarks. This species is closely related to *nigrocephalus*, but is distinguished from it by the reddish legs, shorter second antennal segment, and by the differences in the genitalia (Fig. 161).

Habitat. Unknown.

Distribution. Colorado; now known to occur in the Prairie Provinces (Map 62).

Ceratocapsus nigrocephalus Knight

Fig. 162; Map 62

Ceratocapsus nigrocephalus Knight, 1923*b*:534.

Length 2.94–4.55 mm; width 1.40–1.68 mm. Head, pronotum, and scutellum black. First antennal segment pale, black at base, remaining segments brown. Pronotum and hemelytra impunctate, pilose, other pubescence short, appressed. Legs yellowish brown. Females brachypterous with hemelytra completely pale and wing membrane greatly reduced.

Remarks. This species is closely related to *geminatus*, but is distinguished from it by the pilose hemelytra, by the slightly larger size, by the longer second antennal segment, and by the differences in the genitalia (Fig. 162).

Habitat. Males collected at light, females among grass roots.

Distribution. North Central States, Ontario, Quebec; now known to occur in Manitoba and Saskatchewan (Map 62).

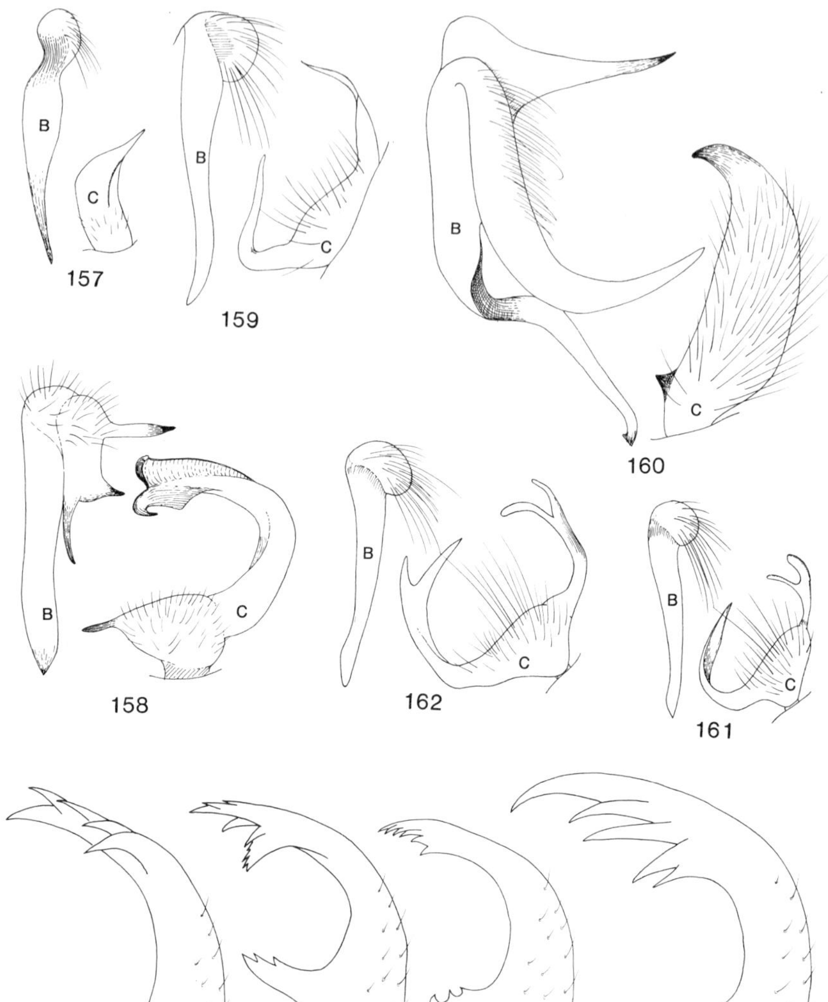

Figs. 157–166. Male clasper of *Ceratocapsus* and *Slaterocoris* spp. 157, *Ceratocapsus pilosulus*; 158, *C. digitulus*; 159, *C. drakei*; 160, *C. modestus*; 161, *C. geminatus*; 162, *C. nigrocephalus*; 163, *Slaterocoris pallidicornis*; 164, *S. atritibialis*; 165, *S. breviatus*; 166, *S. stygicus*.

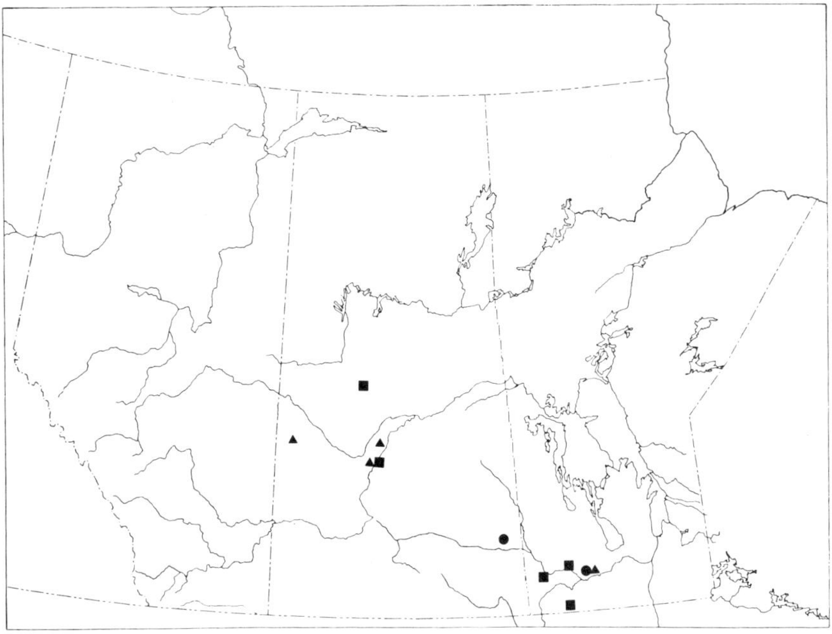

Map 62. Collection localities for *Ceratocapsus modestus* (●), *C. geminatus* (▲), and *C. nigrocephalus* (■).

Genus *Sericophanes* Reuter

Slender, small, antlike species. Head oblique, eyes large, carina between them sharp, prominent. Antennae cylindrical, second and third segments of equal thickness. Pronotum campanulate, strongly declivent, lateral margins rounded, calli obscured. Hemelytra velvety, costal margin broadly concave. Legs brown, hind coxae pale. Brachypterous and macropterous forms.

Nine species have been described from North America, one occurs in the Prairie Provinces.

Sericophanes heidemanni Poppius

Figs. 141, 142; Map 63

Sericophanes heidemanni Poppius, 1914:260.

Length 2.94–3.50 mm; width 1.12–1.26 mm. Head brown, smooth. Rostrum 1.12–1.19 mm long. Antennae brown. Pronotum dark brown. Hemelytra fully developed in male (Fig. 141), reduced in female (Fig. 142), clavus velvety brown with large, pale spot on middle of commissure; corium

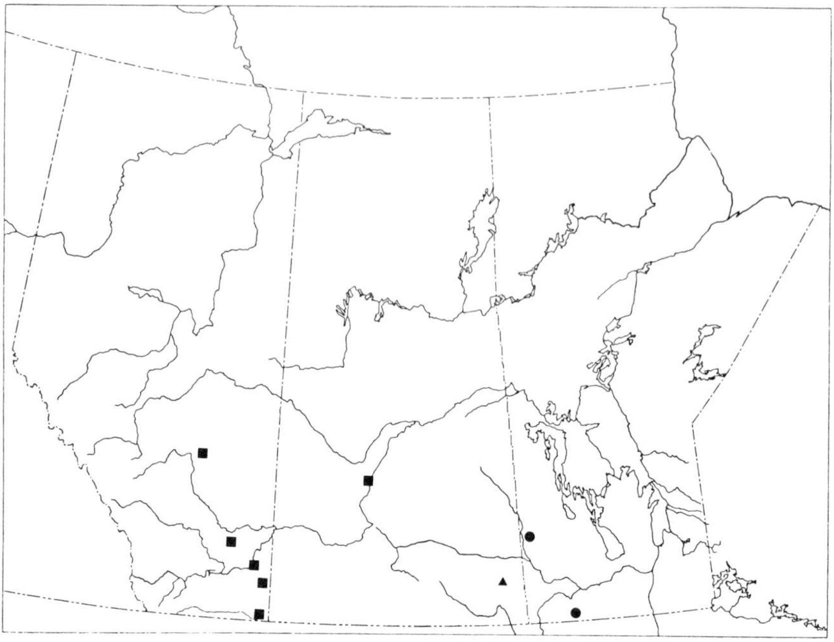

Map 63. Collection localities for *Sericophanes heidemanni* (■), *Schaffneria davisi* (●), and *S. schaffneri* (▲).

with two large, triangular, silvery blotches on each side, extending from costal margin to claval suture; pubescence short, sparse.

Remarks. The females resemble wingless ants, the males are distinguished by the large, pale spot on the clavus.

Habitat. Collected on grass roots, and at light.

Distribution. Widespread in USA, Ontario, Quebec; now known to occur in the Prairie Provinces (Map 63).

Genus *Schaffneria* Knight

Elongate, reddish brown, antlike species. Head oblique, triangular, carina sharp as in *Pilophorus*. Antennal segments thickened similar to those of *Ceratocapsus*. Pronotum campanulate, lateral margins rounded. Scutellum moderately tumid. Hemelytra pruinose on basal half, polished on embolium, apical half of corium, and cuneus; pruinose area behind scutellum with broad and irregular transverse frosted band; pubescence short, dense and appressed on pruinose area; erect, long and sparse on pruinose and polished areas. Osteolar peritreme well developed.

Only five species are known from North America, two occur in the Prairie Provinces. Knight (1966) published a key to the species.

Key to species of *Schaffneria*

1. Larger species, 3.50–3.85 mm long; pronotum strongly convex (Fig. 143) .... ***davisi* (Knight)** (p. 213)
   Smaller species, 2.80–3.01 mm long; pronotum gently convex .... ***schaffneri* Knight** (p. 213)

Schaffneria davisi (Knight)

Fig. 143; Map 63

Pamelia davisi Knight, 1923*b*:535.
Schaffneria davisi: Knight, 1966:2.

Length 3.50–3.85 mm; width 1.33–1.47 mm. Head brown to black. First antennal segment pale to brown with reddish diagonal bar on inner surface; second segment 0.98–1.15 mm long. Rostrum 1.33–1.40 mm long. Pronotum brown to black, shiny, strongly convex (Fig. 130), humpbacked in appearance. Hemelytra dark brown to black. Ventral surface and legs reddish brown.

Remarks. This species is distinguished by the strongly convex pronotum.

Habitat. Collected on *Quercus macrocarpa*.

Distribution. New Jersey; now known to occur in Manitoba (Map 63).

Schaffneria schaffneri Knight

Map 63

Schaffneria schaffneri Knight, 1966:2.

Length 2.80–3.01 mm; width 0.98–1.05 mm. Head brown, clypeus and adjacent area tinged with red. First antennal segment pale; second segment 0.98–1.12 mm long, basal half pale, apical half brown. Rostrum 1.29–1.40 mm long. Pronotum brown, shiny, gently convex. Hemelytra brown, similar to *davisi* in markings and pubescence. Ventral surface and legs reddish brown.

Remarks. This species is smaller than *davisi* and the pronotum is gently convex, not strongly turned down posteriorly as in *davisi*.

Habitat. Collected on *Beckmannia syzigachne.*

Distribution. Texas; now known to occur in Alberta and Saskatchewan (Map 63).

Genus *Slaterocoris* Wagner

Black, shiny species. Head vertical, short, carina sharp, base of head overlaps anterior margin of pronotum. Pronotum rugose, punctate, lateral margins sharply angled, basal margin convex. Hemelytra rugose, punctate, glabrous or pubescent, sericeous hairs absent. Ventral surface black, legs pale or black.

The genus was revised by Kelton (1968) and Knight (1970). Forty-seven species have been described from the Nearctic region, four occur in the Prairie Provinces.

Key to species of *Slaterocoris*

1. Second antennal segment mostly yellow; male claspers (Fig. 163) .... *pallidicornis* (**Knight**) (p. 214)
   Second antennal segment mostly black 2
2. Second antennal segment shorter than head width; rostrum shorter than head width; male claspers (Fig. 164) .... *atritibialis* (**Knight**) (p. 215)
   Second antennal segment longer than head width; rostrum longer than head width 3
3. Hind tibia yellow, often fuscous on basal half; first antennal segment yellow; genital claspers (Fig. 165) .... *breviatus* (**Knight**) (p. 216)
   Hind tibia black; first antennal segment black; genital claspers (Fig. 166) .... *stygicus* (**Say**) (p. 217)

Slaterocoris pallidicornis (Knight)

Fig. 163; Map 64

Strongylocoris pallidicornis Knight, 1938:4.
Slaterocoris pallidicornis: Kelton, 1968:1123.

Length 4.20–4.55 mm; width 1.90–2.20 mm. Rostrum 1.12–2.20 mm. First and second antennal segments yellow. Pubescence on hemelytra short, sparse. Femora black, tibiae yellow.

Remarks. This species is distinguished by the yellow second antennal segment and by the yellow hind tibia. The genital claspers (Fig. 163) are distinctive.

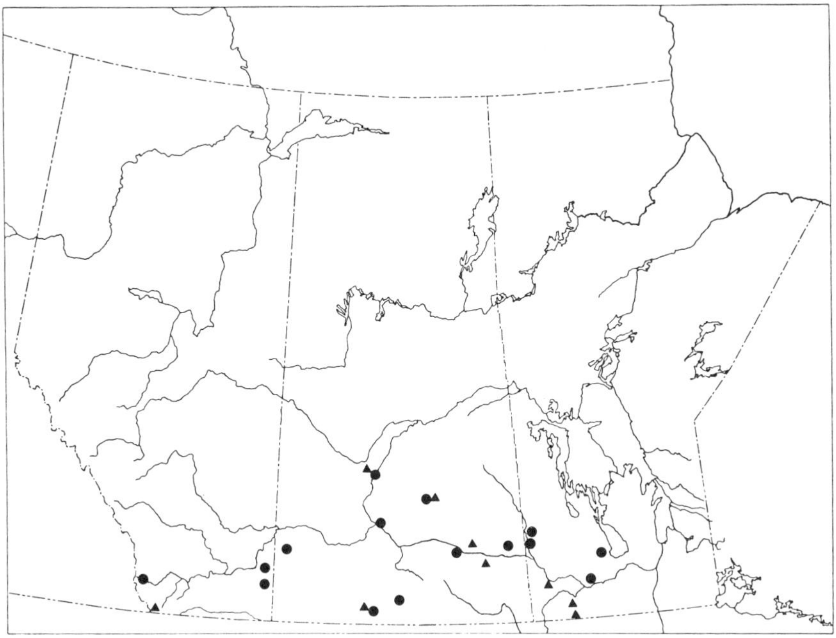

Map 64. Collection localities for *Slaterocoris pallidicornis* (●) and *S. atritibialis* (▲).

Habitat. Collected on *Solidago* spp. and *Symphoricarpos occidentalis*.

Distribution. North Central States, Nova Scotia; Prairie Provinces (Map 64).

Slaterocoris atritibialis (Knight)

Fig. 164; Map 64

Strongylocoris atritibialis Knight, 1938:2.
Slaterocoris atritibialis: Wagner, 1956:280.

Length 4.50–4.60 mm; width 2.40–2.50 mm. Rostrum 1.12–1.26 mm long. Antennae black, second segment in female often pale at middle. Pubescence on hemelytra short, sparse. Femora black, first and second tibiae partly pale, hind tibia black.

Remarks. This species is distinguished by the black hind tibia and by the distinctive genital claspers (Fig. 164).

Habitat. Collected on *Solidago* spp. and *Aster* spp.

Distribution. Widespread in USA, Ontario, Quebec, British Columbia; Alberta, Manitoba, now known to occur in Saskatchewan (Map 64).

Slaterocoris breviatus (Knight)

Fig. 165; Map 65

Strongylocoris breviatus Knight, 1938:1.
Slaterocoris breviatus: Wagner, 1956:280.

Length 4.10–4.30 mm; width 2.03–2.20 mm. Rostrum 1.05–1.12 mm long. First and second antennal segments yellow, second antennal segment often black. Pubescence on hemelytra short, sparse. Femora black, tibiae yellow, hind tibia often fuscous on basal half.

Remarks. This species is distinguished by the male genitalia (Fig. 165).

Habitat. Collected on *Solidago* spp., *Aster* spp., and *Salix* spp.

Distribution. Northeastern and North Central States, Ontario, Quebec, New Brunswick, Nova Scotia; Prairie Provinces (Map 65).

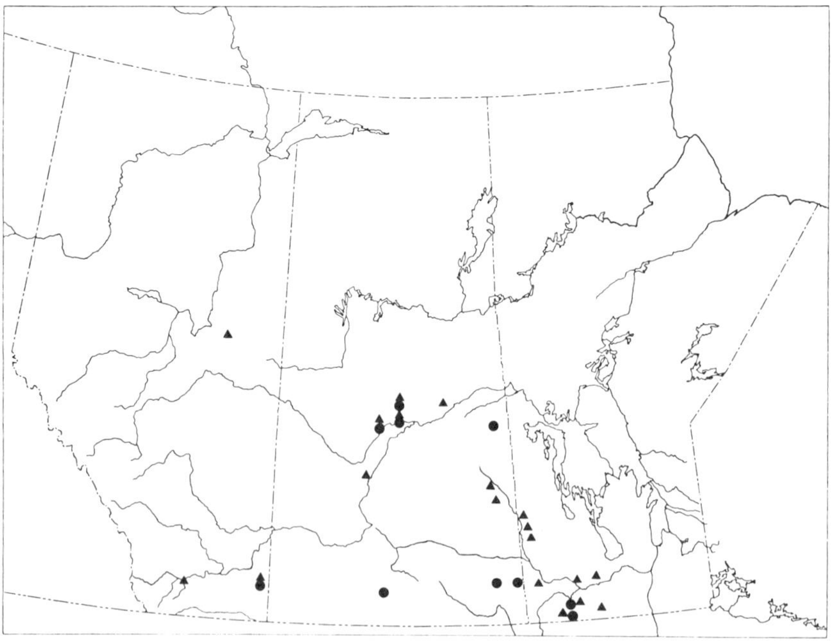

Map 65. Collection localities for *Slaterocoris breviatus* (●) and *S. stygicus* (▲).

Slaterocoris stygicus (Say)

Figs. 144, 166; Map 65

Capsus stygicus Say, 1832:24.
Slaterocoris stygicus: Wagner, 1956:280.

Length 4.20–4.40 mm; width 2.03–2.10 mm. Rostrum 1.05–1.12 mm long. Antennae black, second segment in female often pale. Pubescence on hemelytra short, sparse. Femora black, tibiae pale, in male hind tibia often black, anterior and middle femora partly black (Fig. 144).

Remarks. This species is often difficult to distinguish from *atritibialis* except by the genitalia (Fig. 166).

Habitat. Collected on *Solidago* spp.

Distribution. Widespread in USA, Eastern Canada, British Columbia; Prairie Provinces (Map 65).

Genus *Ilnacora* Reuter

Elongate, slender, green species. Head vertical, vertex flattened, carina between eyes distinct. Second antennal segment longer than width of pronotum at base. Pronotum trapeziform, lateral margins rounded, calli distinct with clump of black scales behind. Scutellum with similar clump at middle of base. Hemelytra smooth, shiny. Pubescence of two types: appressed, black scales, intermixed with erect, pale hairs. Legs long, slender. Dorsal margin of genital cavity with short hook.

Twenty-two species are known from North America, three occur in the Prairie Provinces.

Key to species of *Ilnacora*

1. Face below antennae black; frons with trifurcate black marks; male clasper (Fig. 167) .... ***vittifrons* Knight** (p. 217)
   Face mostly pale 2
2. First antennal segment almost as long as width of vertex; genital clasper (Fig. 168) .... ***stalii* Reuter** (p. 218)
   First antennal segment shorter than width of vertex; genital claspers (Fig. 169) .... ***albifrons* Knight** (p. 220)

Ilnacora vittifrons Knight

Fig. 167; Map 66

Ilnacora vittifrons Knight, 1963:166.

Length 4.80–5.48 mm; width 1.68–1.82 mm. Head mostly black below antennal sockets, pale above with trifurcate black marks. Rostrum

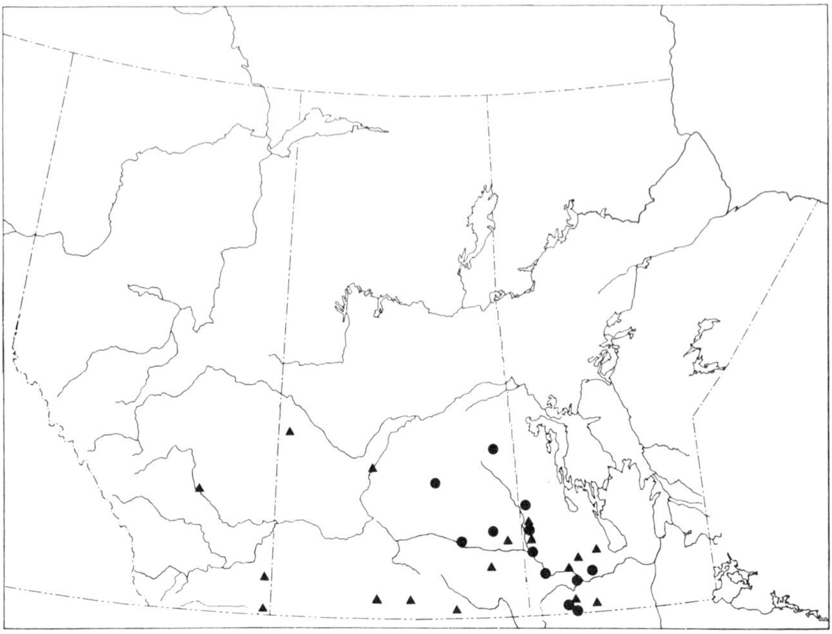

Map 66. Collection localities for *Ilnacora vittifrons* (●) and *I. stalii* (▲).

1.12–1.26 mm long. Pronotum and scutellum with typical black patches of scales. Hemelytra green, black scales uniformly scattered, often rubbed off. Ventral surface black, legs pale green.

Remarks. This species is distinguished by the bicolored head and by the genital claspers (Fig. 167).

Habitat. Collected on *Helianthus* spp.

Distribution. North Central States; now known to occur in Manitoba and Saskatchewan (Map 66).

Ilnacora stalii Reuter

Fig. 168; Map 66

Ilnacora stalii Reuter, 1875*c*:86.
Parthenicus psalliodes Provancher, 1877:146.

Length 5.60–6.16 mm; width 1.82–2.10 mm. Head pale green. Rostrum 1.40–1.54 mm long. First antennal segment almost as long as width of vertex, pale, annulate with black at base and apex. Pronotum and scutellum with typical black patches of scales. Hemelytra green, black scales more or less in clumps. Ventral surface green.

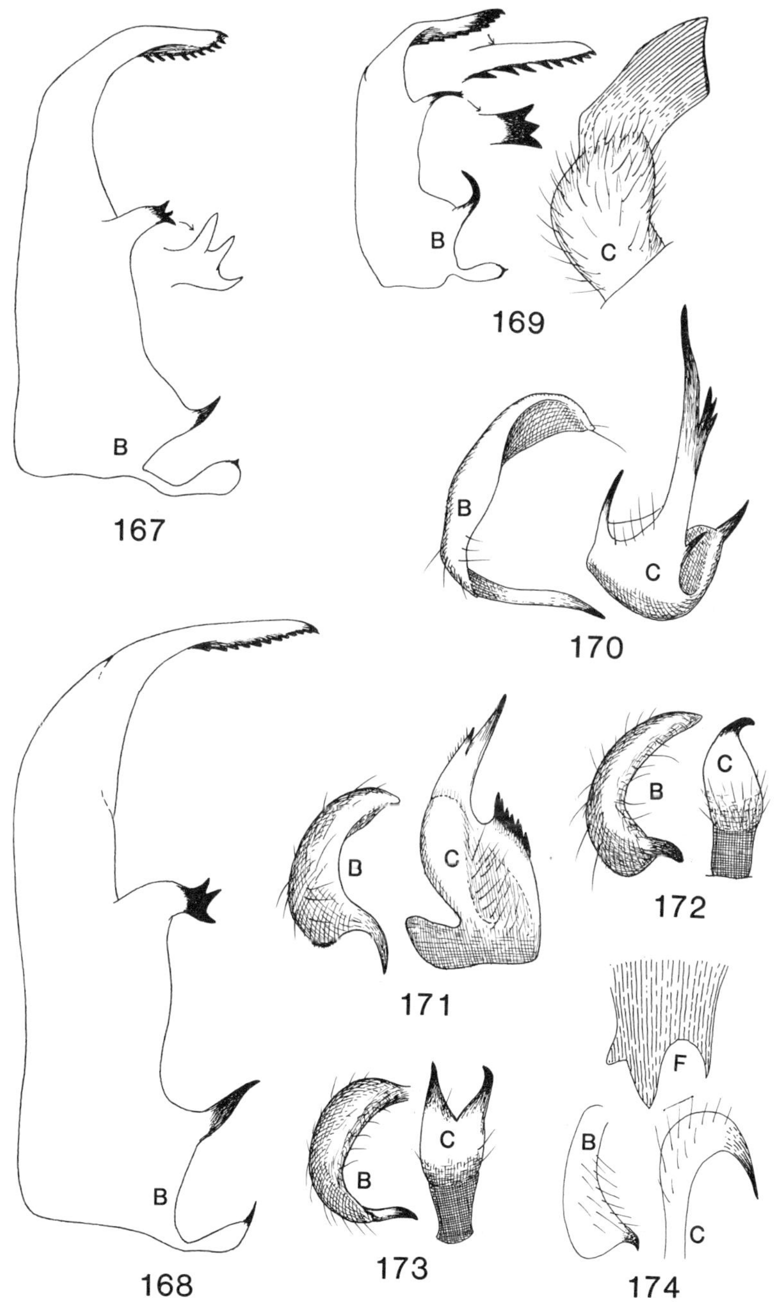

Figs. 167–174. Male claspers of Miridae. 167, *Ilnacora vittifrons*; 168, *I. stalii*; 169, *I. albifrons*; 170, *Ilnacorella sulcata*; 171, *Brooksetta inconspicua*; 172, *B. viridicata*; 173, *B. chelifer*; 174, *B. incurva*.

Remarks. This species is larger than *vittifrons* with longer first antennal segment and longer rostrum. The genital claspers are distinctive (Fig. 168).

Habitat. Collected on *Helianthus* spp.

Distribution. Eastern and Central States, Ontario, Quebec; Alberta, now known to occur in Saskatchewan and Manitoba (Map 66).

Ilnacora albifrons Knight

Figs. 145, 169; Map 67

Ilnacora albifrons Knight, 1963:165.

Length 4.20–4.76 mm; width 1.54–1.68 mm. Head mostly pale, base and apex of clypeus, lorum, and short bar at middle of vertex often black. Rostrum 0.98–1.12 mm long. First antennal segment short and mostly black. Pronotum and scutellum with typical black patches of scales. Hemelytra green, black scales uniformly scattered. Ventral surface black in male, mostly pale in female.

Remarks. This species is distinguished by the nearly white head (Fig. 145) and the genital claspers (Fig. 169).

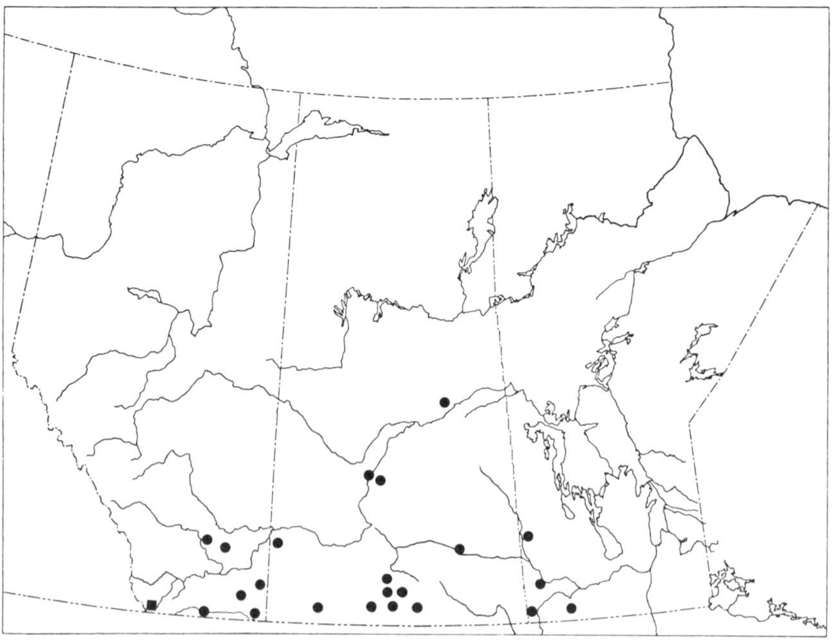

Map 67. Collection localities for *Ilnacora albifrons* (●) and *Ilnacorella sulcata* (■).

Habitat. Collected on *Grindelia perennis*.

Distribution. Central and western States; now known to occur in the Prairie Provinces (Map 67).

Genus *Ilnacorella* Knight

Elongate, slender, green species. Head vertical, vertex flattened, carina between them distinct. First antennal segment almost as long as head width. Pronotum trapeziform, flattened, lateral margins sharply angled. Hemelytra smooth, shiny. Pubescence of two types: black, appressed scales, intermixed with longer, erect, black hairs. Legs long, slender.

Four species are known from North America, one occurs in the Prairie Provinces.

Ilnacorella sulcata Knight

Fig. 170; Map 67

Ilnacorella sulcata Knight, 1925*a*:93.

Length 5.39–5.95 mm; width 1.68–1.96 mm. Head pale yellowish. Rostrum 1.19–1.33 mm long. First antennal segment green, base black. Second antennal segment longer than width of pronotum at base. Pronotum pale green, calli yellowish. Hemelytra green. Ventral surface and legs pale yellow. Dorsal rim of genital cavity without sclerite.

Remarks. This species is distinguished by the long and slender form, by the long antennae and legs, by the pubescence, and by the genital claspers (Fig. 170).

Habitat. Unknown.

Distribution. Alberta (Map 67).

Genus *Brooksetta* Kelton

Oblong, small, green species. Head subvertical, area between eyes flattened; carina between eyes distinct. First antennal segment shorter than width of vertex between eyes. Rostrum relatively short, extending to middle coxae. Pronotum trapeziform, lateral margins angulate; calli distinct. Hemelytra smooth, entire in males, often reduced in females. Pubescence of two types: appressed, black, scaly hairs, intermixed with erect, pale or black, longer simple hairs. Genital claspers variable, highly specific; dorsal margin of genital cavity with slender or flattened sclerite.

Several North American species formerly included in *Melanotrichus* Reuter were transferred to *Brooksetta* by Kelton (1979*c*, *d*). Fifteen species are known from North America, four occur in the Prairie Provinces.

Key to species of *Brooksetta*

1. First antennal segment marked with black; long pubescence on dorsum black; genital structures (Fig. 171) .... ***inconspicua* (Uhler)** (p. 222)
   First antennal segment pale; long pubescence on dorsum pale 2
2. Tibial spines black; genital structures (Fig. 172) ...... ***viridicata* (Uhler)** (p. 223)
   Tibial spines pale 3
3. Rostrum 0.84–0.87 mm long; genital structures (Fig 173) ....
   .... ***chelifer* (Knight)** (p. 223)
   Rostrum 1.05–1.19 mm long; genital structures (Fig. 174) ....
   .... ***incurva* (Knight)** (p. 224)

Brooksetta inconspicua (Uhler)

Figs. 146, 171; Map 68

Asciodema inconspicua Uhler, 1893:376.
Melanotrichus inconspicuus: Knight, 1927*c*:142.
Brooksetta inconspicua: Kelton, 1979*c*:950; 1979*d*:1423.

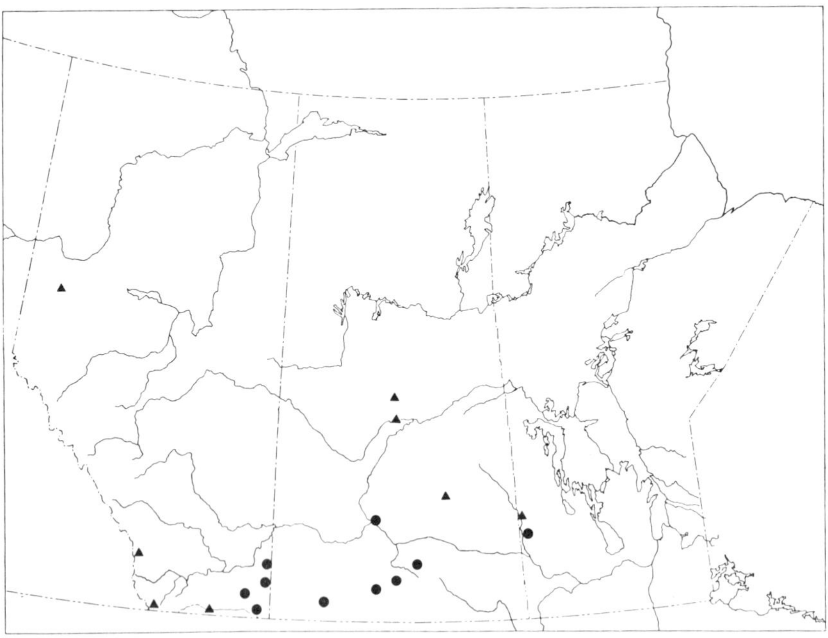

Map 68. Collection localities for *Brooksetta inconspicua* (●) and *B. viridicata* (▲).

Length 3.22–3.64 mm; width 1.26–1.47 mm. Head pale bluish green, carina with row of black bristles each with black spot at base. Rostrum 0.77–0.91 mm long. First antennal segment mostly black in male, with only black line on ventral surface in female. Pronotum, scutellum, and hemelytra bluish green; pubescence consists of black scales, intermixed with erect, black hairs. Ventral surface and legs pale green, tibial spines black. Dorsal margin of genital cavity with tiny sclerite.

Remarks. This species is easily recognized by the black or partly black first antennal segment, by the black hairs on the carina (Fig. 146), and by the genital claspers (Fig. 171).

Habitat. Collected on *Artemisia cana.*

Distribution. California, Utah; Prairie Provinces (Map 68).

Brooksetta viridicata (Uhler)

Fig. 172; Map 68

Orthotylus viridicatus Uhler, 1895:48.
Melanotrichus viridicatus: Knight, 1927:143.
Brooksetta viridicata: Kelton, 1979*c*:952; 1979*d*; 1423.

Length 3.92–4.06 mm; width 1.26–1.40 mm. Head, pronotum, and hemelytra green; pubescence consists of black scales with erect, pale hairs. Rostrum 0.91–1.08 mm long. Tibial spines black. Dorsal margin of genital cavity with short, flattened, bifurcate sclerite.

Remarks. This species is distinguished by the genital structures (Fig. 172).

Habitat. Collected on *Potentilla fruticosa.*

Distribution. Western States; Prairie Provinces (Map 68).

Brooksetta chelifer (Knight)

Fig. 173; Map 69

Melanotrichus chelifer Knight, 1927*c*:144.
Brooksetta chelifer: Kelton, 1979*c*:952; 1979*d*:1423.

Length 3.36–3.64 mm; width 1.26–1.40 mm. Head yellowish green, carina with white bristles and without black spots. Rostrum 0.84–0.87 mm long. First antennal segment pale. Pronotum, scutellum, and hemelytra bluish green; pubescence consists of black scales, intermixed with longer white hairs. Ventral surface and legs pale green, tibial spines pale. Dorsal margin of genital cavity with slender, pointed sclerite.

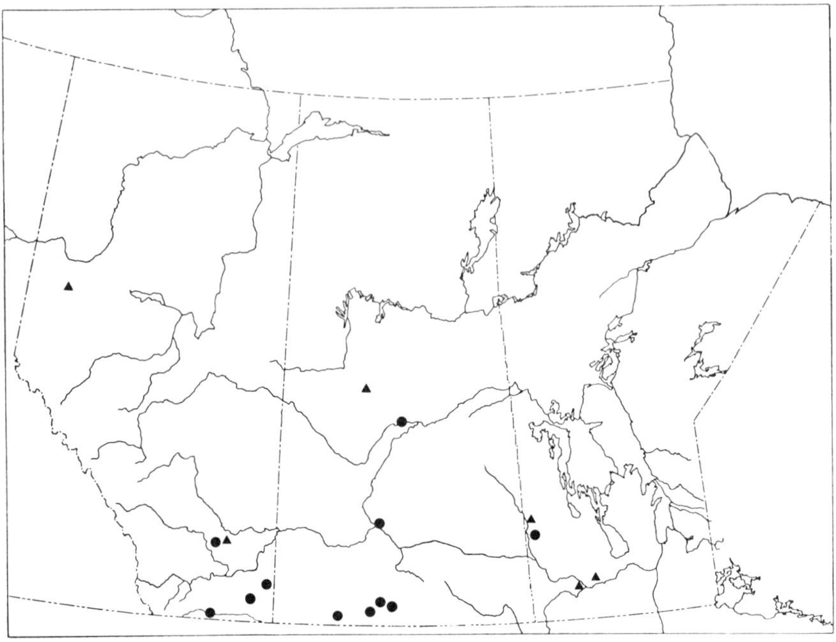

Map 69. Collection localities for *Brooksetta chelifer* (●) and *B. incurva* (▲).

Remarks. This species is similar to *inconspicua* in size and color, but the first antennal segment is pale, the carina has white bristles, the long pubescence on the hemelytra is pale, and the genital claspers (Fig. 173) are different.

Habitat. Collected on *Malvastrum coccineum.*

Distribution. Western States; Prairie Provinces (Map 69).

Brooksetta incurva (Knight)

Fig. 174; Map 69

Melanotrichus incurvus Knight, 1927*c*:143.
Brooksetta incurva: Kelton, 1979*c*:953; 1979*d*:1423.

Length 3.39–3.64 mm; width 1.26–1.40 mm. Head yellowish green, bristles on carina pale. Rostrum 1.05–1.19 mm long. Pronotum pale green, calli yellowish. Scutellum and hemelytra similar to *viridicata* in color and pubescence. Tibial spines pale. Dorsal margin of genital cavity with broadly flattened and divided sclerite.

Remarks. This species is distinguished by the genital structures (Fig. 174).

Habitat. Unknown.

Distribution. Western States; Prairie Provinces (Map 69).

Genus *Hadronema* Uhler

Elongate, black and red, pubescent species. Head vertical, carina between eyes distinct. Pronotum trapeziform, transversely rugose, lateral margins carinate. Hemelytra shiny. Pubescence simple. Front femur of male with or without strongly developed tubercle.

Thirteen species are known from North America, four occur in the Prairie Provinces.

Key to species of *Hadronema*

1. Rostrum extending to hind coxae; pubescence on dorsum pale .... ***princeps* Uhler** (p. 225)
   Rostrum shorter, extending to middle coxae; pubescence on dorsum black 2
2. Femora mostly pale .... ***bispinosum* Knight** (p. 226)
   Femora black 3
3. Smaller species, less than 3.7 mm .... ***simplex* Knight** (p. 226)
   Larger species, more than 4.2 mm .... ***militaris* Uhler** (p. 228)

Hadronema princeps Uhler

Map 70

Hadronema princeps Uhler, 1894:251.

Length 4.20–4.48 mm; width 1.33–1.54 mm. Head black, jugum and frons often reddish. Rostrum 1.68–1.82 mm long. Pronotum red, calli black. Scutellum red. Hemelytra black, base of cuneus pale, claval suture often pale in female. Pubescence on dorsum pale, long, slanting. Ventral surface black marked with red, legs reddish brown or black, front femur in male without tubercle.

Remarks. This species is distinguished by the pale pubescence on the dorsum and by the long rostrum.

Habitat. Collected on *Juniperus horizontalis* and *Pinus contorta*.

Distribution. Western States, British Columbia; Alberta, now known to occur in Saskatchewan (Map 70).

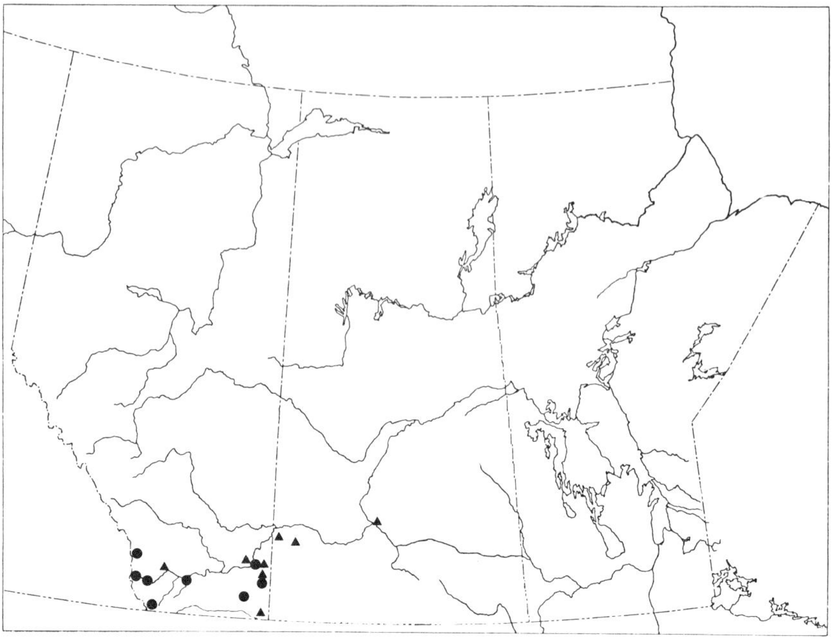

Map 70. Collection localities for *Hadronema princeps* (●) and *H. bispinosum* (▲).

Hadronema bispinosum Knight

Fig. 184; Map 70

Hadronema bispinosum Knight, 1928*b*: 179.

Length 3.36–4.06 mm; width 1.12–1.54 mm. Head almost black, lorum, jugum, and area above antennal socket pale. Pronotum yellow or reddish, calli black. Scutellum partly pale. Hemelytra generally black, wide costal margin and cuneus white; pubescence black, erect (Fig. 184). Ventral surface and femora mostly pale; tibiae black.

Remarks. This species is distinguished by the mostly pale femora.

Habitat. Collected on *Psoralea lanceolata.*

Distribution. South Dakota, Wyoming, Colorado; now known to occur in Alberta and Saskatchewan (Map 70).

Hadronema simplex Knight

Map 71

Hadronema simplex Knight, 1928*b*:178.

Length 2.80–3.64 mm; width 0.98–1.47 mm. Head black. Pronotum mostly black, basal angles sometimes reddish. Scutellum and hemelytra

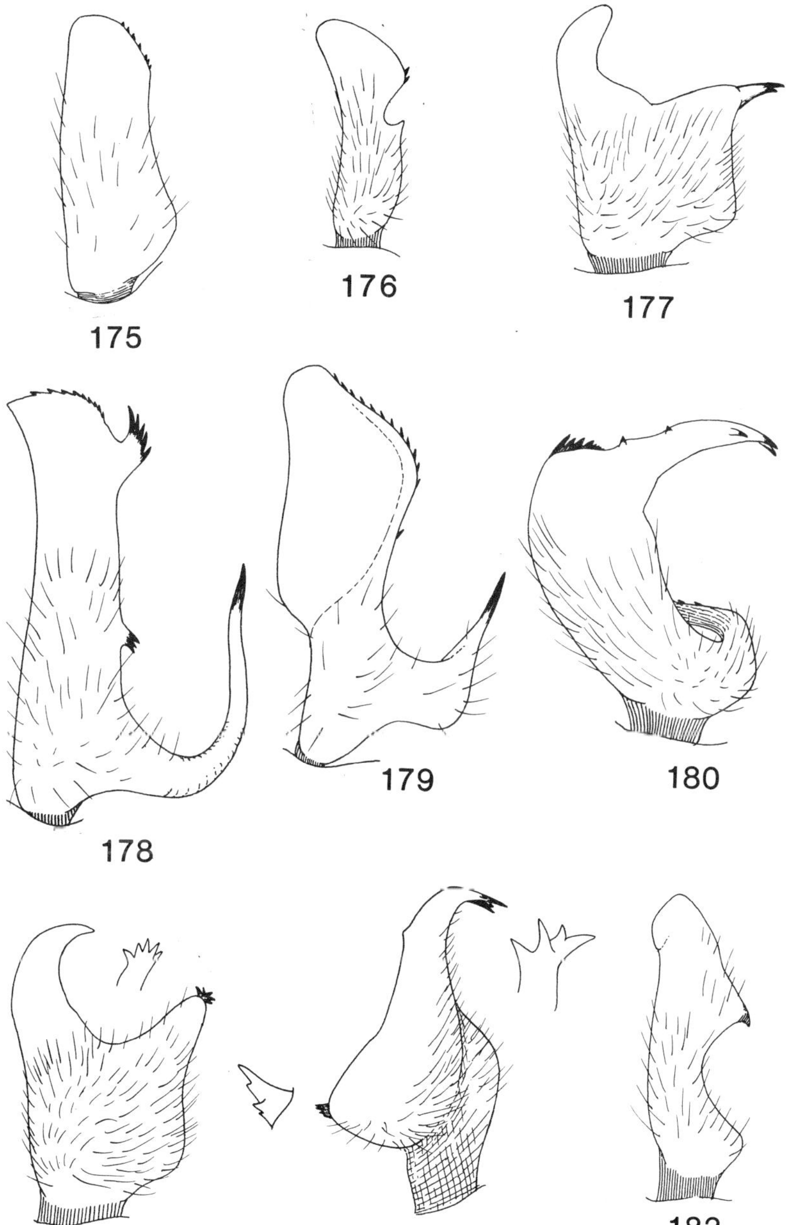

Figs. 175–183. Male claspers of *Lopidea* spp. 175, *yampae*; 176, *minor*; 177, *lathyri*; 178, *media*; 179, *lepidii*; 180, *serica*; 181, *balli*; 182, *dakota*; 183, *teton*.

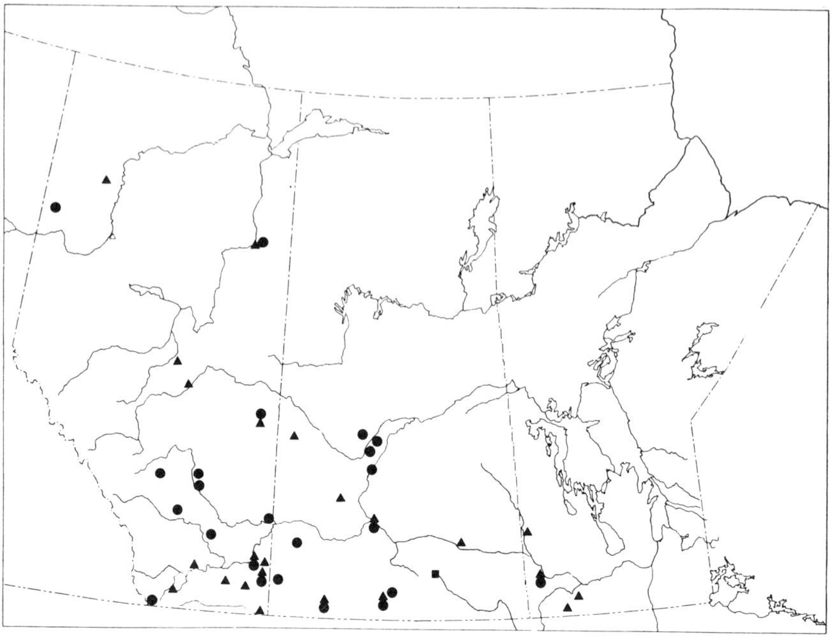

Map 71. Collection localities for *Hadronema simplex* (▲) and *H. militaris* (●).

black, costal margin and cuneus white; pubescence black, erect. Ventral surface and legs black.

Remarks. This species is similar to *bispinosum* in size, but the head and femora are black.

Habitat. Collected on various plants.

Distribution. Wyoming; now known to occur in the Prairie Provinces (Map 71).

Hadronema militaris Uhler

Map 71

Hadronema militaris Uhler, 1872:412.

Length 4.20–5.18 mm; width 1.54–1.82 mm. Head black. Pronotum red, calli black. Scutellum and hemelytra black, costal margin and cuneus white; pubescence black, erect. Ventral surface and legs black.

Remarks. This species is similar to *simplex* in color and appearance, but is much larger in size.

Habitat. Collected on *Thermopsis rhombifolia*.

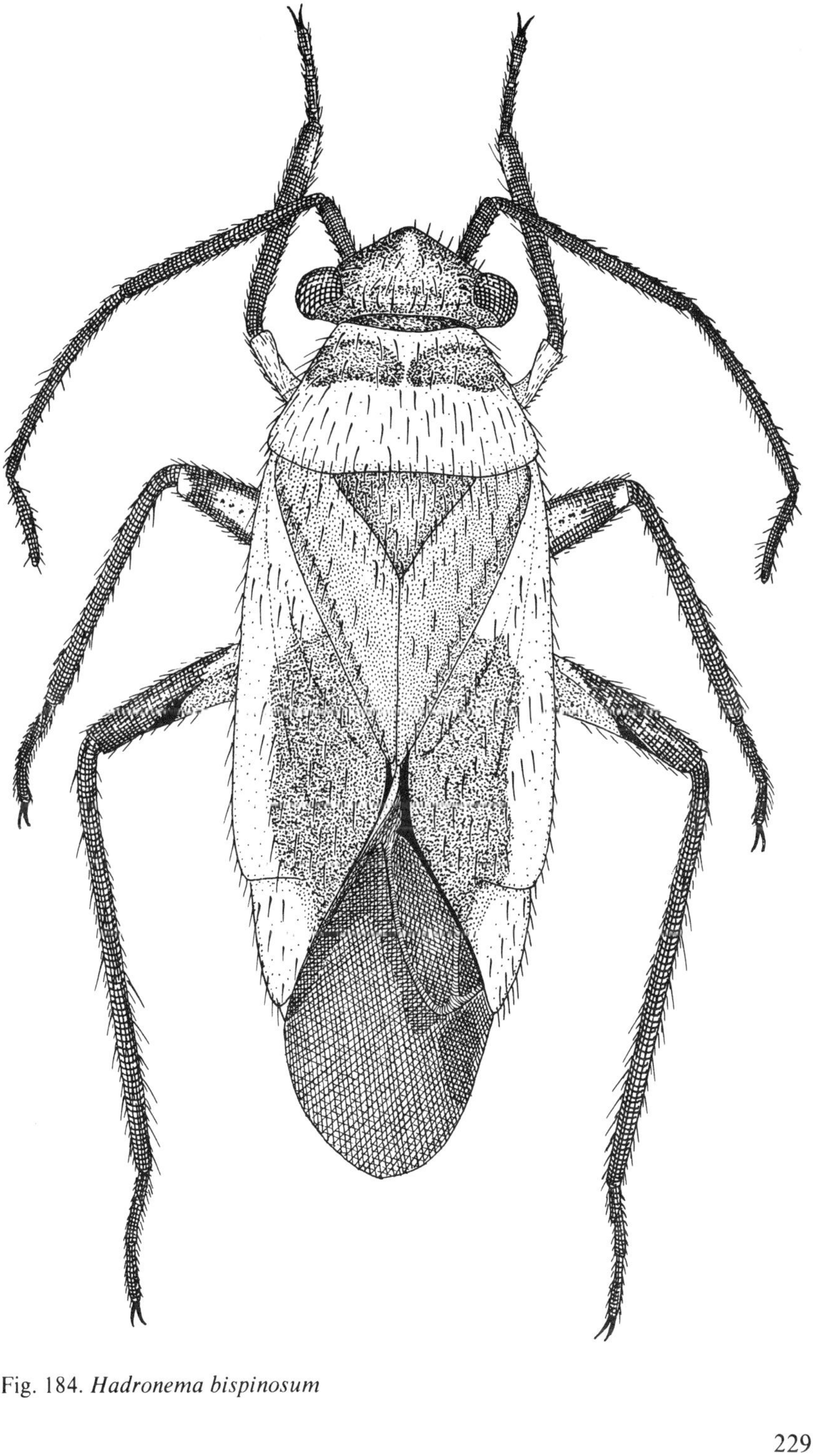

Fig. 184. *Hadronema bispinosum*

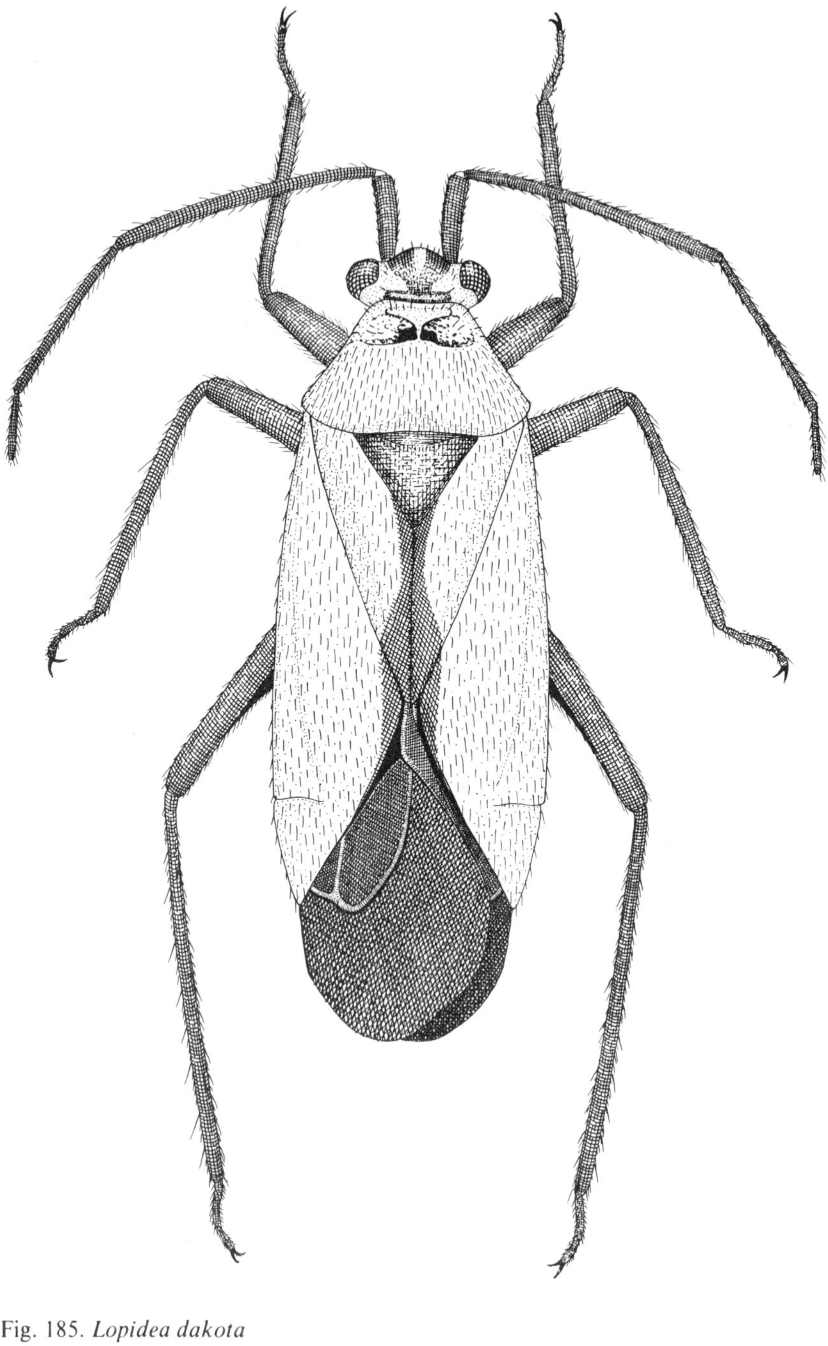

Fig. 185. *Lopidea dakota*

Distribution. New York, Michigan, western States, Mexico; Alberta, now known to occur in Saskatchewan and Manitoba (Map 71).

Genus *Lopidea* Uhler

Elongate, impunctate, black and white, or red and black species. Head vertical, red or white with clypeus, sides of frons and vertex black, frons smooth, eyes spherical, carina between them distinct. Pronotum trapeziform, calli distinct, lateral margins carinate. Scutellum and hemelytra glabrous or pubescent. Legs black.

There are approximately 125 species of *Lopidea* in North America, nine occur in the Prairie Provinces.

Key to species of *Lopidea*

1. Dorsal surface black with costal margin and cuneus pale; scutellum with broad median line pale; head white with black markings; right clasper (Fig. 175) .... *yampae* **Knight** (p. 232)
   Dorsal surface red, callus and broad median line on hemelytra may be black; head red with black markings 2
2. Length 3.9–4.9 mm; right clasper (Fig. 176) .... *minor* **Knight** (p. 233)
   Length 5.0–7.0 mm 3
3. Black median line on hemelytra broad and parallel, extending from pronotum to apex of corium; right clasper (Fig. 177) .... *lathyri* **Knight** (p. 233)
   Black median line behind scutellum narrow or absent (Fig. 185) 4
4. Pubescence on pronotum pale 5
   Pubescence on pronotum black 6
5. Rostrum shorter than second antennal segment; right clasper (Fig. 178) .... *media* **(Say)** (p. 233)
   Rostrum equal to or longer than second antennal segment; right clasper (Fig. 179) .... *lepidii* **Knight** (p. 234)
6. Hemelytra with appressed, sericeous pubescence, intermixed with longer black hairs 7
   Hemelytra with black hairs only 8
7. First antennal segment shorter than vertex between eyes; right clasper (Fig. 180) .... *serica* **Knight** (p. 235)
   First antennal segment equal to or longer than vertex between eyes; right clasper (Fig. 181) .... *balli* **Knight** (p. 236)
8. Rostrum longer than width of pronotum at base; right clasper (Fig. 182) .... *dakota* **Knight** (p. 236)
   Rostrum equal to or shorter than width of pronotum at base; right clasper (Fig. 183) .... *teton* **Knight** (p. 236)

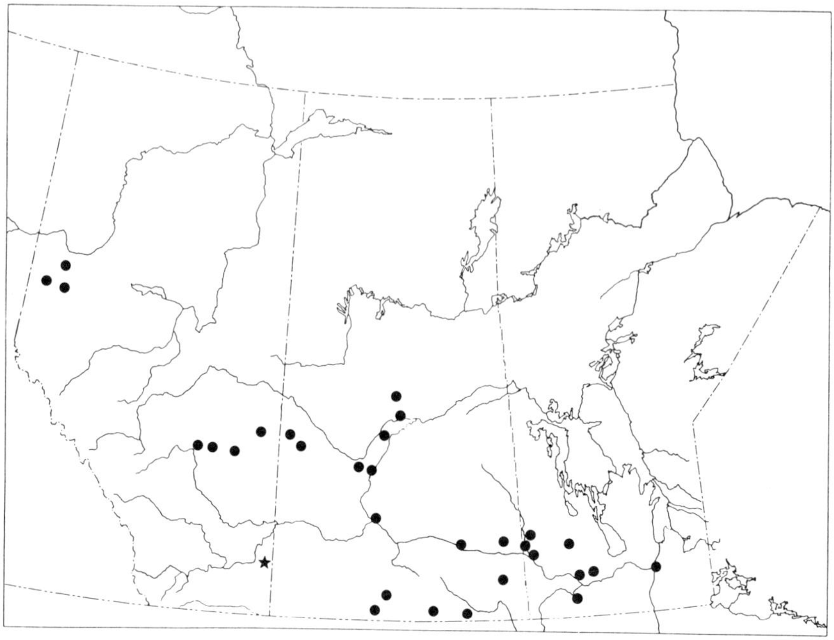

Map 72. Collection localities for *Lopidea yampae* (★) and *L. minor* (●).

Lopidea yampae Knight

Fig. 175; Map 72

Lopidea yampae Knight, 1965:7.

Length 4.30–4.76 mm; width 1.40–1.54 mm. Head mostly white, clypeus, sides of frons black. Pronotum grayish black, calli black, apical margin and longitudinal median line pale. Scutellum black, median line pale. Hemelytra mostly black, claval suture, costal margin, and cuneus pale. Pubescence on dorsum pale. Ventral surface black with pale markings, legs black.

Remarks. This is the only black species in the Prairie Provinces. The right clasper (Fig. 175) is distinctive.

Habitat. Collected on *Artemisia cana*.

Distribution. Colorado; now known to occur in Alberta (Map 72).

Lopidea minor Knight

Fig. 176; Map 72

Lopidea minor Knight, 1918*c*:213.

Length 3.92–4.90 mm; width 1.68–1.82 mm. Head red, usual markings black. Pronotum red, calli black. Scutellum black. Hemelytra red, narrow median area behind scutellum dark; pubescence black. Ventral surface reddish black, legs black.

Remarks. This is the smallest of the reddish species in the Prairie Provinces. The right clasper (Fig. 176) is distinctive.

Habitat. Collected on *Hedysarum* spp.

Distribution. North Dakota, Colorado; Alberta, now known to occur in Saskatchewan and Manitoba (Map 72).

Lopidea lathyri Knight

Fig. 177: Map 73

Lopidea lathyri Knight 1923*a*:66.

Length 6.30–6.58 mm; width 1.89–2.10 mm. Head and pronotum red with black markings. Scutellum black. Hemelytra red, costal margin often pale, broad median line behind scutellum black; pubescence pale on red areas and black on black areas. Ventral surface red along side and black along middle; legs black.

Remarks. This species is distinguished by the broad, black median line on the hemelytra and by the right clasper (Fig. 177).

Habitat. Collected on *Lathyrus* spp.

Distribution. North Central States; Prairie Provinces (Map 73).

Lopidea media (Say)

Fig. 178; Map 73

Capsus media Say, 1832:22.
Lopidea media: Uhler, 1872:412.

Length 5.25–5.74 mm; width 1.70–1.96 mm. Head and pronotum red with black markings; pubescence on pronotum pale. Rostrum shorter than second antennal segment. Hemelytra red.

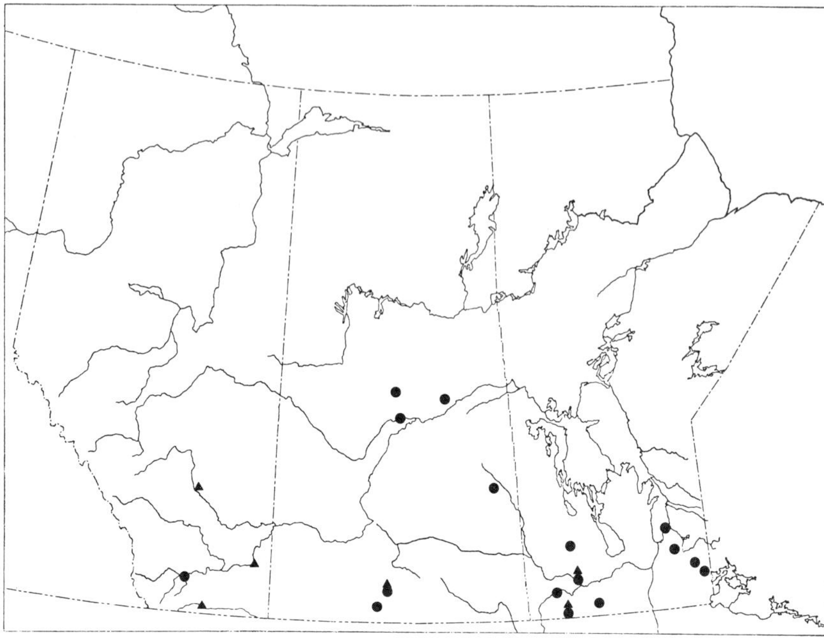

Map 73. Collection localities for *Lopidea lathyri* (●), *L. media* (▲).

Remarks. This species is distinguished by the pale pubescence on the pronotum and by the distinctive right clasper (Fig. 178).

Habitat. Collected on *Solidago* spp.

Distribution. Eastern and Central States, California, Mexico; Alberta, now known to occur in Saskatchewan and Manitoba (Map 73).

Lopidea lepidii Knight

Fig. 179; Map 74

Lopidea lepidii Knight, 1918*b*:175.

Length 5.25–5.46 mm; width 1.75–1.96 mm. Head and pronotum red with black markings. Rostrum equal to or longer than second antennal segment. Pronotum and hemelytra similar to *media* in color and pubescence.

Remarks. This species has pale pubescence on the pronotum as in *media*, but the rostrum is equal to or longer than the second antennal segment. The right clasper (Fig. 179) is distinctive.

Habitat. Collected on *Artemisia* spp.

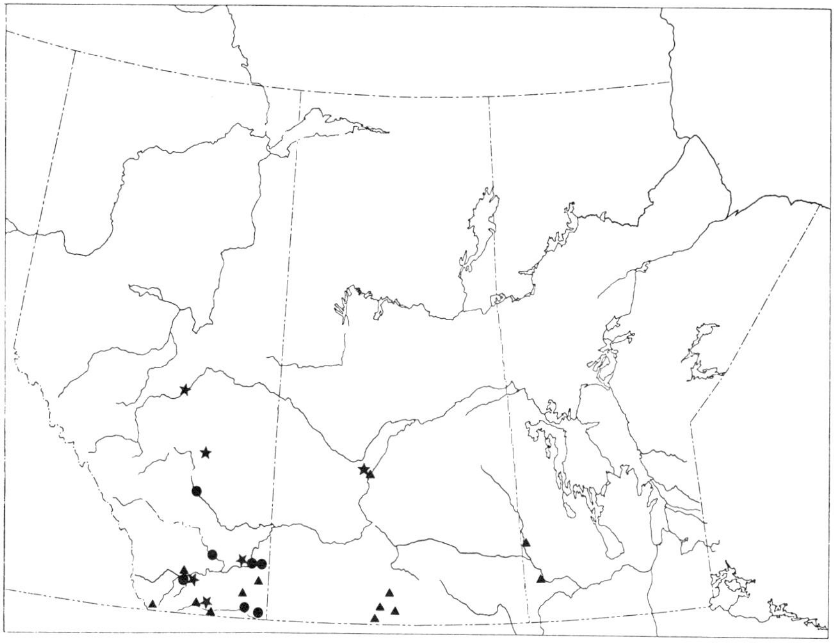

Map 74. Collection localities for *Lopidea lepidii* (●), *L. serica* (▲), and *L. balli* (★).

Distribution. Colorado, Arizona, California; now known to occur in Alberta (Map 74).

Lopidea serica Knight

Fig. 180; Map 74

Lopidea serica Knight, 1923*a*:69.

Length 5.60–6.02 mm; width 2.24–2.38 mm. Head, pronotum, and hemelytra red with black markings. First antennal segment shorter than vertex between eyes. Pronotum and scutellum with black pubescence, hemelytra with black hairs, intermixed with pale, sericeous pubescence.

Remarks. This species is distinguished by the two types of pubescence on the hemelytra and by the right clasper (Fig. 180).

Habitat. Collected on *Lupinus argenteus* and on *Psoralea* spp.

Distribution. Colorado; Alberta, now known to occur in Saskatchewan and Manitoba (Map 74).

Lopidea balli Knight

Fig. 181; Map 74

Lopidea balli Knight, 1923*a*:66.

Length 6.30–6.86 mm; width 1.96–2.24 mm. Head, pronotum, and hemelytra red with black markings. First antennal segment equal to or longer than width of vertex. Pronotum and scutellum with black hairs, hemelytra with black hairs, intermixed with pale, sericeous pubescence.

Remarks. This species is similar to *serica* as both have the same type of pubescence on the hemelytra. It is distinguished from *serica* by the longer first antennal segment and by the right clasper (Fig. 181).

Habitat. Unknown.

Distribution. Colorado, Nebraska; Alberta, now known to occur in Saskatchewan (Map 74).

Lopidea dakota Knight

Figs. 182, 185; Map 75

Lopidea dakota Knight, 1923*a*:67.

Length 6.30–6.58 mm; width 2.24–2.52 mm. Head red, black markings often diffuse. Rostrum longer than width of pronotum at base. Pronotum mostly red, hemelytra red with narrow area along commissure black (Fig. 185).

Remarks. This is a large, robust species in which the rostrum is longer than the width of pronotum at base. The right clasper (Fig. 182) is distinctive.

Habitat. Collected on a great variety of plants; large numbers have been collected on *Caragana arborescens*.

Distribution. North Central States, British Columbia; Prairie Provinces (Map 75).

Lopidea teton Knight

Fig. 183; Map 75

Lopidea teton Knight, 1923*a*:70.

Length 5.60–6.16 mm; width 2.10–2.52 mm. Head red with black markings, median red area often obscured. First antennal segment shorter

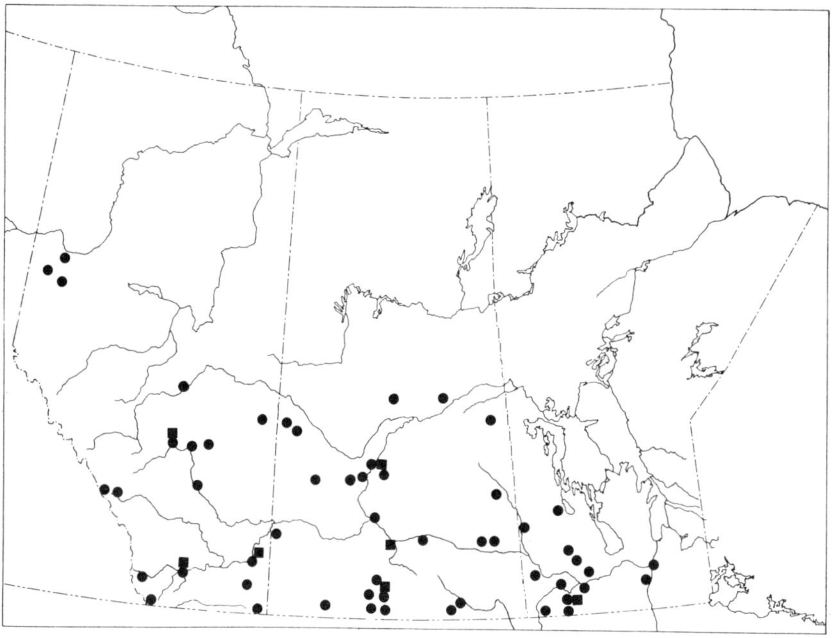

Map 75. Collection localities for *Lopidea dakota* (●) and *L. teton* (■).

than width of vertex. Rostrum equal to or shorter than width of pronotum at base. Pronotum and hemelytra mostly red with black pubescence.

Remarks. This species resembles *dakota*, but has a shorter rostrum and distinctive right clasper (Fig. 183).

Habitat. Collected on *Astragalus caryocarpus*.

Distribution. North Central and western States; now known to occur in the Prairie Provinces (Map 75).

Genus *Labopidea* Uhler

Elongate-oblong, green and black, pubescent species. Head vertical, broad, clypeus prominent, eyes small, far apart; genal suture extending from antennal socket to below eye; vertex wider than dorsal width of eyes. Pronotum trapeziform, twice as wide as long, calli distinct. Hemelytra smooth, pubescence variable. Macropterous and brachypterous forms. Genital claspers distinctive in form.

Kelton (1979*b*) revised the genus and retained seven species in *Labopidea*, five species occur in the Prairie Provinces.

Key to species of *Labopidea*

1. Rostrum extending beyond hind coxae .... ***brooksi*** **Kelton** (p. 238)
   Rostrum extending to middle coxae 2
2. Tibiae mostly black .... ***nigrisetosa*** **Knight** (p. 241)
   Tibiae mostly pale 3
3. Hemelytra green .... ***simplex*** **(Uhler)** (p. 241)
   Hemelytra fuscous or gray 4
4. Pubescence on hemelytra white .... ***pallida*** **Knight** (p. 243)
   Pubescence on hemelytra black .... ***discolor*** **(Sahlberg)** (p. 244)

Labopidea brooksi Kelton

Figs. 186, 187, 188; Map 76

Labopidea brooksi Kelton, 1979*b*:754.

Length 2.66–3.50 mm; width 1.12–1.26 mm. Head oblique, clypeus and frons black. Rostrum 2.10–2.17 mm long, extending beyond hind coxae. Pronotum gray, calli black, median line between them pale; pubescence sericeous, intermixed with longer pale hairs. Osteolar peritreme inconspicuous. Legs brown, tibial spines black. Males macropterous (Fig. 186), females brachypterous (Fig. 187).

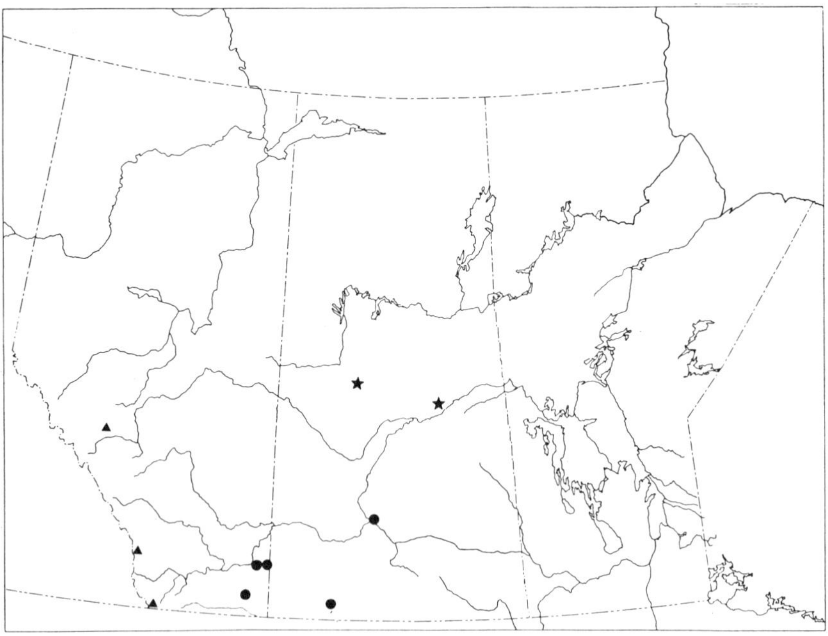

Map 76. Collection localities for *Labopidea brooksi* (●), *L. nigrisetosa* (▲), and *L. simplex* (★).

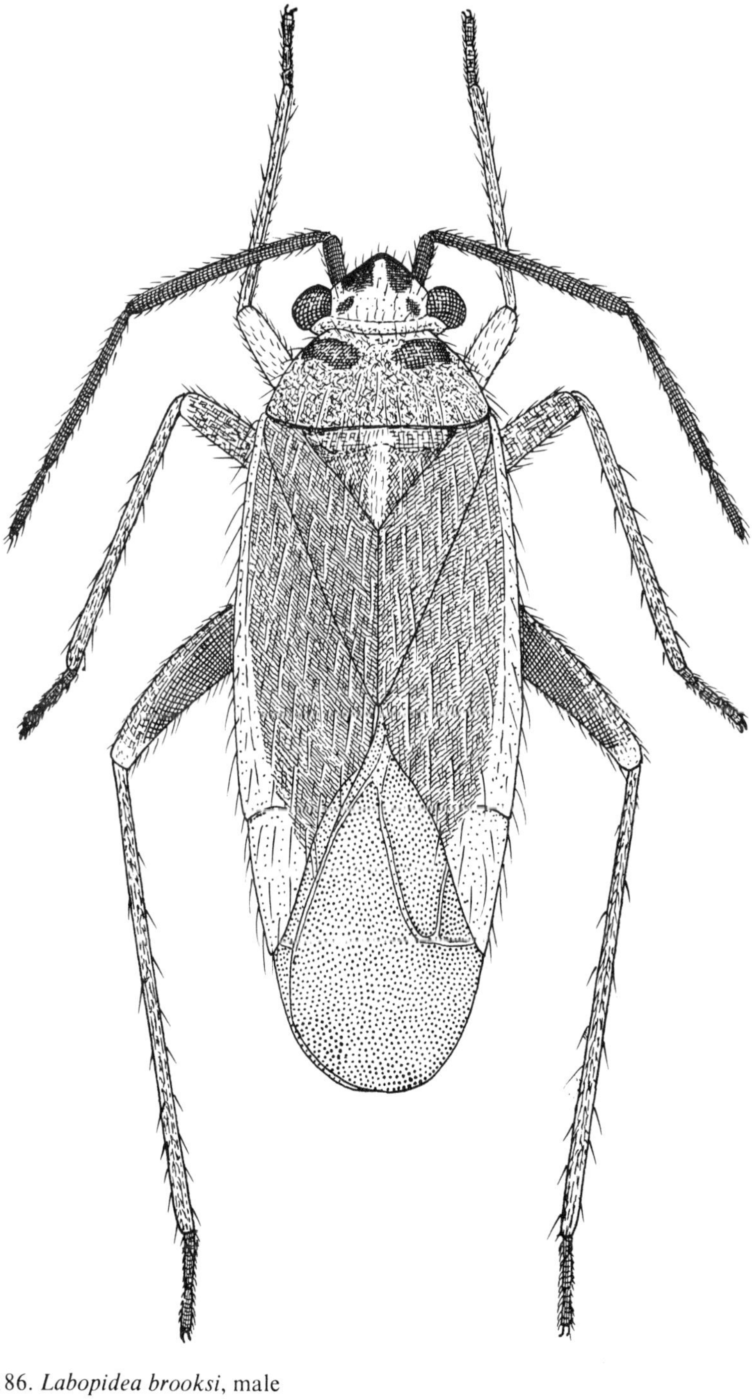

Fig. 186. *Labopidea brooksi*, male

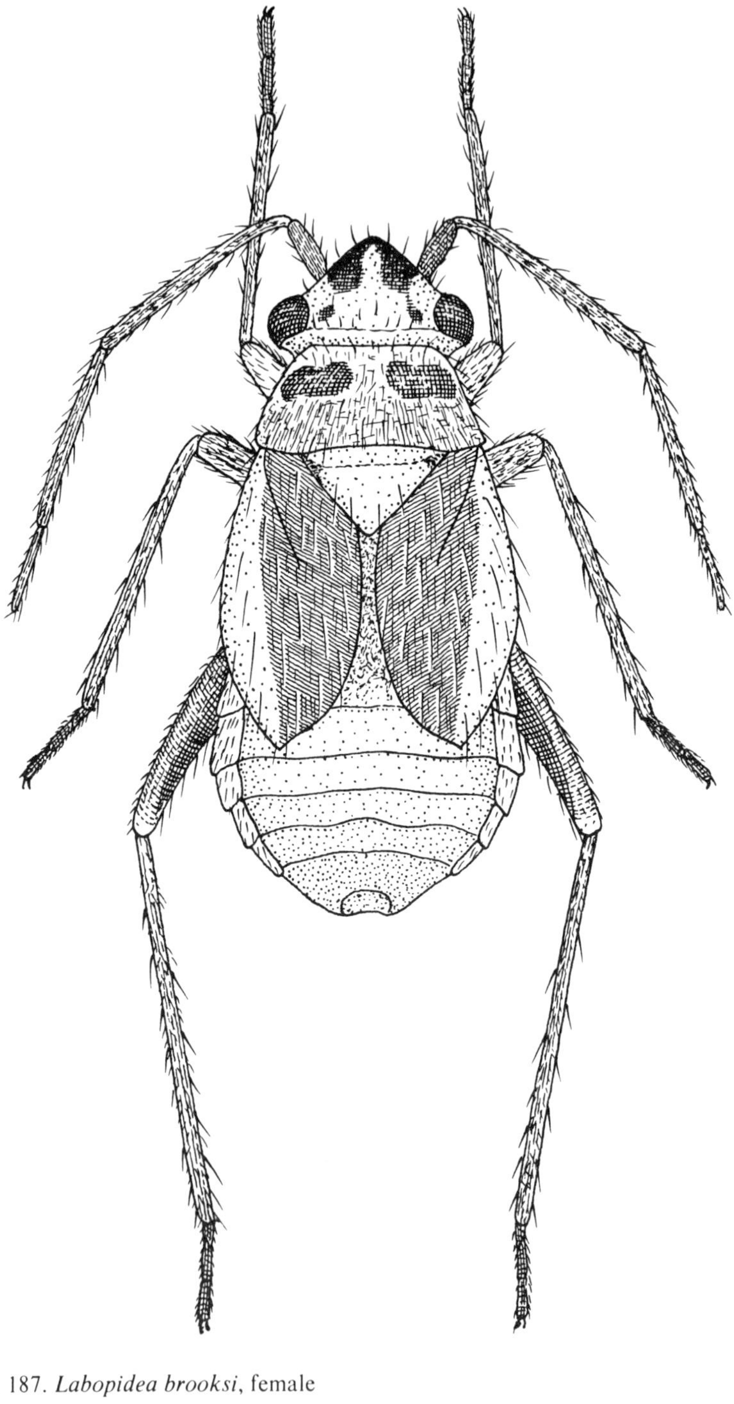

Fig. 187. *Labopidea brooksi*, female

Remarks. This species is distinguished by the small size, gray color, reduced osteolar peritreme, and the genital claspers (Fig. 188).

Habitat. Collected on *Artemisia cana.*

Distribution. Alberta and Saskatchewan (Map 76).

Labopidea nigrisetosa Knight

Fig. 189; Map 76

Labopidea nigrisetosa Knight, 1925*a*:94.

Length 3.08–4.55 mm; width 1.68–2.17 mm. Head oblique, black, area between eyes on top pale yellow. Rostrum 0.98–1.12 mm long, extending to middle coxae. Pronotum yellowish, calli black; pubescence black, erect. Hemelytra green; pubescence black, erect, intermixed with pale hairs. Ventral surface and legs black. Tibial spines black. Macropterous and brachypterous forms.

Remarks. This species is distinguished by the black ventral surface and legs. The genital claspers (Fig. 189) are similar to those of *brooksi*, but larger.

Habitat. Unknown.

Distribution. Wyoming; Alberta (Map 76).

Labopidea simplex (Uhler)

Fig. 190; Map 76

Tinicephalus simplex Uhler, 1872:417.
Hyoidea grisea Reuter, 1909:73.
Labopidea simplex Van Duzee, 1916*a*:220.
Labopidea chloriza Uhler, 1877:416.

Length 4.55–4.90 mm; width 1.82–2.03 mm. Head almost vertical, pale yellowish green, clypeus and frons shaded with fuscous. Rostrum 1.12–1.19 mm long, extending to middle coxae. Pronotum green, calli often shaded with fuscous; pubescence black, erect, intermixed with pale, shorter hairs. Hemelytra green, pubescence silvery, dense, intermixed with blackish hairs. Ventral surface and legs green. Macropterous forms.

Remarks. This species is distinguished by the green ventral surface and legs. The genital claspers (Fig. 190) are similar to those of *pallida.*

Habitat. Collected on *Senecio eremophilus.*

Distribution. Western States; now known to occur in Saskatchewan (Map 76).

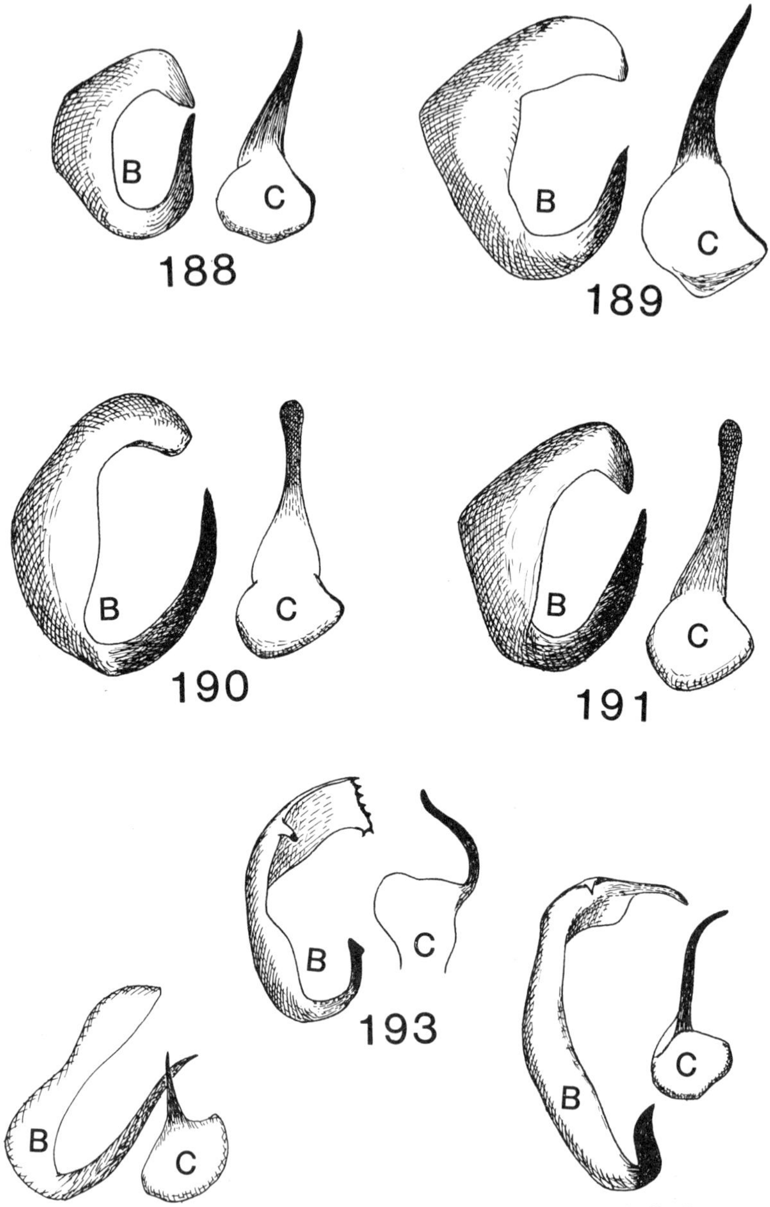

Figs. 188–194. Male claspers of *Labopidea* and *Labopidicola* 188, *Labopidea brooksi*; 189, *Labopidea nigrisetosa*; 190, *Labopidea simplex*; 191, *Labopidea pallida*; 192, *Labopidea discolor*; 193, *Labopidicola idahoensis*; 194, *Labopidicola planifrons*.

Labopidea pallida Knight

Fig. 191; Map 77

Labopidea pallida Knight, 1928*c*:233.

Length 3.92–4.34 mm; width 1.68–2.24 mm. Head shape and color similar to *brooksi*. Rostrum 1.05–1.19 mm long, extending to middle coxae. Pronotum and hemelytra similar to *brooksi* in color, but sericeous pubescence absent. Osteolar peritreme conspicuous. Legs pale. Tibial spines brown. Males macropterous, females brachypterous and macropterous.

Remarks. This species is a larger replica of *brooksi*, but the rostrum is shorter and the osteolar peritreme is prominent. The genital claspers (Fig. 191) are also similar.

Habitat. Collected on *Senecio canus*.

Distribution. Washington; now known to occur in Alberta and Saskatchewan (Map 77).

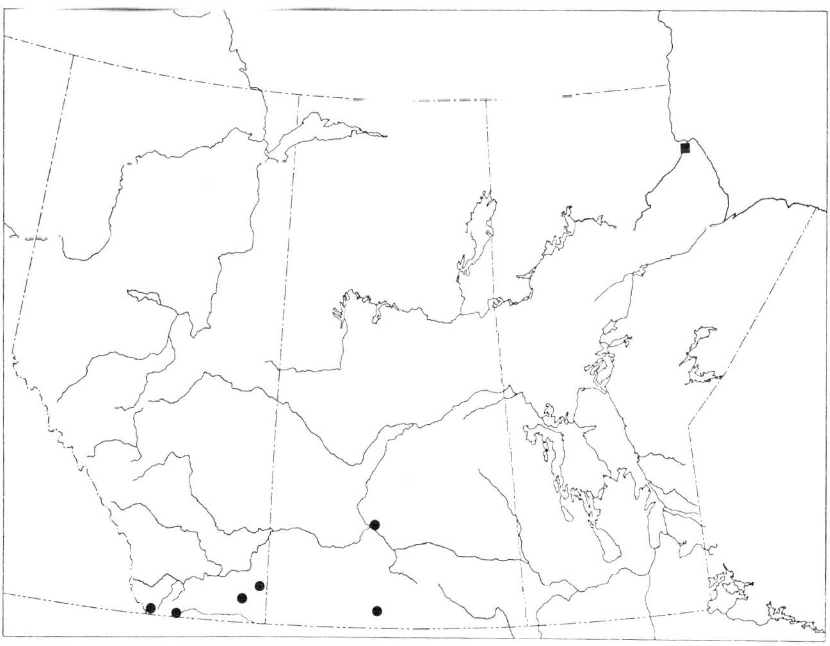

Map 77. Collection localities for *Labopidea pallida* (●) and *L. discolor* (■).

Labopidea discolor (Sahlberg)

Fig. 192; Map 77

Orthotylus discolor Sahlberg, 1878:29.
Labopidea discolor: Kelton, 1979*b*:754.

Length 3.08–3.22 mm; width 1.26–1.40 mm. Head black in male, light gray in female; carina between eyes with a row of black, erect bristles. Rostrum 1.12–1.26 mm long. Pronotum and scutellum black in male, light brown in female. Hemelytra pale gray; pubescence black and erect on head and pronotum, slanting on hemelytra, intermixed with few sericeous hairs. Ventral surface and femora black in male, pale in female; tibiae pale, spines black. Males macropterous, females brachypterous.

Remarks. Kelton (1979*b*) transferred the species to *Labopidea*. It is distinguished by the stiff bristles on the carina and the coloration of the male. The genital claspers (Fig. 192) are distinctive.

Habitat. Unknown.

Distribution. Holarctic; Manitoba (Map 77).

Genus *Labopidicola* Kelton

Elongate, green, pubescent species. Head horizontal, broad, clypeus prominent, eyes small and far apart; genal suture extending from antennal socket to below eye; vertex wider than dorsal width of eyes. Pronotum twice as wide as long; calli prominent. Hemelytra smooth. Pubescence of two types: sericeous and appressed, intermixed with longer semierect hairs. Macropterous and brachypterous forms.

Kelton (1979*b*) erected the genus for the five species that were not congeneric with *Labopidea*. Two species occur in the Prairie Provinces.

Key to species of *Labopidicola*

1. Pronotum and hemelytra with black hairs (Fig. 147) .... ***idahoensis* (Knight)** (p. 245)
   Pronotum and hemelytra with pale hairs .... ***planifrons* (Knight)** (p. 246)

Labopidicola idahoensis (Knight)

Figs. 147, 193; Map 78

Labopidea idahoensis Knight, 1968*a*:97.
Labopidicola idahoensis: Kelton, 1979*b*: 757.

Length 3.64–4.20 mm; width 1.26–1.40 mm. Head green, gently sloping, clypeus sharply angled. Rostrum 0.66–0.77 mm long. Pronotum green, flattened, subrectangular, lateral margins sharply angled, calli distinct; pubescence sericeous, intermixed with erect, black hairs. Hemelytra green, shiny; pubescence sericeous, intermixed with erect, black hairs. Ventral surface and legs green. Tibial spines black. Males macropterous (Fig. 147), females brachypterous and macropterous.

Remarks. This species is distinguished by the short rostrum and by the male claspers (Fig. 193).

Habitat. Collected on *Allium textile.*

Distribution. Colorado, Idaho, British Columbia; Prairie Provinces (Map 78).

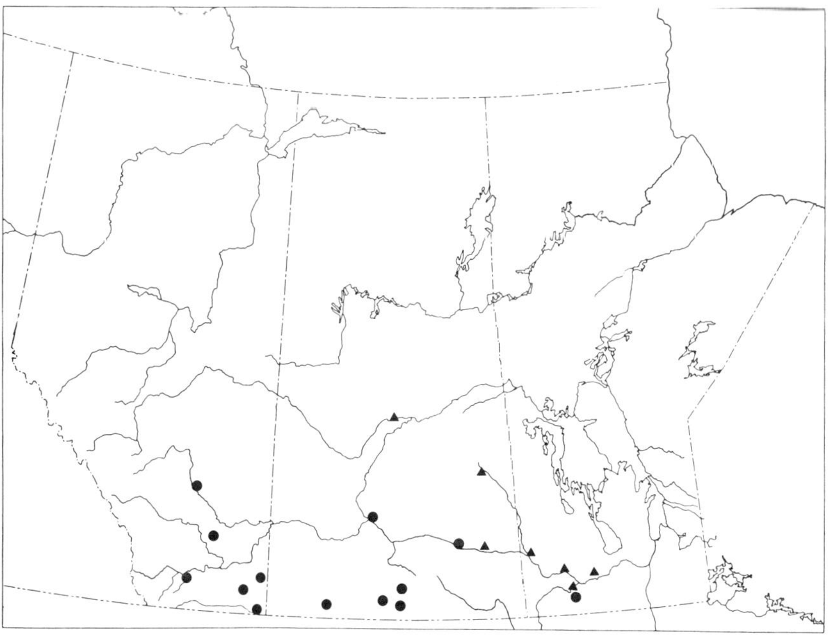

Map 78. Collection localities for *Labopidicola idahoensis* (●) and *L. planifrons* (▲).

Labopidicola planifrons (Knight)

Fig. 194; Map 78

Labopidea planifrons Knight, 1928*c*: 234.
Labopidicola planifrons: Kelton, 1979*b*:757.

Length 3.50–4.20 mm; width 1.33–1.40 mm. Head shape, color, and pubescence similar to *idahoensis*. Rostrum 0.56–0.70 mm long. Pronotum and hemelytra similar to *idahoensis* in shape and color, but erect hairs pale. Ventral surface and legs green. Tibial spines pale. Males macropterous, females macropterous and brachypterous.

Remarks. This species resembles *idahoensis*, but the erect hairs are pale. The genital claspers (Fig. 194) are distinctive.

Habitat. Collected on *Allium stellatum*.

Distribution. South Dakota, Iowa; Manitoba, Saskatchewan (Map 78).

Genus *Noctuocoris* Knight

Elongate, brown, pubescent species. Head oblique, eyes large, carina between them absent. Pronotum trapeziform, lateral margins rounded. Head, pronotum, scutellum, and hemelytra with two types of pubescence: sericeous, intermixed with black, erect, or slanting hairs. Ventral surface and legs brown.

The genus is monobasic and the species occurs in the Prairie Provinces.

Noctuocoris fumidus (Van Duzee)

Map 79

Orthotylus fumidus Van Duzee, 1916:127.
Noctuocoris fumidus: Knight, 1923*b*:510.

Length 5.74–6.16 mm; width 1.68–1.96 mm. Head oblique, margin at base with erect, black bristles. Rostrum 1.40–1.54 mm long. Antennae brown, first segment with black hairs, other segments with short, white hairs. Legs long, slender.

Remarks. Knight (1923*b*) transferred *fumidus* to this genus. The species is distinguished by the large, prominent eyes, by the brown color, and by the two types of pubescence.

Habitat. Collected on *Quercus macrocarpa*, and at light.

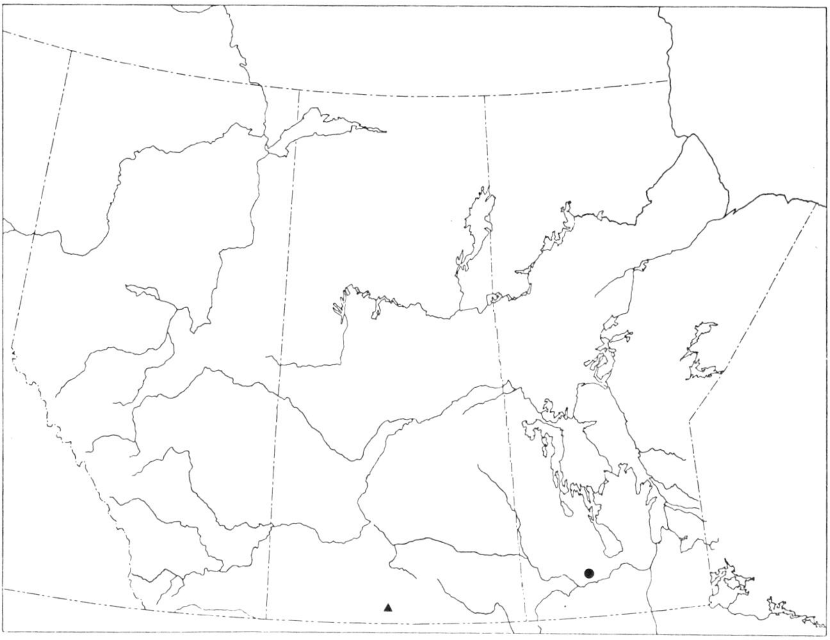

Map 79. Collection localities for *Noctuocoris fumidus* (●) and *Parthenicus brooksi* (▲).

Distribution. Northeastern and North Central States, Ontario, Quebec; now known to occur in Manitoba (Map 78).

Genus *Parthenicus* Reuter

Elongate-oblong, small, brown and gray, pubescent species. Head oblique, eyes large, prominent, hairy, base of head truncate. Rostrum extending beyond hind coxae. Pronotum trapiziform, lateral margins sharply angled, but not carinate; calli inconspicuous. Hemelytra smooth, pubescence of two types: sericeous and appressed, intermixed with pale, erect hairs.

Approximately 71 species are known from North America, one occurs in the Prairie Provinces.

Parthenicus brooksi Kelton

Fig. 195; Map 79

Parthenicus brooksi Kelton, 1980*c*:341.

Length 3.85–3.92 mm; width 1.26–1.33 mm. Head brown. First and terminal antennal segments brown, second segment pale green. Rostrum

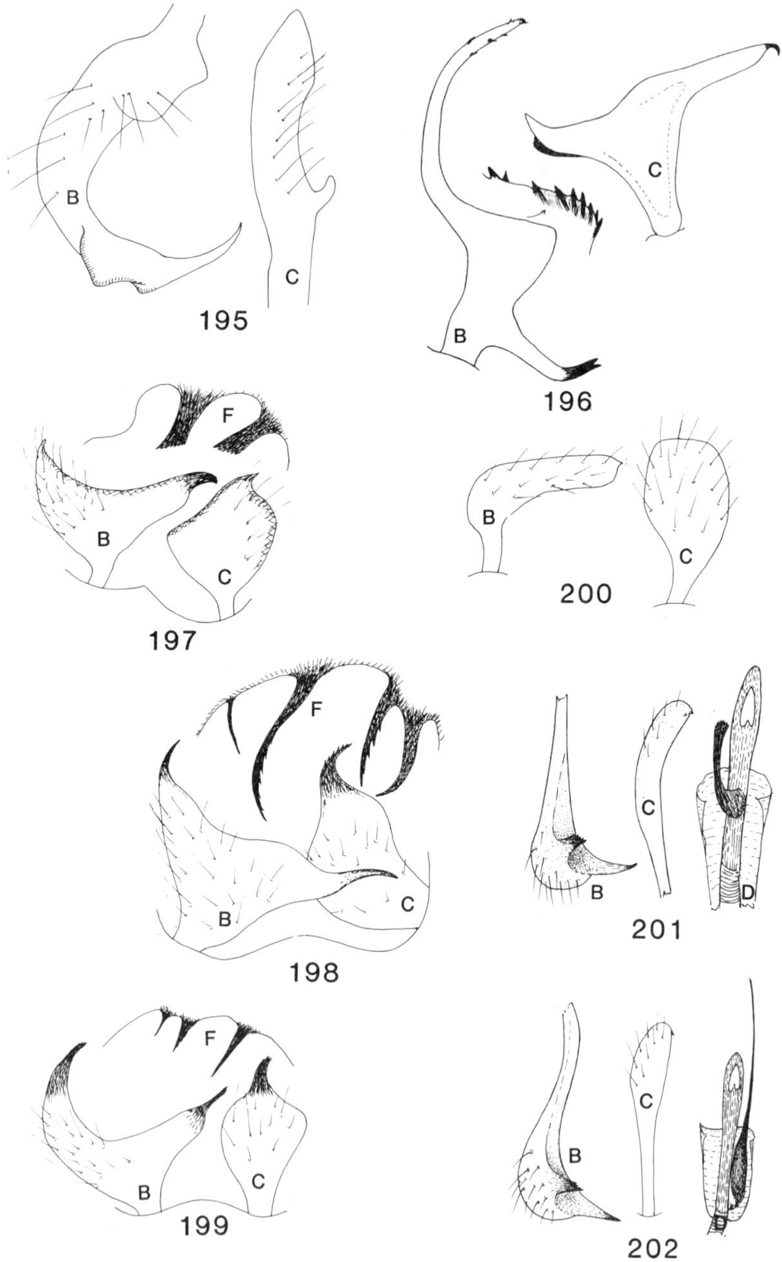

Figs. 195–202. Male genitalia of Miridae. 195, *Parthenicus brooksi*; 196, *Reuteria querci*; 197, *Hesperocapsus demensus*; 198, *H. anograe*; 199, *H. sericatus*; 200, *Fieberocapsus flaveolus*; 201, *Diaphnocoris provancheri*; 202, *D. ulmi*.

1.40–1.43 mm long. Pronotum and scutellum brown. Hemelytra dusky gray with faint fuscous spots. Ventral surface brown, femora brown.

Remarks. This species is easily recognized by the brown first antennal segment, brown head, pronotum, and scutellum, contrasting dusky gray hemelytra, and genital claspers (Fig. 195).

Habitat. Unknown, collected at light.

Distribution. Saskatchewan (Map 79).

Genus *Hesperocapsus* Knight

Elongate, green, pubescent species. Head oblique, eyes large, prominent, carina between them faint. First antennal segment shorter than width of vertex, second segment cylindrical, almost as thick as first. Rostrum variable in length. Pronotum smooth, lateral margins rounded, calli indistinct. Hemelytra smooth, shiny. Pubescence of two types: sericeous and appressed, intermixed with erect, longer hairs. Ventral surface and legs green, tibial spines black. Genital segment of male with several sclerotized projections along dorsal margin.

There are approximately 15 species in North America, three occur in the Prairie Provinces.

The genus was erected by Knight (1968*a*) and contains most of the species formerly included in the genus *Pseudopsallus* Van Duzee.

Key to species of *Hesperocapsus*

1. Rostrum 0.98–1.12 mm long, extending just beyond front coxae; genital claspers (Fig. 197) .... ***demensus* (Van Duzee)** (p. 249)
   Rostrum longer, extending to margin of mesosternum or beyond 2
2. Rostrum 1.26–1.40 mm long, extending to margin of mesosternum; hind femur and tibia without spots; genital claspers (Fig. 198) .... ***anograe* (Knight)** (p. 250)
   Rostrum 1.54 mm or longer, extending to middle coxae; hind femur and tibia spotted with brown; genital claspers (Fig. 199) ...... ***sericatus* (Uhler)** (p. 251)

Hesperocapsus demensus (Van Duzee)

Fig. 197; Map 80

Orthotylus demensus Van Duzee, 1925:398.
Hesperocapsus demensus: Knight, 1968*a*:107.

Length 4.62–5.18 mm; width 1.61–1.82 mm. Head yellowish green, first antennal segment green, other segments brown. Rostrum 0.98–1.12 mm long, extending just beyond front coxae. Pronotum green, calli yellowish. Scutellum and hemelytra green; pubescence sericeous, intermixed with dense, long, slanting hairs. Ventral surface and legs green. Macropterous forms.

Remarks. This species is distinguished by the short rostrum, by the sclerites on the genital capsule, and by the genital claspers (Fig. 197).

Habitat. Collected on *Oenothera serrulata.*

Distribution. Southwestern States, South Dakota; now known to occur in the Prairie Provinces (Map 80).

Hesperocapsus anograe (Knight)

Fig. 198; Map 80

Pseudopsallus anograe Knight, 1930*a*:4.
Hesperocapsus anograe: Knight, 1968*a*:107.

Length 4.76–5.32 mm; width 1.75–1.96 mm. General color and pubescence similar to *demensus.* Rostrum 1.26–1.40 mm long, extending to margin of mesosternum. Macropterous forms.

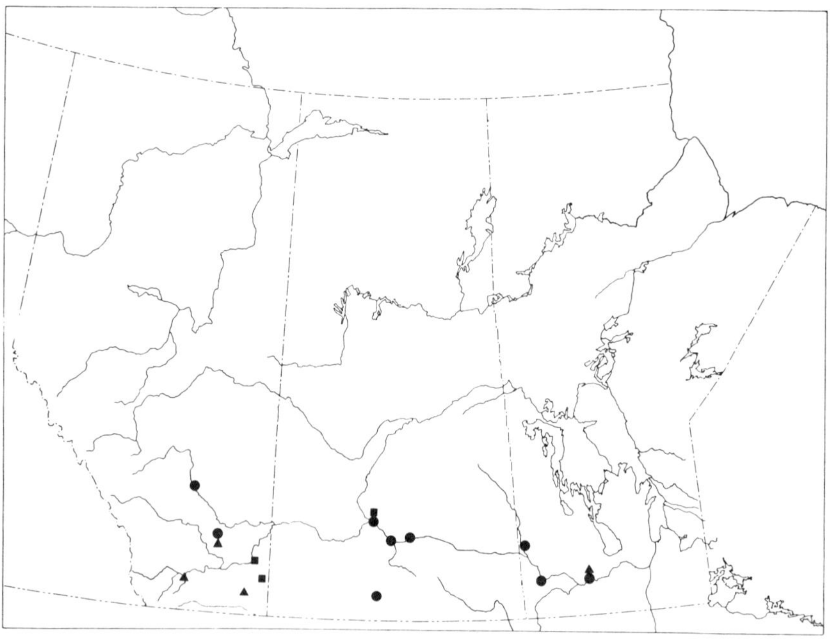

Map 80. Collection localities for *Hesperocapsus demensus* (●), *H. anograe* (▲), and *H. sericatus* (■).

Remarks. This species is readily separated from *demensus* by the longer rostrum, by the differences in the sclerites on the genital segment, and by the genital claspers (Fig. 198).

Habitat. Collected on *Oenothera caespitosa.*

Distribution. Arizona, Colorado; now known to occur in Alberta and Manitoba (Map 80).

Hesperocapsus sericatus (Uhler)

Fig. 199; Map 80

Oncotylus sericatus Uhler, 1895:49.
Oncotylus repertus Uhler, 1895:49.
Hesperocapsus sericatus: Knight, 1968*a*:105.

Length 3.78–5.88 mm; width 1.82–1.96 mm. General appearance, color, and pubescence similar to *demensus* and *anograe*. Rostrum 1.54–1.61 mm long, extending to middle coxae. Hind femur and tibia spotted with brown. Males long, slender, females brachypterous.

Remarks. This species is readily separated from the others by the longer rostrum, by the spotted femur and tibia, by the differences in the projections on the genital segment, and by the genital claspers (Fig. 199).

Habitat. Collected on *Artemisia frigida.*

Distribution. Colorado, Nevada; now known to occur in Alberta and Saskatchewan (Map 80).

Genus *Melanotrichus* Reuter

Elongate, green, pubescent species. Head oblique, eyes large, prominent; carina between them distinct. Pronotum trapeziform, lateral margins rounded; calli distinct, smooth. Hemelytra smooth, shiny. Pubescence of two types: sericeous and appressed, intermixed with long, suberect hairs. Genital segment of male without sclerite.

Kelton (1979*c*) transferred the species with black, scaly pubescence from *Melanotrichus* to *Brooksetta* Kelton.

There are approximately 20 species of *Melanotrichus* in North America, nine occur in the Prairie Provinces.

Key to species of *Melanotrichus*

1. Wing membrane with darkened area behind vein 2
   Wing membrane without darkened area behind vein 3
2. Rostrum 1.22–1.40 mm long, extending behind hind coxae .... ***wallisi* Kelton** (p. 252)
   Rostrum 0.98–1.12 mm long, extending just beyond middle coxae .... ***coagulatus* (Uhler)** (p. 253)
3. Hemelytra with clumps of sericeous pubescence .... ***flavosparsus* (Sahlberg)** (p. 254)
   Hemelytra with sericeous pubescence regularly distributed, not in clumps 4
4. Costal margin of hemelytra pale 5
   Costal margin of hemelytra not pale 6
5. Ventral margin of eye about equal with ventral margin of antennal socket .... ***albocostatus* (Van Duzee)** (p. 254)
   Ventral margin of eye extending below ventral margin of antennal socket .... ***elongatus* Kelton** (p. 256)
6. Larger species, over 4.20 mm in length 7
   Smaller species, less than 3.64 mm in length 8
7. Pronotum and hemelytra mottled with green; abdomen with sericeous pubescence, intermixed with simple hairs .... ***atriplicis* Knight** (p. 257)
   Pronotum and hemelytra not mottled; abdomen with simple hairs .... ***mistus* (Knight)** (p. 257)
8. Rostrum 0.84–0.98 mm, extending to middle coxae .... ***leviculus* Knight** (p. 257)
   Rostrum 1.54–1.68 mm, extending beyond hind coxae .... ***brindleyi* Knight** (p. 258)

Melanotrichus wallisi Kelton

Fig. 203; Map 81

Melanotrichus wallisi Kelton, 1980*b*:337.

Length 3.36–3.64 mm; width 1.12–1.26 mm. Head pale bluish green. Rostrum 1.22–1.40 mm long. Pronotum, scutellum, and hemelytra bluish green; pubescence sericeous, appressed, intermixed with slanting, longer black hairs. Ventral surface and legs pale green, tibial spines brown.

Remarks. This species resembles *coagulatus* in size and color, but has a longer rostrum and differs in the details of genital claspers (Fig. 203).

Habitat. Unknown.

Distribution. Saskatchewan (Map 81).

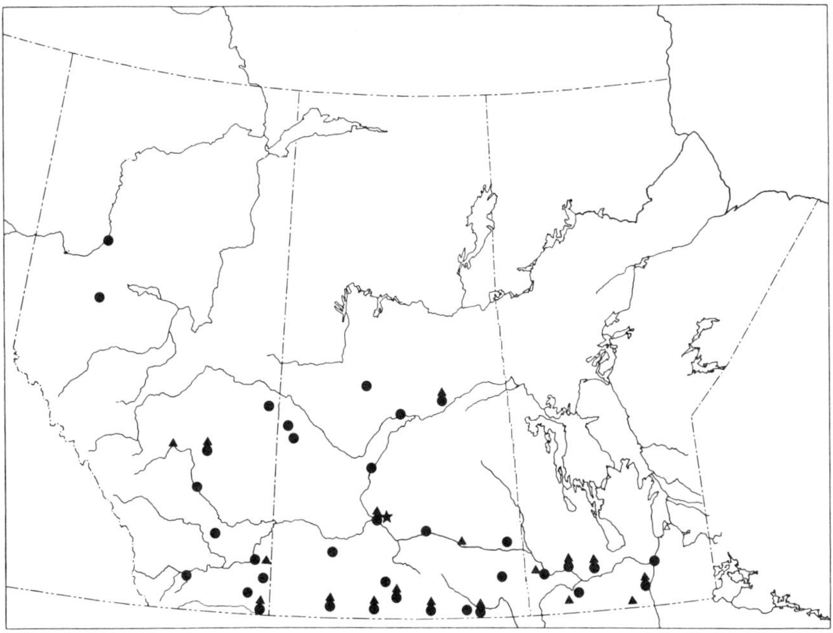

Map 81. Collection localities for *Melanotrichus wallisi* (★), *M. coagulatus* (●), and *M. flavosparsus* (▲).

Melanotrichus coagulatus (Uhler)

Figs. 148, 204; Map 81

Macrocoleus coagulatus Uhler, 1877:417.
Melanotrichus coagulatus: Knight, 1927*c*:142.

Length 3.50–4.06 mm; width 1.12–1.54 mm. Head green marked with yellow. Rostrum 0.98–1.12 mm long. Pronotum green, calli yellowish. Hemelytra pale green mottled with darker green; pubescence sericeous and appressed, usually in small clumps, intermixed with slanting, black hairs. Ventral surface and legs pale green, tibial spines black.

Remarks. This species (Fig. 148) resembles *flavosparsus* in size and appearance, but is distinguished from it by the male claspers (Fig. 204).

Habitat. Collected on *Chenopodium album*.

Distribution. Maine to California; Alberta, now known to occur in Saskatchewan and Manitoba (Map 81).

Melanotrichus flavosparsus (Sahlberg)

Fig. 205; Map 81

Phytocoris flavosparsus Sahlberg, 1842:411.
Lygus unicolor Provancher, 1872:105.
Melanotrichus flavosparsus: Knight, 1927*c*:142.

Length 3.78–4.20 mm; width 1.40–1.54 mm. Head yellowish green. Rostrum 1.12–1.26 mm long. Pronotum green, calli yellowish. Scutellum and hemelytra green, hemelytra appears mottled with paler spots when sericeous clumps absent. Ventral surface and legs pale yellowish, tibial spines pale.

Remarks. The genital claspers (Fig. 205) are distinctive.

Habitat. Collected on *Chenopodium album.*

Distribution. New York to California, Ontario, Quebec; now known to occur in the Prairie Provinces (Map 81).

Melanotrichus albocostatus (Van Duzee)

Fig. 206; Map 82

Orthotylus albocostatus Van Duzee, 1918:299.
Melanotrichus albocostatus: Knight, 1927*c*:147.

Length 4.34–5.18 mm; width 1.40–1.61 mm. Head oblique, yellowish green; eyes prominent, ventral margin of eye about equal with ventral margin of antennal socket. Rostrum 1.19–1.33 mm long. Pronotum pale green, calli yellowish. Scutellum pale green, median longitudinal line pale. Hemelytra uniformly light green, embolium and adjacent corium pale; pubescence sericeous and appressed, intermixed with longer and slanting white and black hairs. Ventral surface and legs pale green, tibial spines fuscous to black.

Remarks. This species is distinguished by the slender, elongate form, by the pale costal margin of the hemelytra, and by the genital claspers (Fig. 206).

Habitat. Collected on *Descurainia sophia.*

Distribution. Western States, British Columbia; now known to occur in Alberta and Saskatchewan (Map 82).

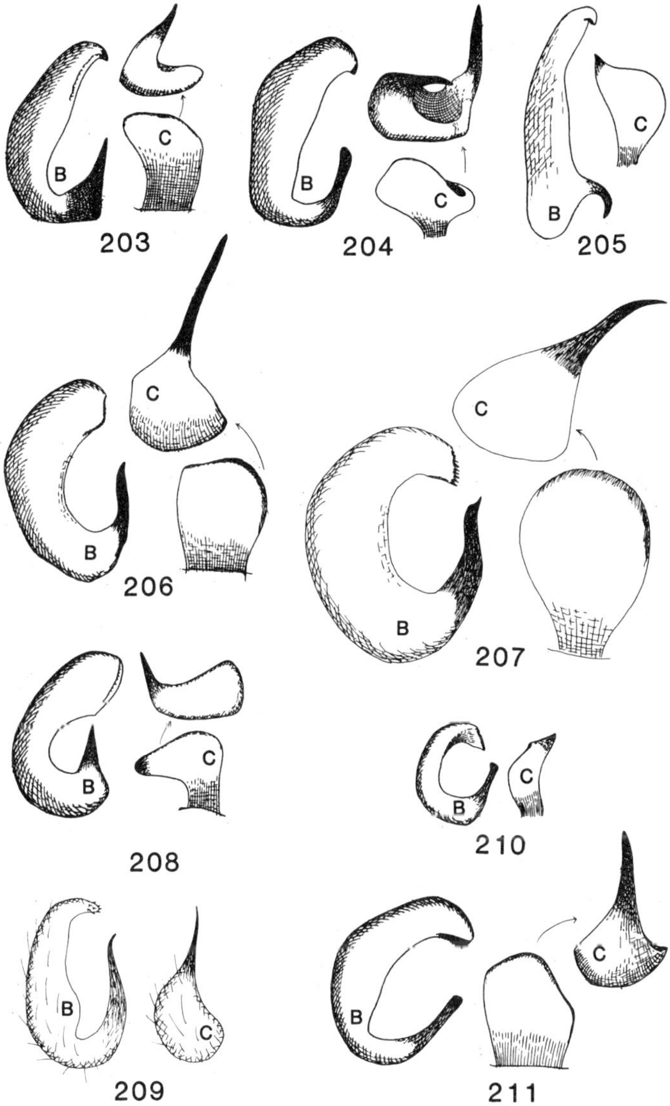

Figs. 203–211. Male claspers of *Melanotrichus* spp. 203, *wallisi*; 204, *coagulatus*; 205, *flavosparsus*; 206, *albocostatus*; 207, *elongatus*; 208, *atriplicis*; 209, *mistus*; 210, *leviculus*; 211, *brindleyi*.

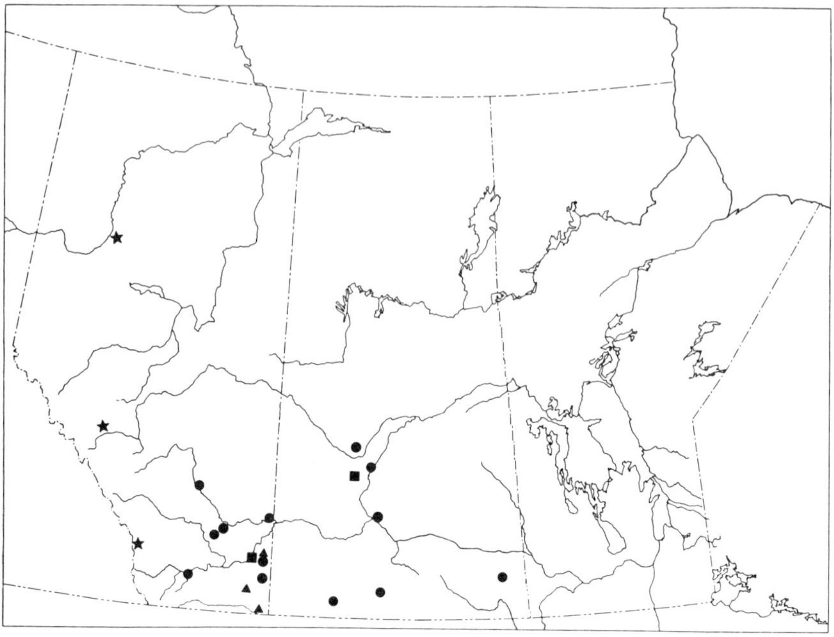

Map 82. Collection localities for *Melanotrichus albocostatus* (●), *M. elongatus* (■), *M. atriplicis* (▲), and *M. mistus* (★).

Melanotrichus elongatus Kelton

Fig. 207; Map 82

Melanotrichus elongatus Kelton, 1980*b*:337.

Length 5.25–6.30 mm; width 1.54–1.82 mm. Head oblique, yellowish green; eyes larger than in *albocostatus,* ventral margin of eye extending below ventral margin of antennal socket. Rostrum 1.19–1.33 mm long. Pronotum pale green, calli yellowish. Scutellum fuscous green, median longitudinal line pale. Hemelytra light green, embolium and adjacent corium pale; cuneus yellowish green; pubescence sericeous and appressed, intermixed with longer and slanting pale and fuscous hairs. Ventral surface and legs pale yellowish green; tibial spines fuscous to black.

Remarks. This species is longer than *albocostatus* and has larger eyes. The genitalia (Fig. 207) are similar.

Habitat. Collected on *Descurainia sophia.*

Distribution. Alberta and Saskatchewan (Map 82).

Melanotrichus atriplicis Knight

Fig. 208; Map 82

Melanotrichus atriplicis Knight, 1968*a*:125.

Length 4.55–5.04 mm; width 1.61–1.96 mm. Head oblique, pale green. Rostrum 1.26–1.40 mm long. Pronotum, scutellum, and hemelytra pale green mottled with darker green; pubescence sericeous and appressed, intermixed with slanting, darker hairs, sericeous hairs also present on pleura and abdomen. Ventral surface and legs pale green, tibial spines pale.

Remarks. This species is readily distinguished by the larger size, by the mottled green hemelytra, and by the genital claspers (Fig. 208).

Habitat. Collected on *Salicornia rubra.*

Distribution. Nevada; now known to occur in Alberta (Map 82).

Melanotrichus mistus (Knight)

Fig. 209; Map 82

Orthotylus mistus Knight, 1925*a*:91.
Melanotrichus mistus: Knight, 1927*c*:142.

Length 4.55–5.11 mm; width 1.82–1.96 mm. Head oblique, pale yellowish. Rostrum 1.12–1.26 mm long. Pronotum and scutellum yellowish green. Hemelytra light green; pubescence sericeous and appressed, intermixed with longer black hairs. Ventral surface and legs pale yellowish green, tibial spines black.

Remarks. This species is large and robust and is uniformly green. The genital claspers (Fig. 209) are similar to those of *elongatus*.

Habitat. Unknown.

Distribution. Northwestern States, British Columbia; Alberta (Map 82).

Melanotrichus leviculus Knight

Fig. 210; Map 83

Melanotrichus leviculus Knight, 1927*a*:146.

Length 3.08–3.64 mm; width 1.19–1.40 mm. Head pale green or yellowish. Rostrum 0.84–0.98 mm long. Pronotum green, calli yellowish. Hemelytra uniformly green; pubescence sericeous and appressed, intermixed

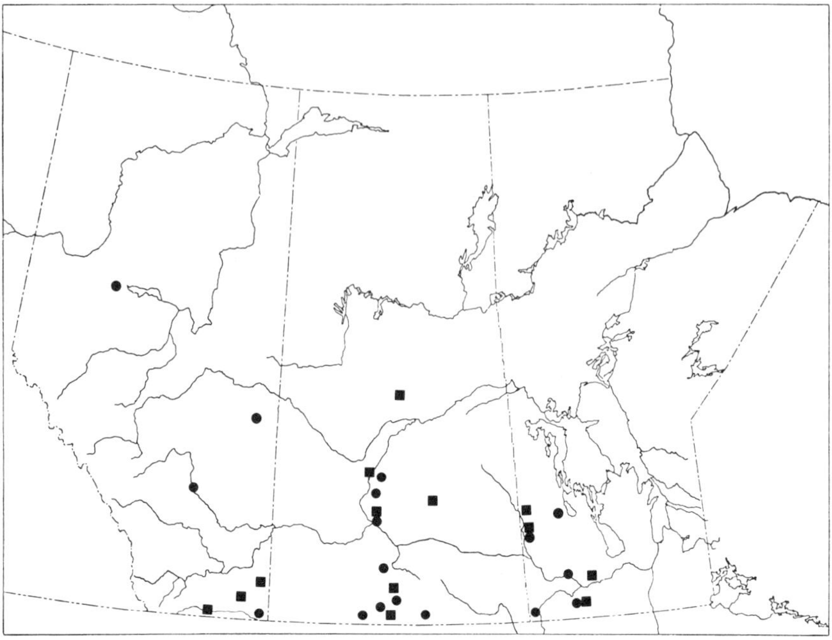

Map 83. Collection localities for *Melanotrichus leviculus* (●) and *M. brindleyi* (■).

with longer and fuscous or blackish hairs. Ventral surface and legs pale green, tibial spines pale.

Remarks. This species resembles *coagulatus,* but the hemelytra is uniformly green, and the tibial spines are pale. The genital claspers (Fig. 210) are distinctive.

Habitat. Collected on *Suaeda depressa* and *Salicornia rubra.*

Distribution. New York; now known to occur in the Prairie Provinces (Map 83).

Melanotrichus brindleyi Knight

Fig. 211; Map 83

Melanotrichus brindleyi Knight, 1968*a*: 127.

Length 3.36–3.50 mm; width 1.26–1.40 mm. Head pale, somewhat yellowish. Rostrum 1.54–1.68 mm long. Pronotum, scutellum, and hemelytra pale green to almost white; pubescence sericeous and appressed, intermixed with slanting, black hairs. Ventral surface and legs pale green, tibial spines pale.

Remarks. This species is distinguished by the long rostrum, by the pale color of the hemelytra and pale tibial spines, and by the genital claspers (Fig. 211).

Habitat. Collected on *Antennaria campestris.*

Distribution. North Central States; now known to occur in the Prairie Provinces (Map 83).

Genus *Reuteria* Puton

Elongate, pale green, delicate species. Head oblique, eyes large, removed from basal margin, carina between them absent. First antennal segment green with two black lines. Pronotum trapeziform, pale green, smooth, lateral margins rounded. Scutellum and hemelytra pale green; pubescence simple, white, long, dense. Ventral surface and legs pale green, femur often with dark green patches.

Henry (1977) reviewed this genus and reported eight species from North America; one occurs in the Prairie Provinces.

Reuteria querci Knight

Fig. 196; Map 84

Reuteria querci Knight, 1939*c*:131.

Length 4.48–4.90 mm; width 1.54–1.68 mm. Head pale green, second antennal segment black at base. Rostrum 1.26–1.33 mm long. Hemelytra pale green, often with several dark green patches, cuneus with apex green.

Remarks. This species is readily distinguished by the black lines on the first antennal segment (Fig. 149), and by the distinctive genitalia (Fig. 196).

Habitat. Collected on *Quercus macrocarpa.*

Distribution. Eastern and North Central States; now known to occur in Manitoba (Map 84).

Genus *Fieberocapsus* Carvalho & Southwood

Oval, compact, black and green species. Head vertical, black, margin along eye and area along base pale, basal margin of head truncate. Eyes large, carina between them present. Pronotum trapeziform, calli slightly raised, pubescent. Scutellum black. Hemelytra green, dimorphic in both sexes; pubescence simple and appressed, moderately long and dense. Legs pale.

This genus is monobasic.

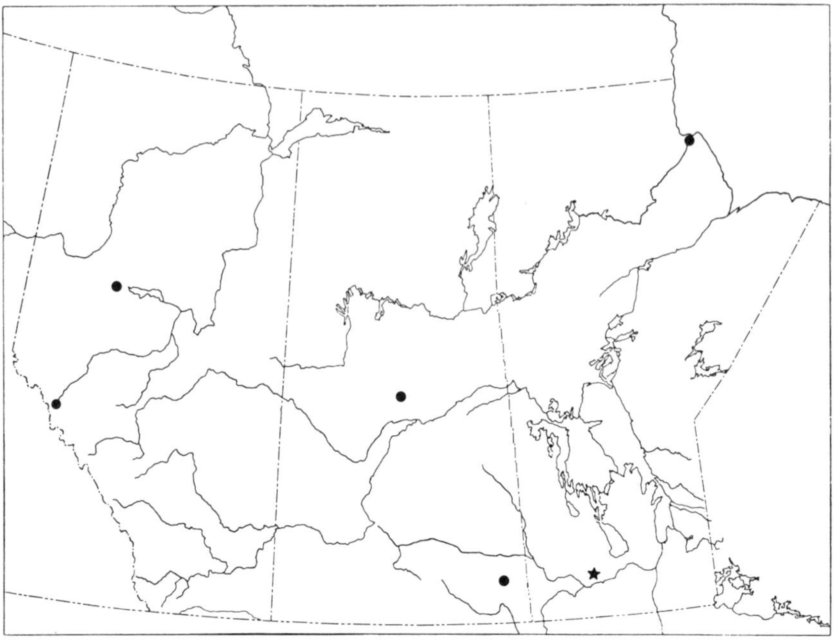

Map 84. Collection localities for *Reuteria querci* (★) and *Fieberocapsus flaveolus* (●).

Fieberocapsus flaveolus (Reuter)

Fig. 200; Map 84

Tytthus flaveolus Reuter, 1870:323.
Fieberocapsus flaveolus: Carvalho & Southwood, 1955:34.

Length 2.10–3.22 mm; width 1.12–1.40 mm. Head smooth and shiny, eyes hairy. First antennal segment black, other segments brown. Pronotum mostly pale, calli and lateral margins often fuscous. Hemelytra pale, wing membrane and cuneus absent in brachypterous forms.

Remarks. The brachypterous females of this species resemble those of *Mecomma gilvipes* in shape, but differ in color. The genitalia (Fig. 200) are distinctive. The species was first reported from North America by Kelton (1980*c*).

Habitat. Collected on *Carex* spp.

Distribution. Holarctic; Prairie Provinces (Map 84).

Genus *Mecomma* Fieber

Elongate, dimorphic species. Head vertical, black, spot near eye pale; eyes large. Pronotum campaniform, black. Scutellum black. Hemelytra pale with fuscous, in brachypterous forms wing membrane absent; pubescence simple, moderately long, dense. Ventral surface black; legs pale.

Five species are known from North America, three occur in the Prairie Provinces.

Key to species of *Mecomma*

1. Calli of pronotum raised as two distinct humps; second antennal segment in males linear (Fig. 150), in females clavate at apex (Fig. 151) .... ***angustatum*** **(Uhler)** (p. 261)
   Calli of pronotum not raised; second antennal segment linear in both sexes (Figs. 152, 153) .... ***gilvipes*** **(Stål)** (p. 261)

Mecomma angustatum (Uhler)

Figs. 150, 151; Map 85

Globiceps angustatus Uhler, 1895:42.
Mecomma angustata: Kelton & Knight, 1962:1300.

Length 3.22–4.06 mm; width 0.98–1.54 mm. Head transversely oval. Rostrum 1.30–1.40 mm long. Second antennal segment linear in male, clavate at apex in female. Pronotum with calli raised as two distinct humps. Hemelytra black, basal half of corium and cuneus pale; in brachypterous female hemelytra reduced to short pads, pale.

Remarks. Kelton and Knight (1962) revised the genus for North America. The shape of the head and the tumid calli readily distinguish the species (Figs. 150, 151).

Habitat. Collected close to the ground on grasses in damp situations.

Distribution. Alaska, northwestern States, New Hampshire, northern Canada, Quebec, Ontario; Manitoba, Alberta, now known to occur in Saskatchewan (Map 85).

Mecomma gilvipes (Stål)

Figs. 152, 153; Map 85

Leptomecoris gilvipes Stål, 1858:187.
Mecomma gilvipes: Reuter, 1883:386.
Chlamydatus luctuosus Provancher, 1887:137.

Length 3.85–4.62 mm; width 1.26–1.40 mm. Head triangular. Rostrum 1.33–1.47 mm long. Second antennal segment linear in both sexes, first

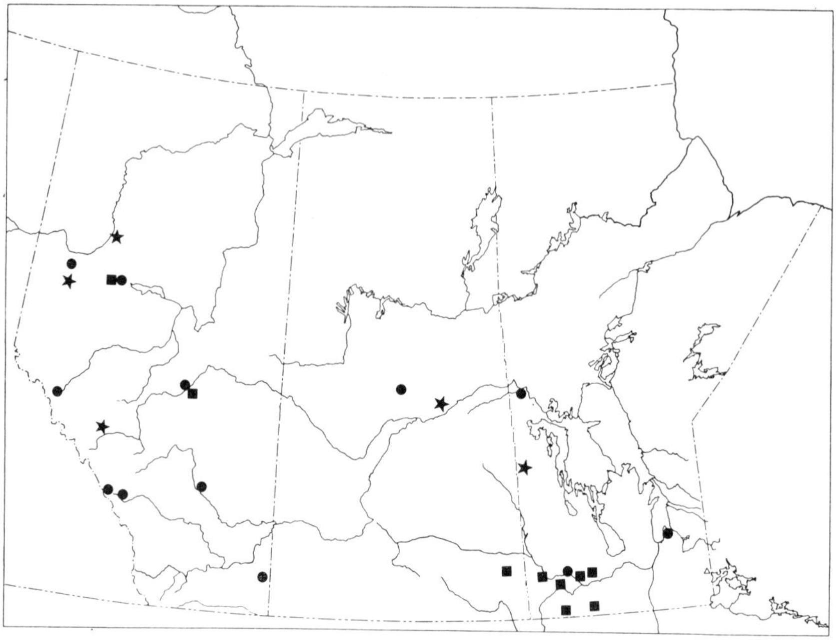

Map 85. Collection localities for *Mecomma angustatum* (●), *M. gilvipes* (■), and *Cyrtorhinus caricis* (★).

segment pale. Calli of pronotum distinct, but not tumid. Hemelytra mostly pale, males macropterous; females brachypterous, cuneus and wing membrane absent.

Remarks. This species differs from *angustata* in the shape of the head and in the normal calli (Figs. 152, 153).

Habitat. Collected on undergrowth in damp situations.

Distribution. Alaska, western and northeastern States, South Carolina, northern and Eastern Canada, British Columbia; Alberta, Manitoba, now known to occur in Saskatchewan (Map 85).

Genus *Cyrtorhinus* Fieber

Elongate, black and green species. Head vertical, black, spot near eye pale; eyes large. Pronotum campaniform, black. Scutellum black. Hemelytra pale with fuscous; pubescence simple, moderately long, dense. Ventral surface black, abdomen in female often pale; legs pale.

One Holarctic species occurs in North America.

Cyrtorhinus caricis (Fallén)

Fig. 154; Map 85

Capsus caricis Fallén, 1807:102
Cyrtorhinus caricis: Sahlberg, 1875:176.

Length 3.92–4.62 mm; width 1.19–1.40 mm. Head, pronotum, and scutellum similar to *Mecomma gilvipes*. Rostrum 1.12–1.26 mm long. Antennae black, linear. Hemelytra mostly pale, clavus, corium, and cuneus fuscous; both sexes macropterous.

Remarks. This species is similar to *M. gilvipes* in appearance (Fig. 154), but is easily separated from it by the black first antennal segment and by the shorter rostrum.

Habitat. Collected on *Carex* spp.

Distribution. Alaska, Colorado; Prairie Provinces (Map 85).

Genus *Diaphnocoris* Kelton

Pale green, slender, delicate species. Head short, oblique, clypeus prominent, eyes large, carina between them indistinct. Pronotum trapeziform, basal margin concave or sinuate; lateral margins rounded, smooth, shiny; calli slightly raised. Scutellum smooth, shiny. Hemelytra pale green, smooth, shiny; pubescence simple, pale, semierect, moderately long. Ventral surface pale green; legs green.

Three species are known from North America, two occur in the Prairie Provinces.

Key to species of *Diaphnocoris*

1. Rostrum 1.26–1.40 mm long .... ***provancheri* (Burque)** (p. 263)
   Rostrum 1.05–1.19 mm long .... ***ulmi* (Knight)** (p. 264)

Diaphnocoris provancheri (Burque)

Figs. 155, 201; Map 86

Melacocoris provancheri Burque, (*in* Provancher, 1887: 114).
Diaphnidea pellucida Uhler, 1895:44.
Orthotylus translucens Tucker, 1907:58.
Diaphnocoris pellucida: Kelton, 1961*b*: 566.
Diaphnocoris provancheri: Kelton, 1961*b*:566.

Length 4.20–4.76 mm; width 1.40–1.54 mm. Head pale green, eyes situated forward from posterior margin of head. Rostrum 1.26–1.40 mm

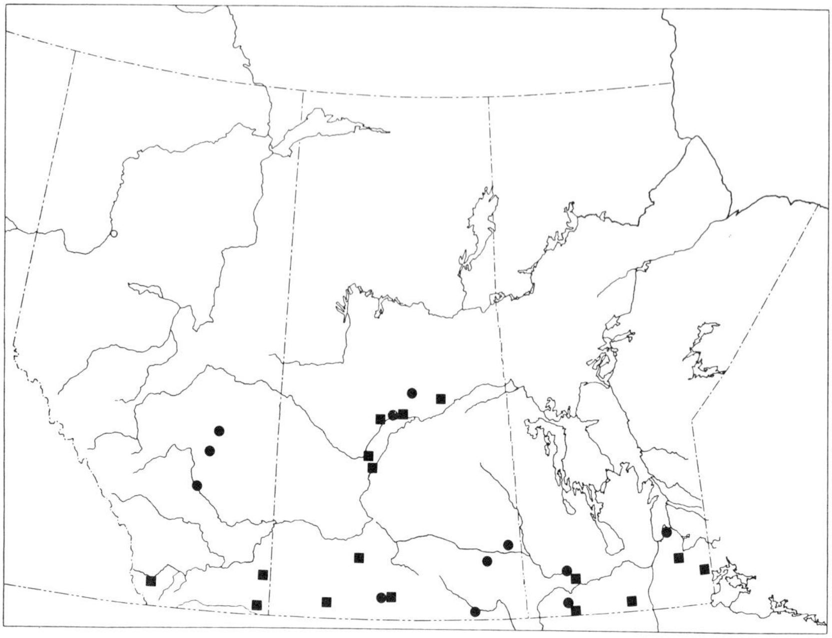

Map 86. Collection localities for *Diaphnocoris provancheri* (■) and *D. ulmi* (●).

long. Second antennal segment 1.57–1.75 mm long, green, fuscous, or black. Pronotum, scutellum, and hemelytra pale green. Hind tibia often fuscous.

Remarks. Kelton (1980*c*) synonymized *pellucida* with *provancheri.* The second antennal segment is pale green in teneral specimens and fuscous to black in mature specimens. The eyes in this species are located on the middle of the head (Fig. 155). The genital claspers (Fig. 201) are distinctive.

Habitat. Collected on *Alnus* spp., *Betula* spp., *Shepherdia canadensis*, *Crataegus* spp., *Corylus* spp., *Acer negundo*, and *Quercus macrocarpa.*

Distribution. Widespread in USA, Eastern Canada, British Columbia; Prairie Provinces (Map 86).

Diaphnocoris ulmi (Knight)

Fig. 202; Map 86

Orthotylus ulmi Knight, 1927*c*:179.
Diaphnocoris ulmi: Kelton, 1965*a*:1027.

Length 3.50–4.20 mm; width 1.19–1.33 mm. Head pale green, eyes closer to basal margin of head than in *provancheri.* Rostrum 1.05–1.19 mm long. Second antennal segment 1.19–1.40 mm long, green to fuscous. Pronotum, scutellum, and hemelytra green.

Remarks. In this species the eyes are closer to the basal margin of the head than in *provancheri*, and the rostrum and second antennal segment are shorter. The claspers are similar to those of *provancheri*, but the vesica is different (Fig. 202).

Habitat. Collected on *Ulmus americana* and *Salix interior*.

Distribution. Widespread in USA, Quebec, Ontario; Prairie Provinces (Map 86).

Genus *Orthotylus* Fieber

Elongate-oval, green or black, pubescent species. Head oblique, frons smooth, carina between eyes usually present. Pronotum trapeziform, smooth, often finely transversely rugose, lateral margins angulate; calli distinct, smooth. Hemelytra impunctate; pubescence simple.

Approximately 43 species are known from North America, 13 occur in the Prairie Provinces.

Key to species of *Orthotylus*

1. Species green, dark markings not extensive 2
   Species greenish brown to black, dark markings extensive 4
2. Lateral margins of mesoscutum black; genital claspers (Fig. 212) ..... ***fuscicornis* Knight** (p. 266)
   Lateral margins of mesoscutum green 3
3. Larger species 6.30–7.00 mm; genital claspers (Fig. 213) ..... ***notabilis* Knight** (p. 268)
   Smaller species 5.60–5.88 mm; male claspers (Fig. 214) ..... ***basicornis* Knight** (p. 268)
4. Median line on scutellum brown or black; male claspers (Fig. 215) ..... ***candidatus* Van Duzee** (p. 269)
   Scutellum black or median line pale 5
5. Wide costal margin pale, not interrupted by brown or black 6
   Costal margin beyond embolium mostly brown or black 9
6. Dark area on corium not extending outside radial vein; genital claspers (Fig. 216) ..... ***neglectus* Knight** (p. 270)
   Dark area on corium extending outside radial vein 7
7. Rostrum 1.54 mm or shorter; genital claspers (Fig. 217) ..... ***dorsalis* (Provancher)** (p. 270)
   Rostrum longer than 1.54 mm 8
8. Lorum and jugum black; genital claspers (Fig. 218) ..... ***pacificus* Van Duzee** (p. 271)
   Lorum and jugum mostly green; genital claspers (Fig. 219) .. ***ute* Knight** (p. 272)

9. Scutellum partly pale; genital claspers (Fig. 220) .. ***ornatus* Van Duzee** (p. 273)
Scutellum black 10
10. Claval suture pale along its full length; cuneus completely pale green; pronotum black; genital claspers (Fig. 221) .... ***angulatus* (Uhler)** (p. 273)
Claval suture pale only at base of corium; cuneus partly black; pronotum often pale at middle 11
11. Second antennal segment about one third longer than width of pronotum at base; genital clasper (Fig. 222) .... ***alni* Knight** (p. 273)
Second antennal segment only slightly longer than width of pronotum at base; genital clasper (Fig. 223) .... ***katmai* (Knight)** (p. 274)

Orthotylus fuscicornis Knight

Fig. 212; Map 87

Orthotylus fuscicornis Knight, 1927*c*:177.

Length 5.46–5.88 mm; width 1.75–2.10 mm. Head, pronotum, and hemelytra green, spot at base of clypeus, posterior margins of calli, and lateral margins of mesoscutum black. First antennal segment black. Carina between eyes distinct. Pubescence silvery, short, appressed. Ventral surface and legs green.

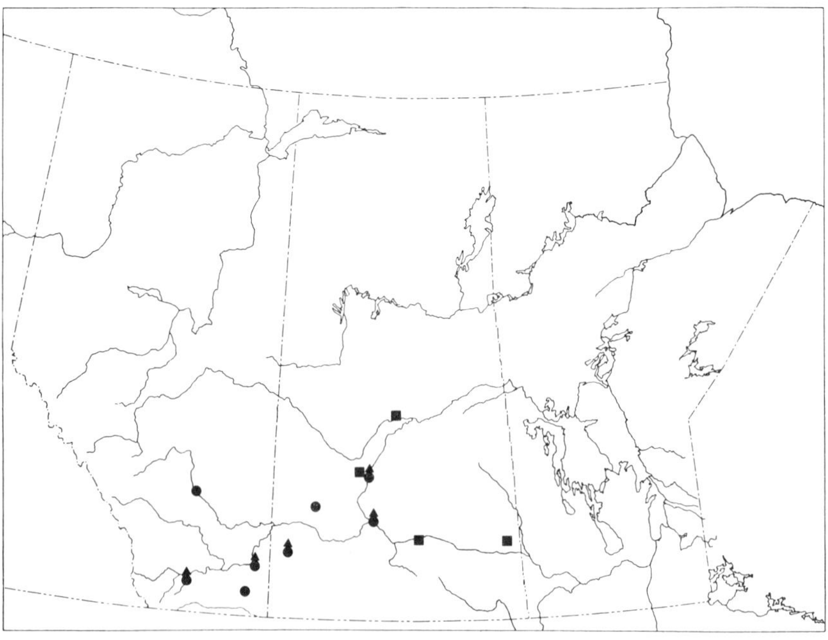

Map 87. Collection localities for *Orthotylus fuscicornis* (●), *O. notabilis* (▲), and *O. basicornis* (■).

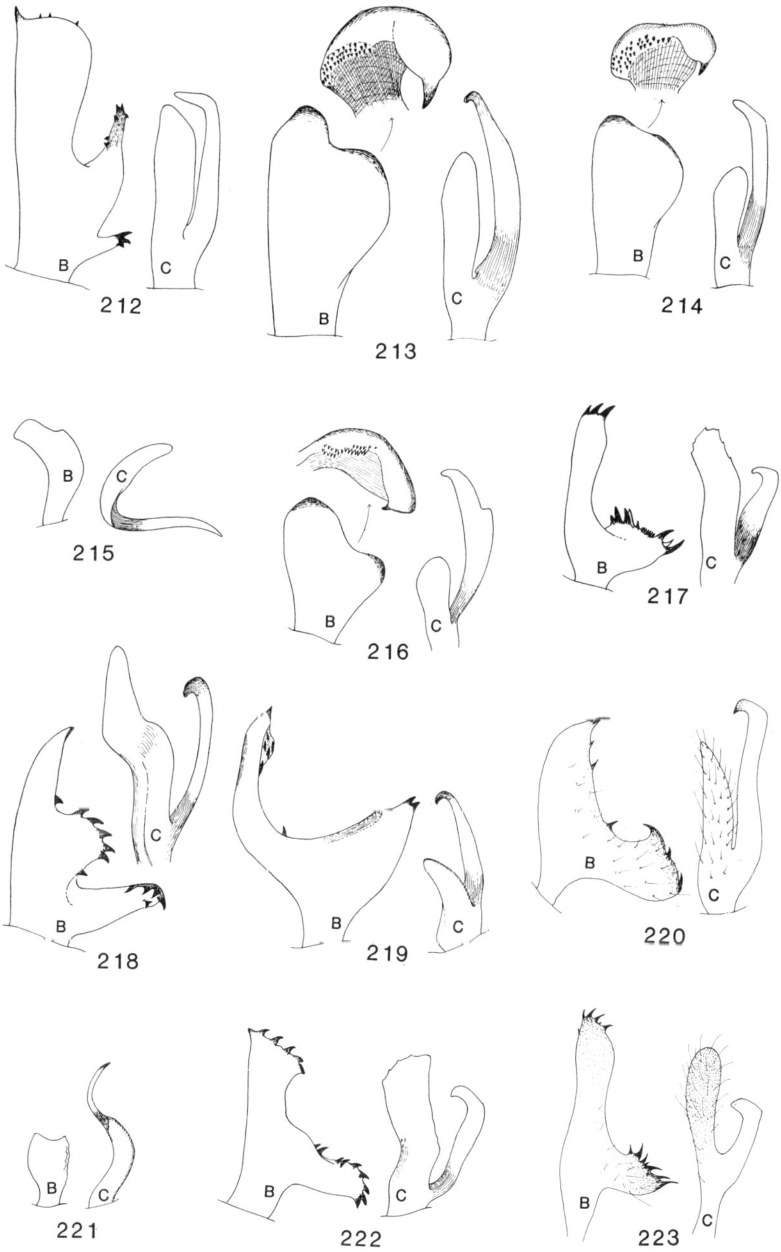

Figs. 212–223. Male claspers of *Orthotylus* spp. 212, *fuscicornis*; 213, *notabilis*; 214, *basicornis*; 215, *candidatus*; 216, *neglectus*; 217, *dorsalis*; 218, *pacificus*; 219, *ute*; 220, *ornatus*; 221, *angulatus*; 222, *alni*; 223, *katmai*.

Remarks. The black markings in the female are often absent except for the lateral margins on the mesoscutum. The male claspers (Fig. 212) are distinctive.

Habitat. Collected on *Salix interior*.

Distribution. Colorado, Utah; now known to occur in the Prairie Provinces (Map 87).

Orthotylus notabilis Knight

Fig. 213; Map 87

Orthotylus notabilis Knight, 1927*d*:176.

Length 6.30–7.00 mm; width 2.10–2.38 mm. Head green, clypeus and frons black. First antennal segment black in male, green in female; carina between eyes distinct. Pronotum green, dorsal margins of calli and two large spots behind black. Scutellum green, basal angles black. Hemelytra green, middle of clavus and inner apical margin of corium often black in male, green in female. Pubescence pale, long, dense, slanting. Ventral surface green, often black in male, legs green.

Remarks. This is the largest of the green species in the Prairie Provinces. The black markings are mostly in the male, in the female the first antennal segment may be black. The genital claspers (Fig. 213) are distinctive.

Habitat. Collected on *Salix interior*.

Distribution. North Central States; Saskatchewan, now known to occur in Alberta and Manitoba (Map 87).

Orthotylus basicornis Knight

Fig. 214; Map 87

Orthotylus basicornis Knight, 1923*b*:515.

Length 5.60–5.88 mm; width 1.68–1.82 mm. Head, pronotum, and hemelytra pale green. First antennal segment black, carina between eyes distinct. Pubescence silvery, long, semierect. Ventral surface and legs pale green.

Remarks. This species resembles *fuscicornis* in size and in the black first antennal segments, but the black marks at the base of clypeus, on the calli, and on the mesoscutum are absent. The genital claspers (Fig. 214) are distinctive.

Habitat. Collected on *Salix interior.*

Distribution. Northeastern and North Central States, Quebec, Ontario; now known to occur in Saskatchewan (Map 87).

Orthotylus candidatus Van Duzee

Fig. 215; Map 88

Orthotylus candidatus Van Duzee, 1916:124.

Length 4.55–5.32 mm; width 1.40–1.68 mm. Head black, two spots on vertex pale. Antennae black in male, only first segment black in female. Pronotum black, large area behind calli pale. Scutellum pale, median line brown or black. Hemelytra mostly pale, inner clavus and outer half of corium fuscous to black; apex of cuneus black. Ventral surface black, legs pale.

Remarks. The brown or black median line on the scutellum, and the genital claspers (Fig. 215) readily distinguish this species.

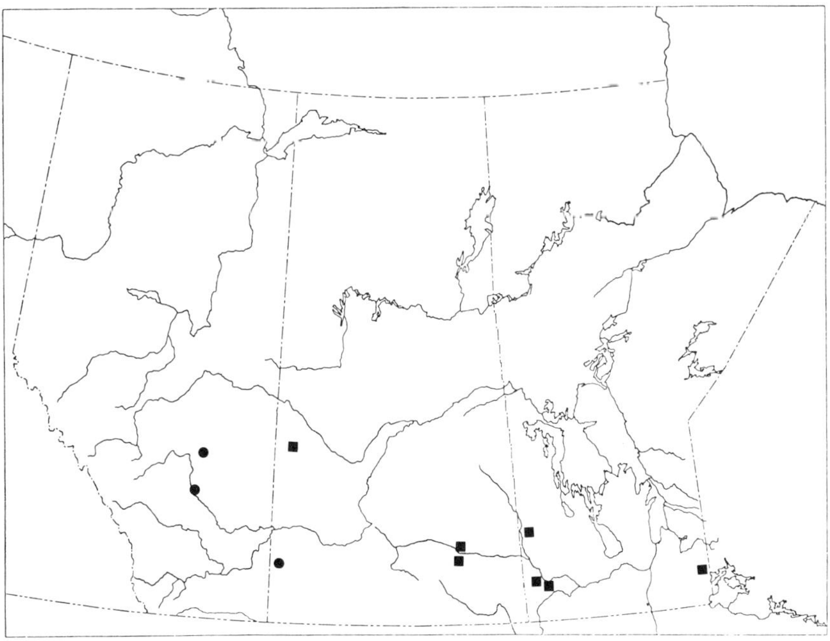

Map 88. Collection localities for *Orthotylus candidatus* (●) and *O. neglectus* (■).

Habitat. Collected on *Populus tremuloides.*

Distribution. Northeastern States, Ontario; Alberta, now known to occur in Saskatchewan (Map 88).

Orthotylus neglectus Knight

Fig. 216; Map 88

Orthotylus neglectus Knight, 1923*b*:515.

Length 5.74–6.30 mm; width 1.68–2.10 mm. Head black in male, green marked with black on frons and vertex in female. First antennal segment black in male, yellowish green in female. Pronotum green, calli and broad rays behind black. Scutellum black in male, longitudinal median line pale in female. Hemelytra black, wide costal margin and cuneus green; dark area on corium not extending outside radial vein; pubescence pale, long, dense. Ventral surface black in male, green in female; legs green.

Remarks. This species is distinguished by the wide green costal margin on the hemelytra. The genital claspers (Fig. 216) are distinctive.

Habitat. Collected on *Salix* spp.

Distribution. Eastern and Central States, Nova Scotia; now known to occur in the Prairie Provinces (Map 88).

Orthotylus dorsalis (Provancher)

Figs. 156, 217; Map 89

Lygus dorsalis Provancher, 1872:104.
Orthotylus dorsalis Provancher, 1887:138.
Diommatus congress Uhler, 1887:33.

Length 5.32–5.60 mm; width 1.54–1.82 mm. Head black, narrow area adjacent to eyes pale, more extensively pale in female. Antennae black in male, pale in female. Pronotum black, median line and lateral margins pale in male; pale in female except for two black rays behind calli. Scutellum black in male, median line pale in female. Hemelytra black, costal margin pale, dark area on corium extending beyond radial vein. Pubescence silvery, long, dense. Ventral surface black in male, only lateral margins black in female, legs pale.

Remarks. This species is similar to *neglectus* in color, but the black markings on the corium extend beyond the radial vein. The genital claspers (Fig. 217) are distinctive.

Habitat. Collected on *Salix bebbiana.*

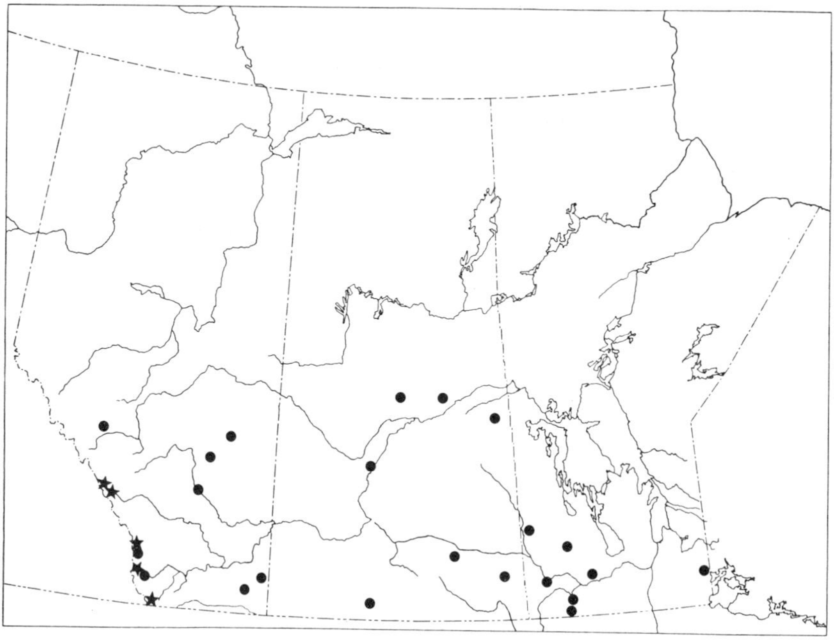

Map 89. Collection localities for *Orthotylus dorsalis* (●) and *O. pacificus* (★).

Distribution. Northeastern and Central States, Quebec, Ontario, British Columbia; Alberta, now known to occur in Saskatchewan and Manitoba (Map 89).

Orthotylus pacificus Van Duzee

Fig. 218; Map 89

Orthotylus pacificus Van Duzee, 1919:33.

Length 4.90–6.30 mm; width 1.68–1.89 mm. Head black, margin next to eye pale. Antennae black. Pronotum black, median area and lateral margins pale. Scutellum black, median line often pale in female. Hemelytra black, costal margin pale; pubescence pale, appressed. Ventral surface black in male, partly green in female; legs pale green.

Remarks. This species is darker than *dorsalis*. The genital claspers (Fig. 218) are distinctive.

Habitat. Collected on *Salix* spp.

Distribution. British Columbia; now known to occur in Alberta (Map 89).

Orthotylus ute Knight

Fig. 219; Map 90

Orthotylus ute Knight, 1927*d*:179.

Length 5.88–6.86 mm; width 1.89–2.03 mm. Head green, clypeus, sides of frons, and vertex black. Antennae black in male, brown in female. Pronotum pale green, calli and basal margin black. Scutellum black in male, median line pale; all pale in female. Hemelytra black, basal half of corium, cuneus, and costal margin pale green; pubescence pale, long, dense. Ventral surface and legs green.

Remarks. This species is not as dark as *pacificus,* and the genital claspers (Fig. 219) are distinctive.

Habitat. Collected on *Betula occidentalis.*

Distribution. Northwestern States; now known to occur in Alberta (Map 90).

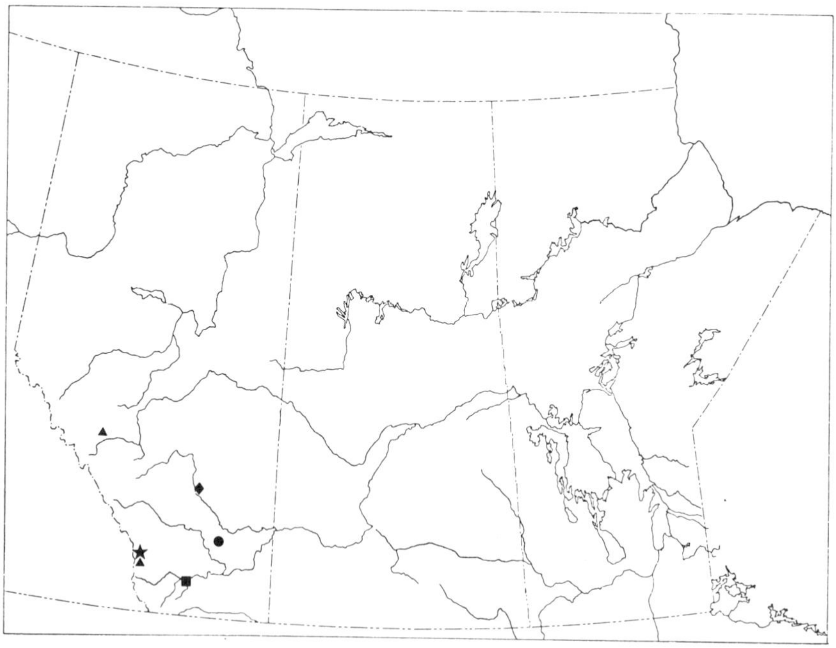

Map 90. Collection localities for *Orthotylus ute* (◆), *O. ornatus* (●), *O. angulatus* (■), *O. alni* (▲), and *O. katmai* (★).

Orthotylus ornatus Van Duzee

Fig. 220; Map 90

Orthotylus ornatus Van Duzee, 1916:122.

Length 5.60–5.88 mm; width 1.96–2.10 mm. Head pale yellow, clypeus and frons black. Antennae black in male, pale green in female. Pronotum black, rugose. Scutellum pale on apical half, black at base. Hemelytra black, basal half of corium and base of cuneus pale; pubescence pale, appressed. Ventral surface pale green marked with black; legs pale, femora black on apical half.

Remarks. This species is distinguished by the rugose pronotum, by the partly pale scutellum, and by the genital claspers (Fig. 220).

Habitat. Collected on *Salix amygdaloides*.

Distribution. Northwestern and Central States, Texas, Ontario; now known to occur in Alberta (Map 90).

Orthotylus angulatus (Uhler)

Fig. 221; Map 90

Diommatus angulatus Uhler, 1895:44.
Orthotylus angulatus: Van Duzee, 1916*b*:115.

Length 4.69–4.90 mm; width 1.26–1.40 mm. Head black. Antennae black. Pronotum black, finely rugose, not as prominent as in *ornatus*. Scutellum black. Hemelytra black, commissure, base of corium, and most of cuneus pale green; pubescence pale, short, appressed. Ventral surface black, femora pale, apices brown, hind and middle tibiae brown to black.

Remarks. This species is distinguished by the pale claval suture and by the genital claspers (Fig. 221).

Habitat. Collected on *Salix amygdaloides*.

Distribution. Colorado; now known to occur in Alberta (Map 90).

Orthotylus alni Knight

Fig. 222; Map 90

Orthotylus alni Knight, 1923*b*:521.

Length 4.90–6.30 mm; width 1.68–1.82 mm. Head black, line adjacent to eye pale. Antennae black. Pronotum black, rugose, longitudinal median line pale. Scutellum black. Hemelytra black, base of corium and base of cuneus often pale; pubescence pale, long, dense. Ventral surface black, legs pale.

Remarks. This species is distinguished by the long second antennal segment and by the genital claspers (Fig. 222).

Habitat. Collected on *Alnus rugosa.*

Distribution. Northeastern States, Ontario; now known to occur in Manitoba (Map 90).

Orthotylus katmai (Knight)

Fig. 223; Map 90

Tichorhinus katmai Knight, 1921*a*:108.
Orthotylus katmai: Downes, 1924:14.

Length 4.90–5.32 mm; width 1.68–1.82 mm. Head black, line adjacent to eye pale. Antennae black. Pronotum black, finely rugose, longitudinal median line pale. Scutellum black. Hemelytra black, base of corium and base of cuneus often pale; pubescence pale, long, dense. Ventral surface and legs black, front femur greenish.

Remarks. This species is similar to *alni* in appearance, but the second antennal segment is shorter and the genital claspers (Fig. 223) are different.

Habitat. Collected on *Salix* spp.

Distribution. Alaska, British Columbia; now known to occur in Alberta and Manitoba (Map 90).

Tribe Pilophorini

In the Prairie Provinces the tribe is represented by one genus and seven species.

Genus *Pilophorus* Hahn

Elongate, brown or black, antlike species. Head oblique, strongly triangular, base of head convex, carina sharp, overlapping apex of pronotum. Pronotum subcampanulate, lateral margins rounded. Scutellum tumid with clumps of sericeous pubescence. Hemelytra pruinose and polished, banded with transverse, silvery bars of sericeous pubescence; costal margin concave; cuneus at inner basal angle with sericeous pubescence. Ventral surface with patches of sericeous pubescence on pleura and abdomen. Osteolar peritreme well developed.

Approximately 55 species are known from North America, seven occur in the Prairie Provinces. Knight (1973) published a key to the Nearctic species.

Key to species of *Pilophorus*

1. Hemelytra polished over entire width behind posterior silvery line (Fig. 138) .. 2
   Hemelytra polished on outer half of corium only 4
2. Third antennal segment reddish brown, or white only on basal half .... *uhleri* **Knight** (p. 275)
   Third antennal segment white 3
3. Base of second antennal segment pale .... *piceicola* **Knight** (p. 276)
   Base of second antennal segment red or brown .... *pinicola* **Knight** (p. 277)
4. Small species, not over 3.50 mm .... *geminus* **Knight** (p. 277)
   Larger species over 3.90 mm 5
5. Light yellowish brown species .... *salicis* **Knight** (p. 277)
   Black or dark brown species 6
6. Species 4.20 mm or less; second antennal segment 1.47 mm or less; rostrum 1.61 mm or less .... *setiger* **Knight** (p. 278)
   Species 5.25 mm or more; second antennal segment 1.68 mm or more; rostrum 1.75 mm or more .... *clavatus* (**Linnaeus**) (p. 279)

Pilophorus uhleri Knight

Map 91

Pilophorus uhleri Knight, 1923*b*:538.

Length 5.25–5.60 mm; width 1.75–1.96 mm. Head reddish brown to black. Second antennal segment 1.82–1.96 mm long, clavate; third segment reddish brown, often white on basal half. Rostrum 1.82–1.96 mm long. Pronotum and scutellum black. Hemelytra black and shiny beyond transverse, silvery band, brown and pruinose in front; wing membrane fuscous with large, velvety spot at base. Pubescence on head, pronotum, and hemelytra yellowish and appressed, intermixed with longer erect, black hairs. Ventral surface black, legs reddish brown.

Remarks. This species is distinguished by the large size and by the mostly black third antennal segment. The rostrum and the second antennal segment are almost equal in length.

Habitat. Collected on *Pinus banksiana* and *P. contorta*.

Distribution. Eastern and North Central States, Ontario; now known to occur in the Prairie Provinces (Map 91).

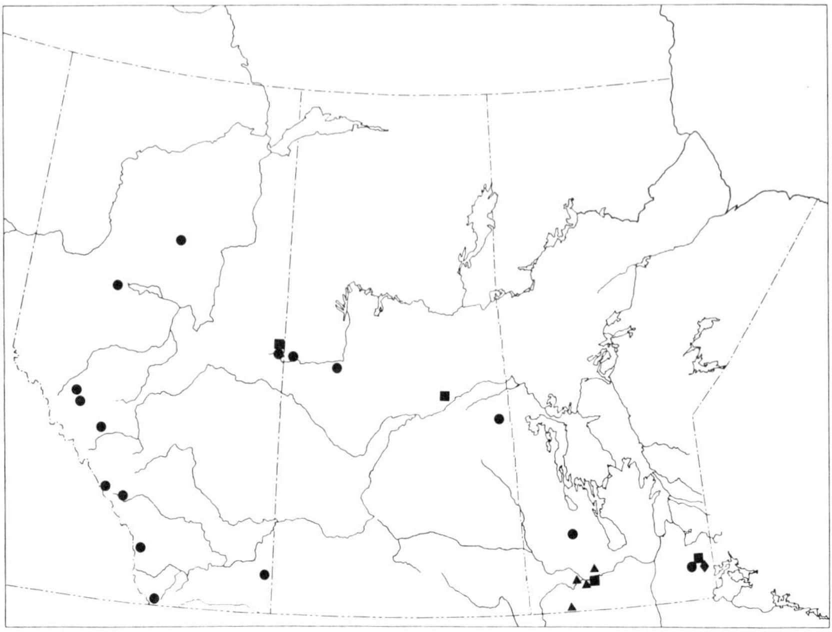

Map 91. Collection localities for *Pilophorus uhleri* (●), *P. piceicola* (▲), *P. pinicola* (■), and *P. geminus* (♦).

Pilophorus piceicola Knight

Fig. 138; Map 91

Pilophorus piceicola Knight, 1926*a*:19.

Length 4.41–4.62 mm; width 1.40–1.61 mm. Head dark brown, area below antennal sockets yellowish brown. Second antennal segment 1.61–1.75 mm long, strongly clavate, basal half yellow and apical half reddish brown. Rostrum 1.75–1.89 mm long. Pronotum and scutellum black, shiny. Hemelytra black and shiny behind transverse, silvery band, yellowish brown and pruinose in front; pubescence black, short, appressed. Ventral surface reddish brown; legs reddish brown, front and hind coxae mostly white.

Remarks. This species is distinguished by the strongly clavate and bicolored second antennal segment, and by the white terminal segments (Fig. 138).

Habitat. Collected on *Picea glauca*.

Distribution. New York, Iowa; now known to occur in Manitoba (Map 91).

Pilophorus pinicola Knight

Map 91

Pilophorus pinicola Knight, 1973:138.

Length 4.76–4.90 mm; width 1.61–1.68 mm. Head similar to *piceicola* in color. Second antennal segment 1.82–1.96 mm long, clavate, reddish brown, terminal segments white. Rostrum 1.75–1.82 mm long. Pronotum, scutellum, and hemelytra similar to *piceicola* in color and pubescence.

Remarks. This species is similar to *piceicola* in appearance, but is slightly larger, the second antennal segment is reddish on basal half, not as strongly clavate, and longer than the rostrum.

Habitat. Collected on *Pinus banksiana.*

Distribution. Minnesota; now known to occur in the Prairie Provinces (Map 91).

Pilophorus geminus Knight

Map 91

Pilophorus geminus Knight, 1926*a*:22.

Length 3.22–3.50 mm; width 1.08–1.26 mm. Head dark brown, clypeus, jugum, and lorum lighter. Second antennal segment 0.84–0.89 mm long, nearly cylindrical, yellowish brown, apex black; terminal segments black. Rostrum 1.19 1.26 mm long. Pronotum and scutellum brown. Hemelytra mostly light brown and pruinose; dark brown and shiny only at base and apex of clavus and on outer half of corium beyond silvery band; cuneus dark brown, shiny; pubescence fuscous, recumbent; transverse band widely dislocated at radial vein. Ventral surface brown, legs yellowish brown.

Remarks. This is the smallest species encountered with widely dislocated transverse band.

Habitat. Collected on *Pinus banksiana.*

Distribution. Minnesota, Wisconsin; now known to occur in Manitoba (Map 91).

Pilophorus salicis Knight

Map 92

Pilophorus salicis Knight, 1968*a*:173.

Length 4.06–4.34 mm; width 1.40–1.61 mm. Head brown. First antennal segment pale, second segment 1.43–1.61 mm long, reddish brown, apex

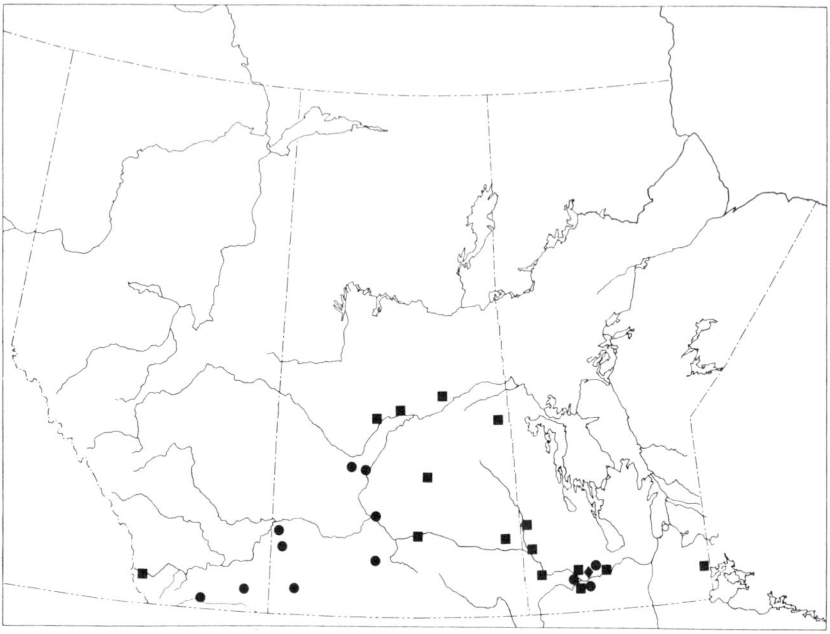

Map 92. Collection localities for *Pilophorus salicis* (●), *P. setiger* (◆), and *P. clavatus* (■).

darker, incrassate. Rostrum 1.49–1.61 mm long. Pronotum and scutellum yellowish brown. Hemelytra yellowish brown, pruinose; base and apex of clavus shiny, outer half of corium beyond transverse, silvery band brown, shiny; cuneus brown, shiny; pubescence silvery and appressed, intermixed with longer, erect hairs; transverse band only slightly dislocated at radial vein. Ventral surface and legs yellowish brown.

Remarks. This species is distinguished by the yellowish brown color and is the lightest species encountered in the Prairie Provinces.

Habitat. Collected on *Elaeagnus commutata*, probably feeding on the psyllids that are common on the shrub.

Distribution. Colorado, Nevada; now known to occur in the Prairie Provinces (Map 92).

Pilophorus setiger Knight

Map 92

Pilophorus setiger Knight, 1941*c*:124.

Length 3.90–4.20 mm; width 1.50–1.68 mm. Head brown with vertex dark brown. Second antennal segment 1.34–1.47 mm long, reddish, apex black, incrassate. Rostrum 1.51–1.61 mm long. Pronotum and scutellum brown. Hemelytra mostly brown, pruinose; base and apex of clavus shiny, outer half of corium beyond transverse, silvery band and cuneus shiny; pubescence golden and appressed, intermixed with longer erect hairs; transverse band widely dislocated at radial vein. Ventral surface and legs brown.

Remarks. This species is darker than *salicis*, and the transverse band is further dislocated at the radial vein.

Habitat. Unknown.

Distribution. North Central States; now known to occur in Manitoba (Map 92).

Pilophorus clavatus (Linnaeus)

Map 92

Cimex clavatus Linnaeus, 1767:729.
Pilophorus clavatus: Westwood, 1840:121.

Length 5.25–5.46 mm; width 1.54–1.68 mm. Head black, area below antennal sockets light brown. Second antennal segment 1.68–1.96 mm long, red, apex black, incrassate. Rostrum and scutellum black. Hemelytra brown to dark brown; shiny and pruinose patterns; pubescence and silvery bands similar to those of *setiger*. Ventral surface and legs reddish brown.

Remarks. This species is similar to *uhleri* in size. It resembles *setiger* in color and in the patterns on the hemelytra, but is separated from it by the larger size, longer rostrum, and longer second antennal segment.

Habitat. Collected on *Quercus macrocarpa* and *Salix* spp.

Distribution. Holarctic; northeastern and North Central States, Nova Scotia, Ontario, British Columbia; now known to occur in the Prairie Provinces (Map 92).

Subfamily Phylinae Douglas & Scott

In North America this is the third largest subfamily with 56 genera and approximately 401 species. Most of the species are phytophagous.

The following are the subfamily characteristics: 1) straight, hairlike parempodia between the claws; 2) pulvilli present; 3) male genitalia with rigid ductus seminis, and 4) distinctive left clasper.

In the Prairie Provinces the subfamily is represented by 2 tribes, 20 genera and 62 species.

Key to tribes of Phylinae

1. Antlike species; pronotum subtriangular; hemelytra complete in males (Fig. 224), absent or reduced in females (Fig. 225) .... **Hallodapini** (p. 280)
   Species not antlike; pronotum trapeziform or subrectangular (Fig. 227) .... **Phylini** (p. 289)

Tribe Hallodapini

In the Prairie Provinces this tribe of antlike species is represented by two genera and four species.

Key to genera of Hallodapini

1. Second antennal segment linear (Fig. 229) .... ***Coquillettia* Uhler** (p. 280)
   Second antennal incrassate or clavate (Fig. 230) ...... ***Orectoderus* Uhler** (p. 282)

Genus *Coquillettia* Uhler

Males elongate, macropterous. Head oblique, base constricted into a short neck; eyes large; second antennal segment linear. Pronotum subtriangular. Hemelytra pruinose and shiny, almost glabrous with characteristic black or brown and white color pattern. Ventral surface and legs reddish brown.

Females apterous, antlike. Pronotum subquadrate, tumid at middle. Abdomen globose.

Approximately 25 species are known from North America, two occur in the Prairie Provinces.

Key to species of *Coquillettia*

1. Eyes bulging and almost vertical on head, vertex concave between eyes (Fig. 231); hemelytra diffuse grayish brown; cuneus pale yellow, apex black .... ***albertae* Kelton** (p. 281)
   Eyes not bulging, oblique on head; vertex convex between eyes (Fig. 232); hemelytra with distinct brown or black markings; cuneus white on basal half, black on apical half .... ***insignis* Uhler** (p. 282)

Coquillettia albertae Kelton

Figs. 226, 231; Map 93

Coquillettia albertae Kelton, 1980*a*:285.

Length 5.95–6.16 mm; width 1.40–1.54 mm. Head, pronotum, and scutellum dark brown. Rostrum 1.61–1.68 mm long. Eyes large and bulging, extending above vertex of head. Hemelytra with diffuse color pattern, clavus and corium lightly fuscous, cuneus pale yellow with apex black; wing membrane lightly fuscous. Ventral surface and legs brown.

Remarks. This species is distinguished by the large and bulging eyes (Fig. 231), and by the diffuse color pattern on the hemelytra. The females are unknown.

Habitat. Unknown, collected at light.

Distribution. Alberta (Map 93).

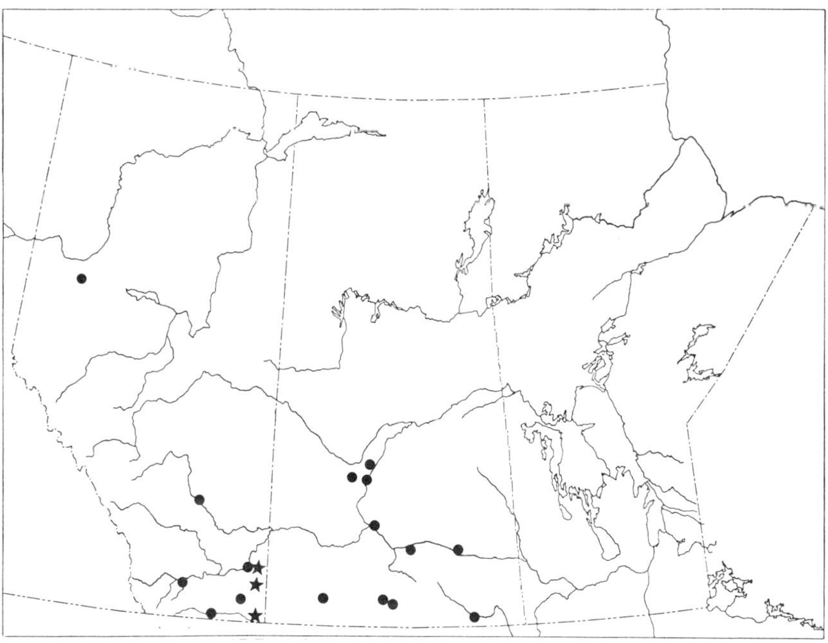

Map 93. Collection localities for *Coquillettia albertae* (★) and *C. insignis* (●).

Coquillettia insignis Uhler

Figs. 224, 225, 229, 232; Map 93

Coquillettia insignis Uhler, 1890:79.

Male. Length 6.16–6.72 mm; width 1.54–1.68 mm. Head reddish brown, eyes not bulging. Rostrum 1.68–1.82 mm. Pronotum and scutellum reddish brown, shiny. Hemelytra with distinctive color pattern; clavus, apical half of corium and apical half of cuneus black, basal half of corium and basal half of cuneus white; dark areas pruinose, white areas shiny; wing membrane fuscous, base white; pubescence pale, short, sparse. Legs long, slender.

Female. Length 4.76–5.60 mm. Antlike, apterous, abdomen globose (Fig. 225).

Remarks. The males are readily distinguished by the color pattern on the hemelytra (Fig. 224), and the females by association with the males. The species is readily separated from *albertae* by the shape of the eyes and head (Fig. 232).

Habitat. Collected on grasses.

Distribution. Western States, British Columbia; Alberta, now known to occur in Saskatchewan (Map 93).

Genus *Orectoderus* Uhler

Males elongate, macropterous. Head oblique, base constricted into short neck as in *Coquillettia*; second antennal segment clavate. Pronotum black, smooth, shiny, subcampanulate. Hemelytra black or with white pattern. Ventral surface black; legs reddish brown.

Females brachypterous, antlike. Pronotum subquadrate, tumid at middle. Hemelytra reduced, padlike, tips upturned, pointed. Abdomen globose. Legs reddish brown.

Ten species are known from North America, two occur in the Prairie Provinces.

Key to species of *Orectoderus*

1. Pubescence on hemelytra black, short, sparse; pronotum shiny .... ***obliquus* Uhler** (p. 288)
   Pubescence on hemelytra silvery, long, dense; pronotum dull .... ***montanus* Knight** (p. 289)

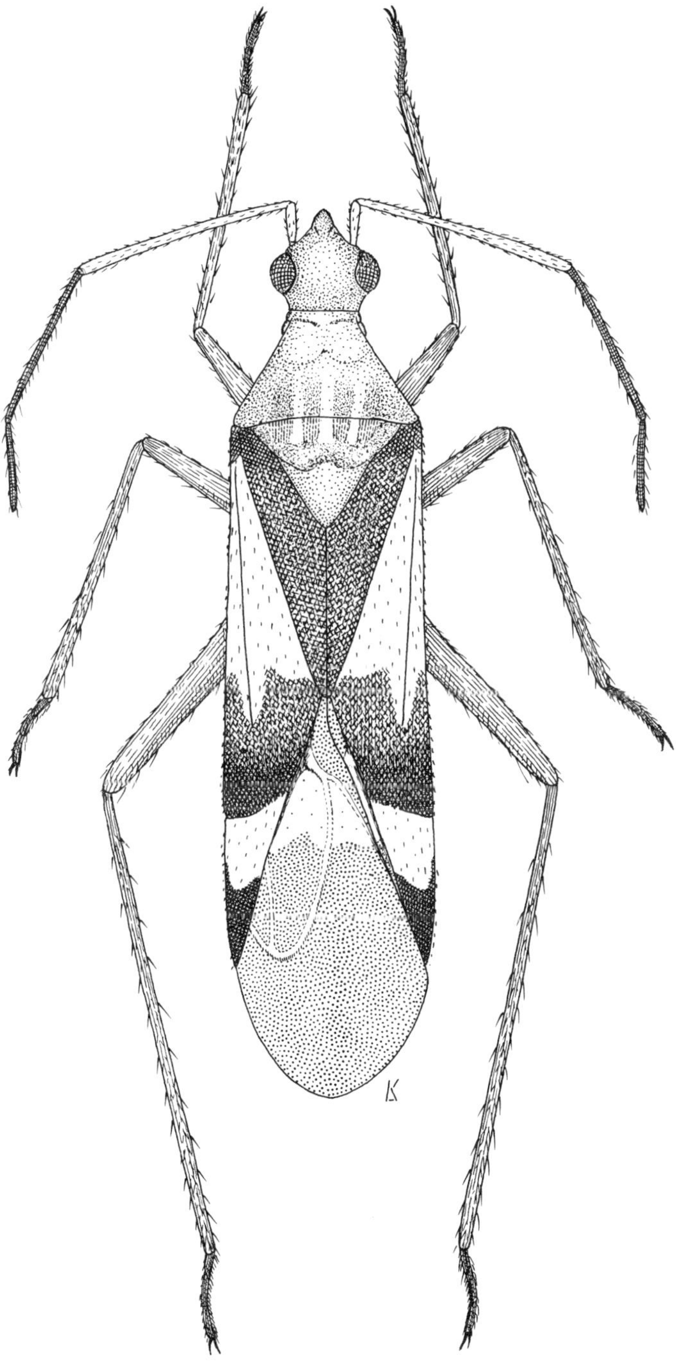

Fig. 224. *Coquillettia insignis*, male

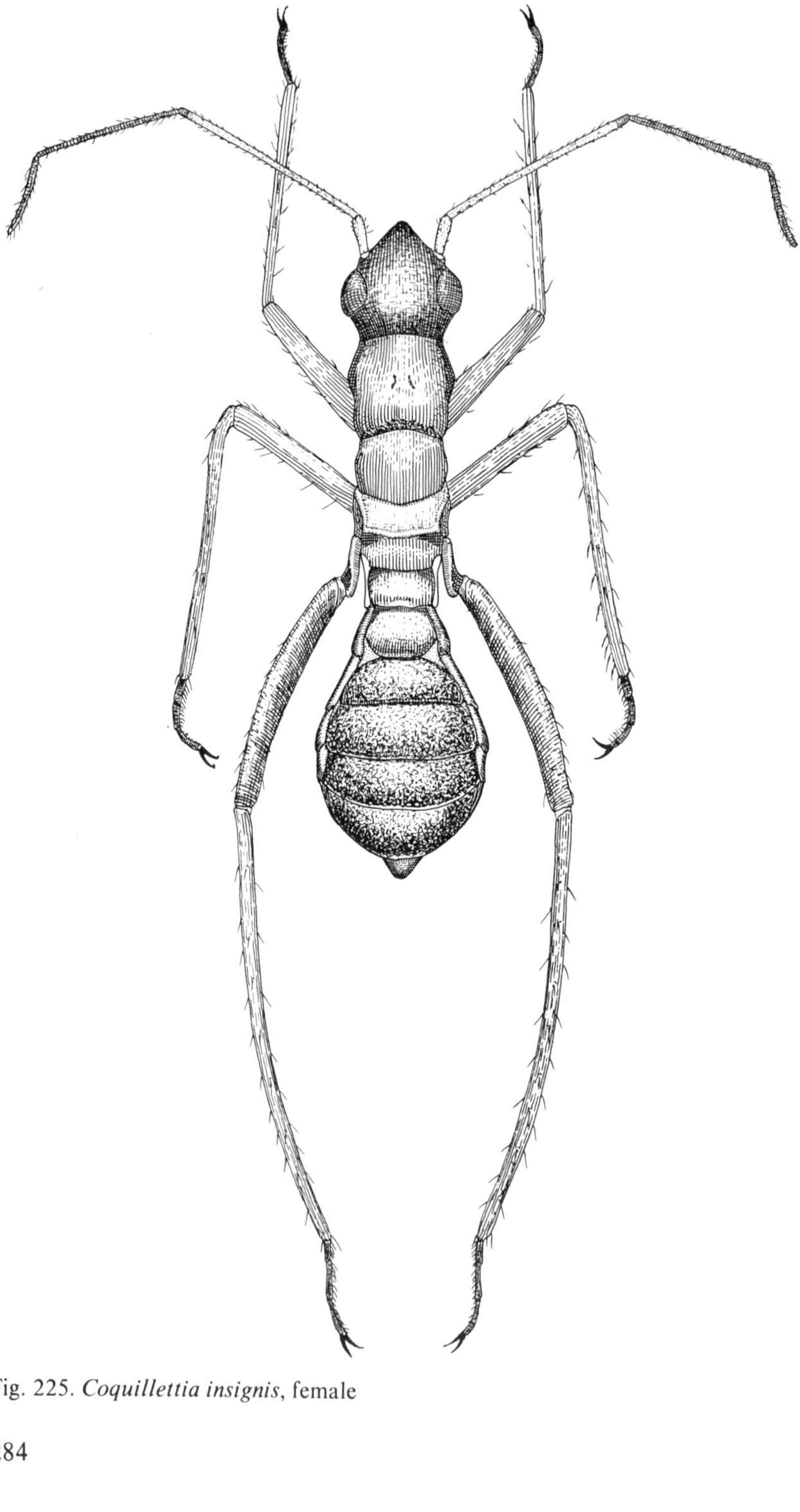

Fig. 225. *Coquillettia insignis*, female

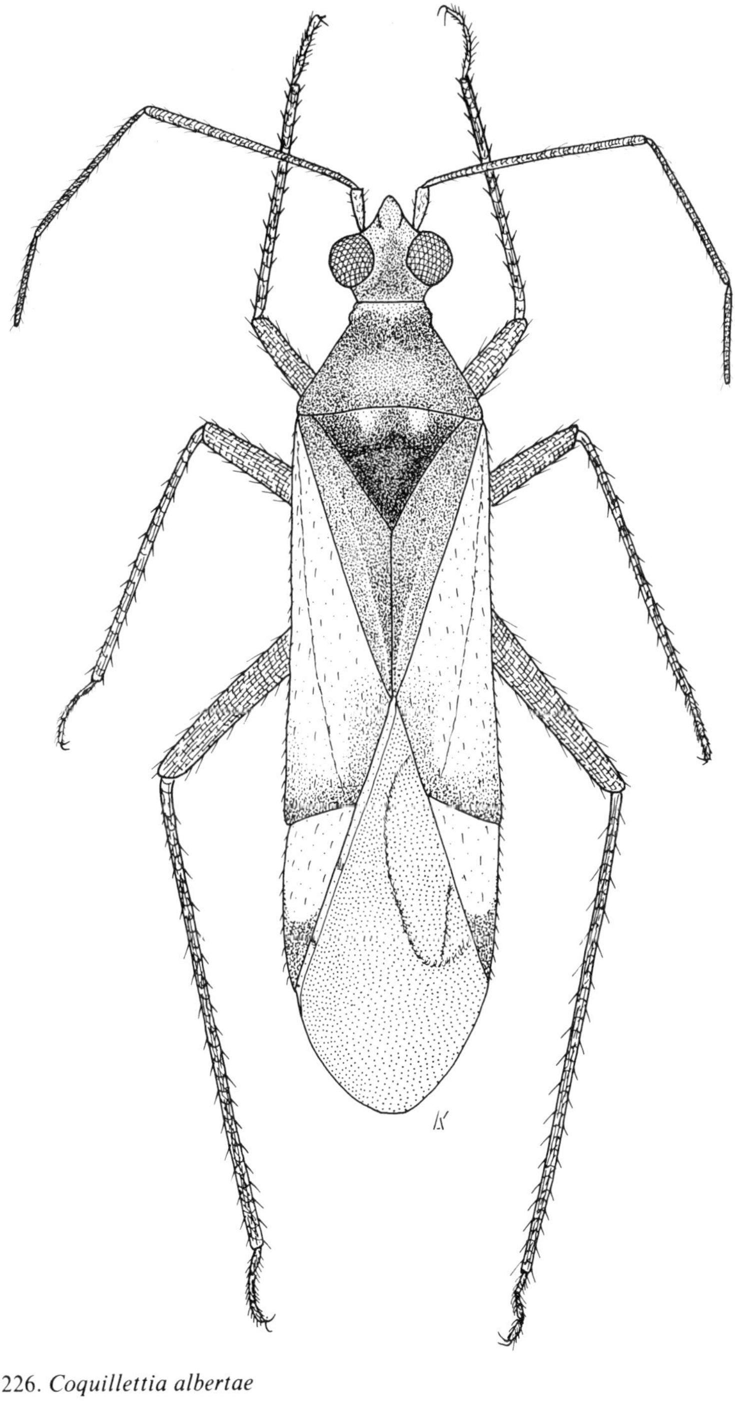

Fig. 226. *Coquillettia albertae*

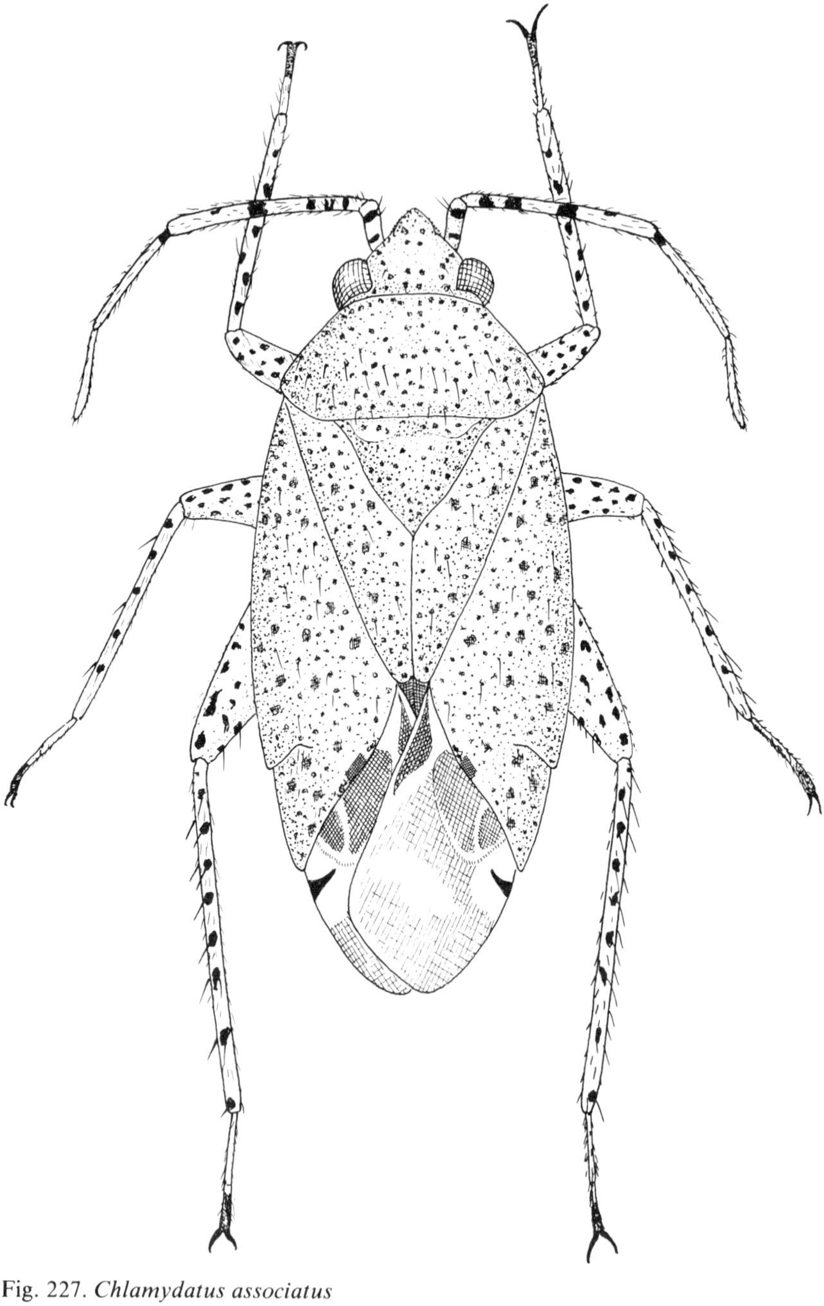

Fig. 227. *Chlamydatus associatus*

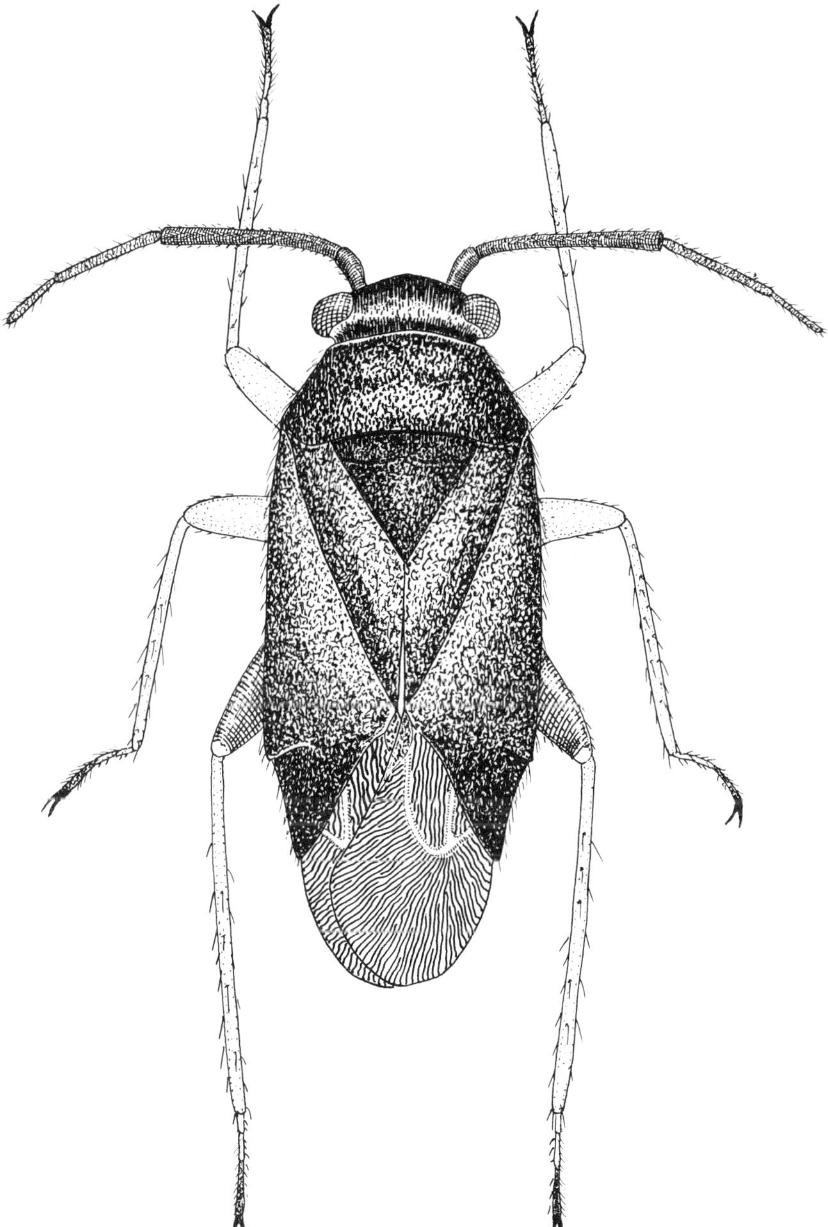

Fig. 228. *Pseudatomoscelis seriatus*

Orectoderus obliquus Uhler

Fig. 230; Map 94

Orectoderus obliquus Uhler, 1876:320.

Male. Length 7.70–7.98 mm; width 2.10–2.24 mm. Head black. Second antennal segment 2.31–2.52 mm long, basal half pale, apical half black. Pronotum and scutellum shiny black, calli flattened. Hemelytra shiny, often all black, but usually corium with wedge-shaped area along claval suture, and basal half of cuneus white; pubescence black, short, sparse.

Female. Length 5.25–5.95 mm. Antlike. Head, pronotum, and scutellum reddish brown. Wing pads upturned, abdomen globose.

Remarks. This species is distinguished by the shiny, polished appearance, and the short, sparse, black pubescence on hemelytra.

Habitat. Collected on prairie grasses, *Rosa acicularis*.

Distribution. Widespread in USA, Eastern Canada; Manitoba, Alberta, now known to occur in Saskatchewan (Map 94).

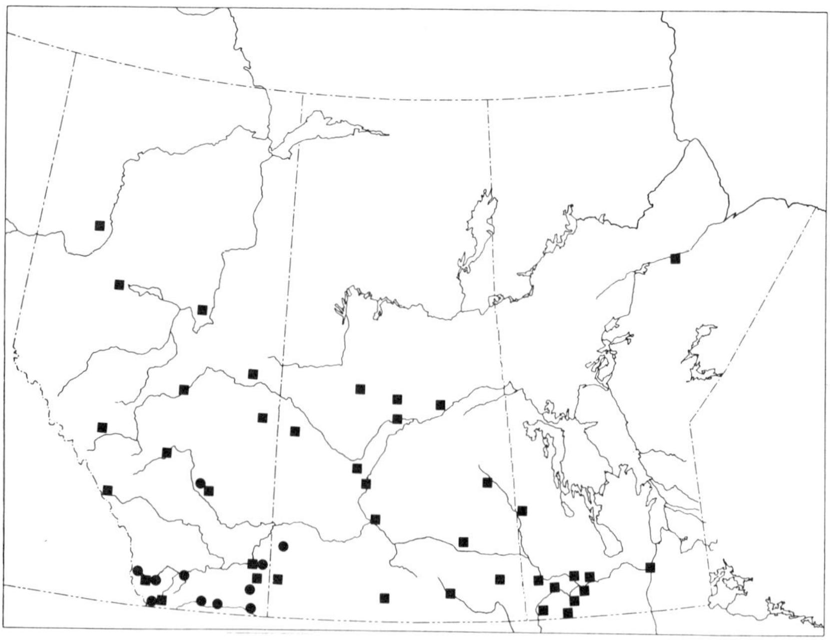

Map 94. Collection localities for *Orectoderus obliquus* (■) and *O. montanus* (●).

Orectoderus montanus Knight

Map 94

Orectoderus montanus Knight, 1968*b*:315.

Male. Length 6.30–6.79 mm; width 1.82–1.96 mm. Head black, second antennal segment 2.10–2.24 mm long, incrassate, reddish brown, apical one third black. Pronotum and scutellum black, not as polished as in *obliquus*; calli distinctly raised. Hemelytra with characteristic white pattern, not as shiny as in *obliquus*; pubescence silvery, long, dense, more distinctive than in *obliquus*.

Female. Length 5.39–5.81 mm. Antlike, similar to *obliquus* in color and appearance.

Remarks. This species is not as polished and shiny as *obliquus* and the calli on the pronotum in the male are distinctly raised. The females are similar to *obliquus* and difficult to identify.

Habitat. Collected on prairie grasses, *Potentilla fruticosa*.

Distribution. Idaho, Wyoming; now known to occur in Alberta and Saskatchewan (Map 94).

Tribe Phylini

In the Prairie Provinces the tribe is represented by 18 genera and 58 species.

Key to genera of Phylini

1. Dorsal surface with one type of pubescence, simple hairs 2
   Dorsal surface with two types of pubescence, scaly or sericeous hairs, intermixed with simple hairs 11
2. First and second antennal segments green with longitudinal black lines on first segment and often on second (Fig. 235) .... ***Icodema* Reuter** (p. 291)
   First and second antennal segments black, if green without longitudinal black lines 3
3. Front tibiae strongly bowed; third antennal segment in male strongly pilose on one side (Fig. 236) .... ***Conostethus* Fieber** (p. 293)
   Front tibiae straight; third antennal segment in male not pilose on one side 4
4. Second antennal segment equal to or shorter than head width .... ***Chlamydatus* Curtis** (p. 294)
   Second antennal segment longer than head width 5
5. Hind tibia without black spots at bases of spines 6
   Hind tibia with black spots at bases of spines 8

229

230

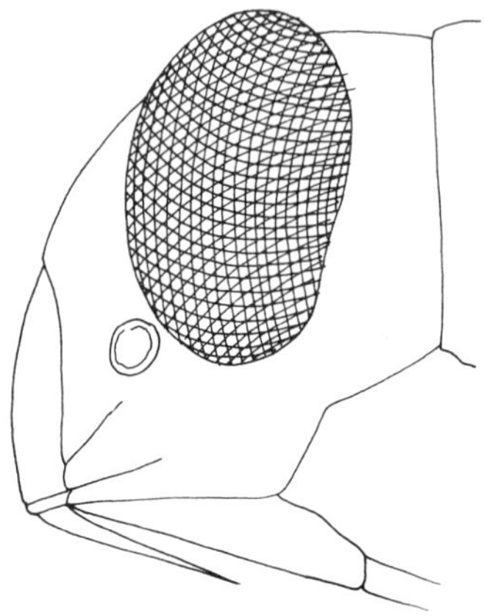

231

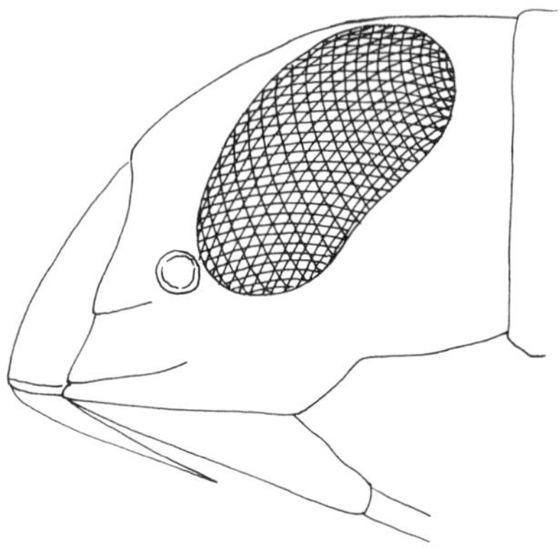

232

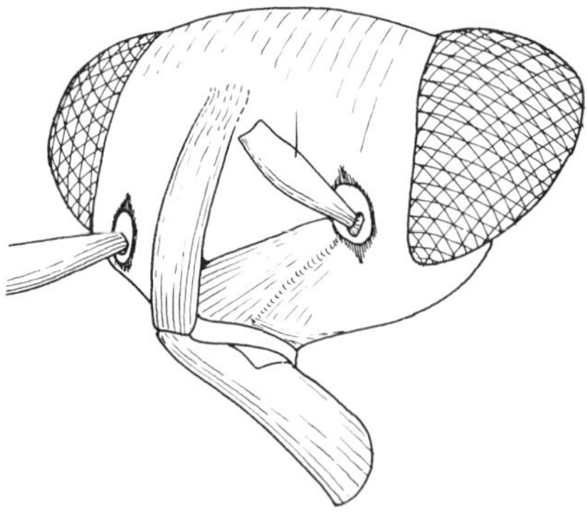

233

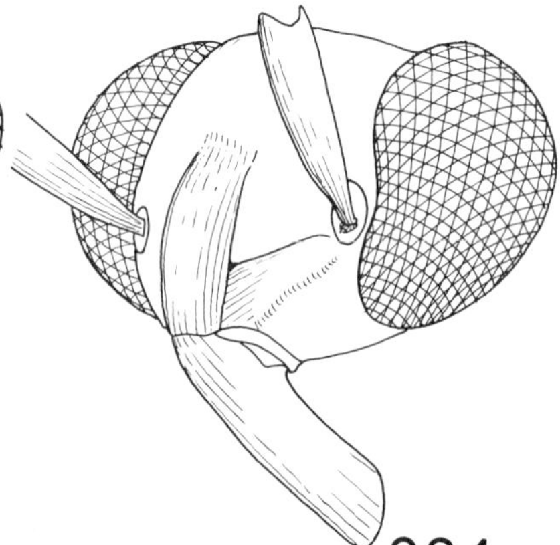

234

Figs. 229–234. Antennae and heads of Miridae. 229, second antennal segment of *Coquillettia* spp.; 230, second antennal segment of *Orectoderus* spp.; 231, head of *Coquillettia albertae*; 232, head of *C. insignis*; 233, head of *Monosynamma bohemanni*; 234, head of *Rhinocapsus* spp.

6. Antennal socket separated from eye by width of third antennal segment; inner margin of eye nearly straight (Fig. 233); species with pale hemelytra (Fig. 237) .... ***Tytthus*** **Fieber** (p. 302)
Antennal socket close to eye (Fig. 234); species with dark hemelytra 7
7. Dorsal surface highly polished, dark reddish species (Fig. 238) .... ***Rhinocapsus*** **Uhler** (p. 305)
Dorsal surface not highly polished, brown to fuscous species .... ***Microphylellus*** **Reuter** (p. 308)
8. Pubescence on dorsal surface black 9
Pubescence on dorsal surface yellow 10
9. Head vertical (Fig. 239) .... ***Phyllopidea*** **Knight** (p. 311)
Head oblique (Fig. 240) .... ***Hoplomachus*** **Fieber** (p. 313)
10. Antennal socket separated from eye by width of third antennal segment; inner margin of eye nearly straight (Fig. 233), pattern on hemelytra distinctive (Fig. 241) .... ***Monosynamma*** **Scott** (p. 315)
Antennal socket close to eye; inner margin of eye emarginate (Fig. 234) .... ***Plagiognathus*** **Fieber** (p. 315)
11. Second antennal segment green with large black spots (Fig. 228) .... ***Pseudatomoscelis*** **Poppius** (p. 330)
Second antennal segment black, if green without black spots 12
12. Hemelytra with flat white scales in addition to normal pubescence (Fig. 242) .... ***Lepidopsallus*** **Knight** (p. 331)
Hemelytra with sericeous pubescence, but without white scales 13
13. Pulvilli reaching tips of claws, claws sharply bent at tips (Fig. 4); large black species with orange legs .... ***Atractotomus*** **Fieber** (p. 334)
Pulvilli much shorter, claws not sharply bent at tips (Fig. 5) 14
14. Hemelytra and legs black; first and second antennal segments of male enlarged (Fig. 243) .... ***Criocoris*** **Fieber** (p. 335)
Hemelytra and legs not completely black; first and second antennal segments linear 15
15. Hemelytra pale green with 6–10 large, diffuse, brownish spots (Fig. 244) .... ***Atomoscelis*** **Reuter** (p. 335)
Hemelytra without diffuse brownish spots 16
16. Head, pronotum, abdomen, and femora strongly marked with red (Figs. 245, 246) .... ***Megalopsallus*** **Knight** (p. 337)
Head, pronotum, abdomen, and femora not marked with red 17
17. Second antennal segment shorter than head width (Fig. 247) .... ***Europiella*** **Reuter** (p. 341)
Second antennal segment longer than head width (Fig. 248) .... ***Psallus*** **Fieber** (p. 345)

Genus *Icodema* Reuter

Elongate, pale green species with simple type of pubescence. Head oblique. First and second antennal segments green with longitudinal black lines. Pronotum trapeziform, lateral margins angulate. Hemelytra shiny; pubescence pale. Femora and tibiae with longitudinal black lines; tibiae with black knee spots.

One species is found in North America. It was originally described as *Plagiognathus*, but Carvalho (1955*b*) transferred it to *Icodema*.

Icodema nigrolineatum (Knight)

Fig. 235; Map 95

Plagiognathus nigrolineatus Knight, 1923*b*:443.
Icodema nigrolineata: Carvalho, 1955*b*:226.

Length 4.34–4.55 mm; width 1.54–1.68 mm. Head pale green, tip of clypeus black. Antennae pale green, first segment with two longitudinal black lines, second segment often with longitudinal black lines on basal half. Rostrum 1.47–1.61 mm long. Pronotum, scutellum, and hemelytra pale green; pubescence pale, long, dense. Ventral surface and legs pale green; femora with black longitudinal lines near apices; tibiae with black knee spots and often with black longitudinal lines near bases.

Remarks. This species is easily distinguished by the pale green color and by the black lines on the antennae and legs (Fig. 235).

Habitat. Collected on *Quercus macrocarpa.*

Distribution. Eastern USA, Texas; now known to occur in Manitoba (Map 95).

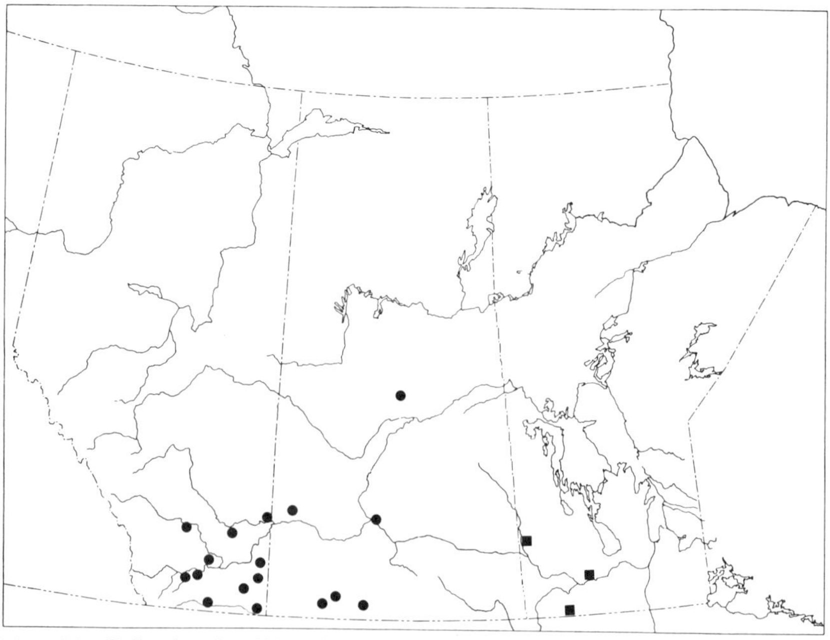

Map 95. Collection localities for *Icodema nigrolineatum* (■) and *Conostethus americanus* (●).

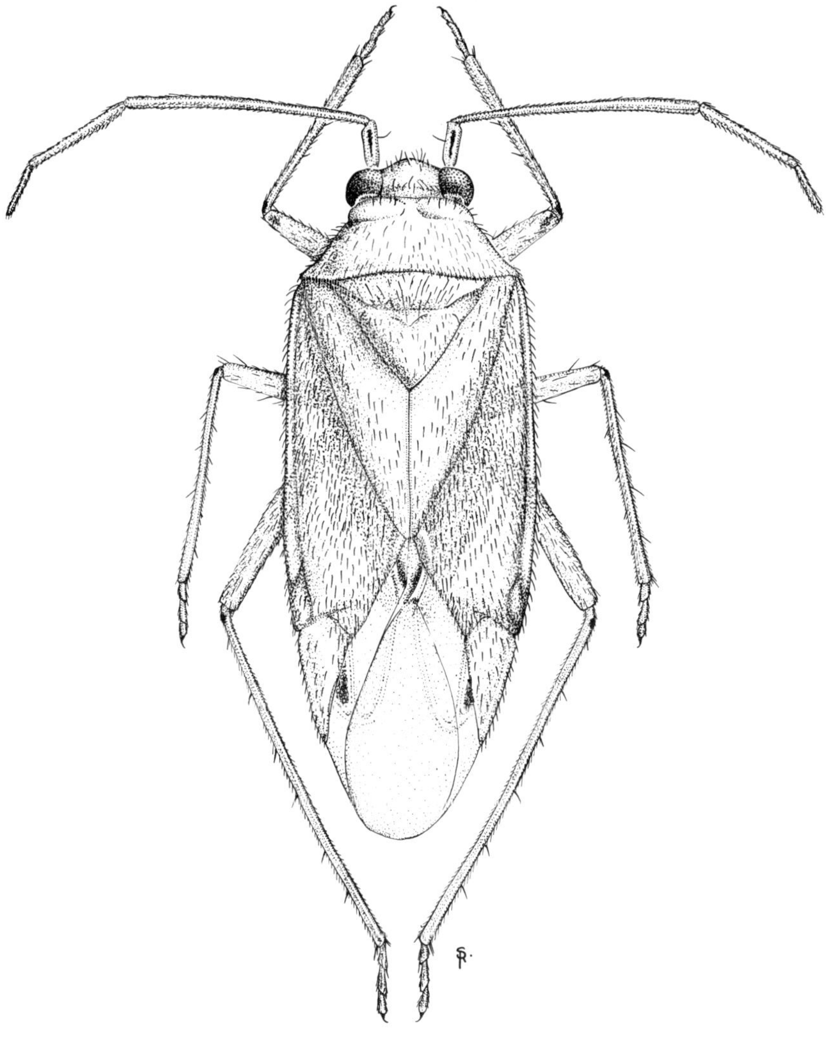

Fig. 235. *Icodema nigrolineatum*

Genus *Conostethus* Fieber

Elongate, pale green and black species. Head vertical, eyes large and protruding. Antennae black, third segment in male curved, strongly pilose on one side. Pronotum broadly campanulate, anterior angle carinate, lateral margins broadly angulate; calli distinct. Hemelytra smooth, pubescence simple, short, appressed. Front tibiae bowed anteriorly.

One species is known from North America.

Conostethus americanus Knight

Fig. 236; Map 95

Conostethus americanus Knight, 1939*b*:132.

Length 3.78–4.06 mm; width 1.12–1.26 mm. Head, antennae, pronotum, and scutellum mostly black in male, fuscous green in female. Rostrum 1.12–1.26 mm long. Hemelytra fuscous, costal margin green. Ventral surface black in male, green in female; legs fuscous.

Remarks. This species is distinguished by the elongate form, by the vertical head, by the bowed front tibiae, and by the curved and densely pilose third antennal segment in the male (Fig. 236).

Habitat. Collected on prairie grasses.

Distribution. North Central States; now known to occur in Alberta and Saskatchewan (Map 95).

Genus *Chlamydatus* Curtis

Small, oblong-oval, generally black species. Head almost vertical; eyes contiguous with pronotum, carina between them distinct. Second antennal segment equal to or shorter than head width. Pronotum trapeziform, lateral margins angulate. Hemelytra brachypterous or macropterous; pubescence simple, black, brown, or silvery, long and dense. Ventral surface black, hind femora enlarged.

Twenty-one species have been reported from North America. Nine occur in the Prairie Provinces, two of these are Holarctic in distribution. Reference to *pulicarius* by Van Duzee (1917), Knight (1923*b*, 1941*c*), Blatchley (1926), Walley (1932), and Kelton (1965*b*) should probably refer to *pullus*.

Key to species of *Chlamydatus*

1. Hemelytra mostly yellowish green, macropterous ...... ***obliquus* (Uhler)** (p. 296)
   Hemelytra black or brown, macropterous or brachypterous 2
2. First and second antennal segments, and legs reddish ...... ***ruficornis* Knight** (p. 296)
   Antennal segments and legs not reddish 3
3. Wing membrane not extending beyond cuneus .... ***opacus* (Zetterstedt)** (p. 297)
   Wing membrane extending beyond cuneus 4
4. Pubescence on dorsum silvery or golden brown 5
   Pubescence on dorsum black 8
5. Pubescence on dorsum golden brown ...... ***auratus* Kelton** (p. 298)
   Pubescence on dorsum silvery 6

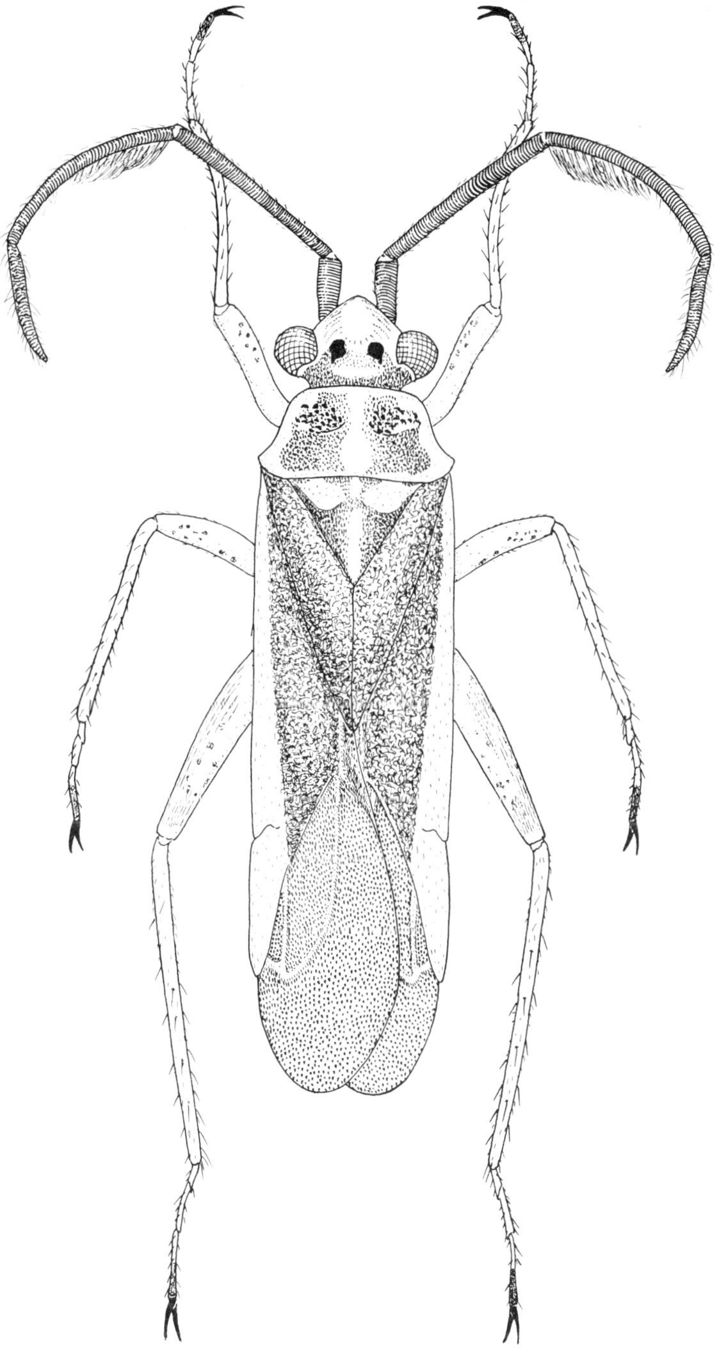

Fig. 236. *Conostethus americanus*

6. Posterior tibia black or fuscous .... ***montanus* Knight** (p. 299)
Posterior tibia pallid 7
7. Second antennal segment as long as head width ...... ***associatus* (Uhler)** (p. 299)
Second antennal segment much shorter than head width .... ***artemisiae* Kelton** (p. 300)
8. First and second femora yellow; second antennal segment pallid .... ***pallidicornis* Knight** (p. 300)
First and second femora black; second antennal segment usually black .... ***pullus* (Reuter)** (p. 301)

Chlamydatus obliquus (Uhler)

Map 96

Agalliastes obliquus Uhler, 1893:378.
Chlamydatus obliquus: Reuter, 1909:83.

Length 2.80–3.22 mm; width 1.12–1.33 mm. Second antennal segment black. Rostrum 1.10–1.22 mm long. Pronotum black, central area at base often pale green. Hemelytra shiny; corium black; clavus, embolium, and cuneus pale green; pubescence silvery, long, dense. First and second femora yellow, third mostly black; tibiae yellow with faint black spots at bases of spines.

Remarks. This species is distinguished by the bicolored hemelytra.

Habitat. Collected on *Lupinus argenteus*.

Distribution. Western USA, British Columbia; Alberta (Map 96).

Chlamydatus ruficornis Knight

Map 96

Chlamydatus ruficornis Knight, 1959:424.

Length 1.90–2.10 mm; width 0.98–1.12 mm. Head black. First and second antennal segments reddish. Rostrum 1.05–1.15 mm long. Pronotum and scutellum black, pruinose; pubescence silvery, appressed. Brachypterous or macropterous. Legs reddish.

Remarks. Brachypterous adults are the most common form of this species and are easily distinguished from all others by the absence of a wing membrane. The macropterous forms are rare, the author has seen only one female with fully developed wings. The adults are also distinguished by the reddish first and second antennal segments, and by the reddish legs.

Habitat. Collected on *Trifolium repens*.

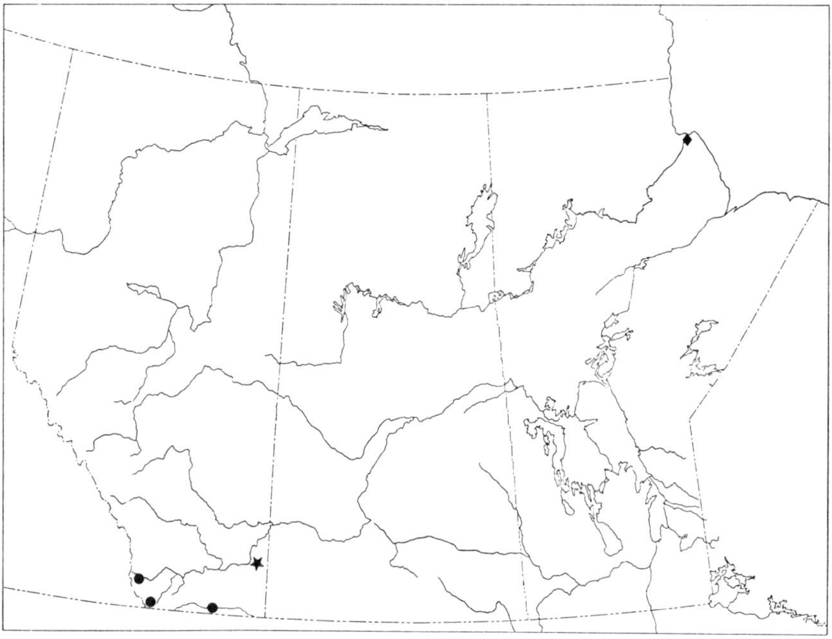

Map 96. Collection localities for *Chlamydatus obliquus* (●), *C. ruficornis* (★), and *C. opacus* (◆).

Distribution. Colorado, Iowa; now known to occur in Alberta and Saskatchewan (Map 96).

Chlamydatus opacus (Zetterstedt)

Map 96

Capsus opacus Zetterstedt, 1840:279.
Chlamydatus opacus: Wagner, 1955:208.

Length 1.96–2.66 mm; width 1.05–1.33 mm. Head generally yellow, black below antennal sockets, light brown above; frons with transverse brown bars each side of median line. First and second antennal segments black. Rostrum 1.12–1.29 mm long. Pronotum and scutellum yellowish brown marked with black. Hemelytra reduced in both sexes, grayish brown with yellow areas, pruinose; pubescence silvery, appressed. Femora black, tibiae yellow.

Remarks. This species has a very short wing membrane. It is also distinguished by the yellow areas on the head, pronotum, scutellum, and hemelytra.

Habitat. Unknown.

Distribution. Holarctic; Northern Canada, Quebec; Manitoba (Map 96).

Chlamydatus auratus Kelton

Map 97

Chlamydatus auratus Kelton, 1965*b*:1133.

Length 2.59–3.08 mm; width 1.26–1.54 mm. Head black, often yellowish along carina. First antennal segment yellow or partly black. Rostrum 1.10–1.19 mm long. Pronotum, scutellum, and hemelytra black, shiny; pubescence golden, appressed; wing membrane fully developed or reduced. Front and middle femora yellow, hind femur fuscous with several black spots; tibiae yellow with black spots at bases of spines.

Remarks. This species is distinguished by the golden pubescence on the hemelytra.

Habitat. Collected on *Anemone canadensis* and *Medicago sativa.*

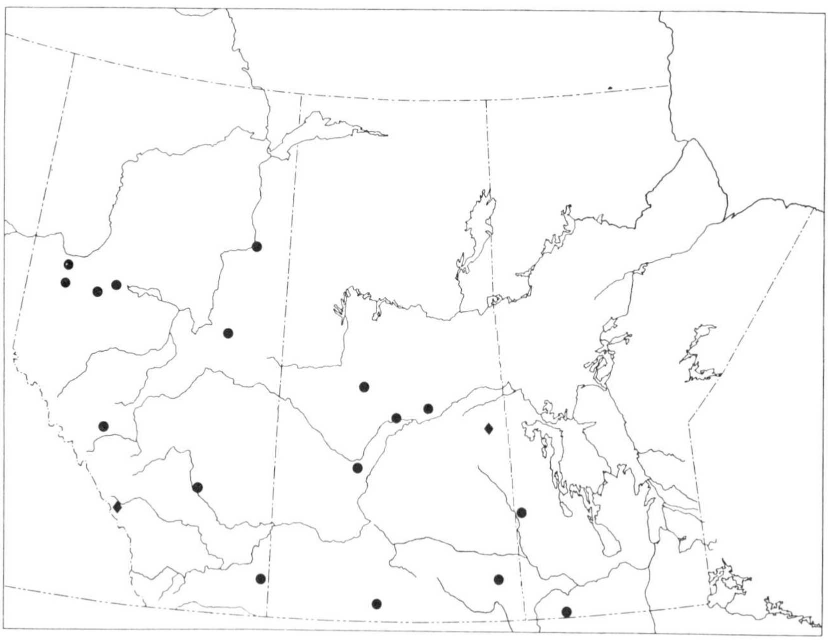

Map 97. Collection localities for *Chlamydatus auratus* (●) and *C. montanus* (◆).

Distribution. New York, British Columbia; Prairie Provinces (Map 97).

Chlamydatus montanus Knight

Map 97

Chlamydatus montanus Knight, 1964:140.

Length 2.31–2.50 mm; width 1.05–1.19 mm. Head black. First and second antennal segments black; second segment shorter than head width. Rostrum 0.98–1.15 mm long. Pronotum, scutellum, and hemelytra black; pubescence silvery, long, dense. Front and middle femora and tibiae orange; hind femur black, tibia fuscous to black.

Remarks. This species resembles *associatus* by the silvery pubescence on the hemelytra. It is separated from it by the short second antennal segment and by the blackish hind tibia.

Habitat. Collected on *Pinus banksiana* and *P. contorta.*

Distribution. Montana; Alberta, Saskatchewan (Map 97).

Chlamydatus associatus (Uhler)

Fig. 227; Map 98

Agalliastes associatus Uhler, 1872:419.
Chlamydatus associatus: Atkinson, 1890:173.

Length 2.66–2.82 mm; width 1.12–1.26 mm. Head black; first and second antennal segments black, second segment as long as head width. Rostrum 0.96–1.05 mm long. Pronotum, scutellum, and hemelytra black; pubescence silvery, long, dense. Front femora and tibiae yellow; hind femur black, tibia yellow.

Remarks. This species is common on the prairies. It is distinguished by the silvery pubescence on the hemelytra, by the second antennal segment that is as long as width of head, by the yellow first and second femora, and by the spotless tibiae (Fig. 227).

Habitat. Collected on a great variety of plants, including alfalfa, cultivated sunflower, and potato.

Distribution. Mexico, widespread in USA, Ontario, Quebec, British Columbia; Prairie Provinces (Map 98).

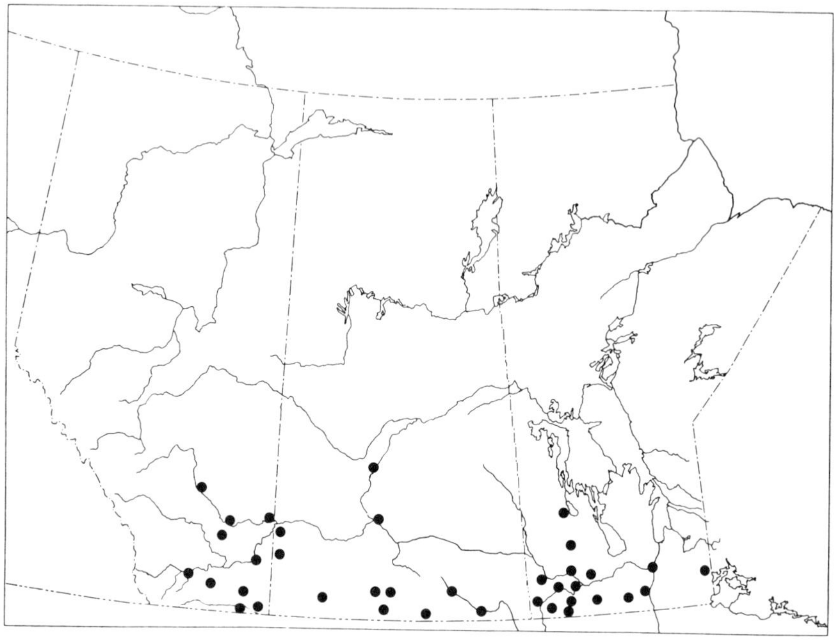

Map 98. Collection localities for *Chlamydatus associatus*.

Chlamydatus artemisiae Kelton

Map 99

Chlamydatus artemisiae Kelton, 1965*b*:1138.

Length 2.31–2.38 mm; width 1.01–1.12 mm. Head black, often pale along carina. First and second antennal segments black. Rostrum 0.80–0.84 mm long. Pronotum, scutellum, and hemelytra black; pubescence silvery, long, dense. Front and middle femora yellow, hind femur black; tibiae yellow.

Remarks. This species is similar in appearance to *associatus* and *montanus*. It is separated from them by the shorter rostrum.

Habitat. Collected on *Artemisia cana*.

Distribution. Alberta (Map 99).

Chlamydatus pallidicornis Knight

Map 99

Chlamydatus pallidicornis Knight, 1964:138.

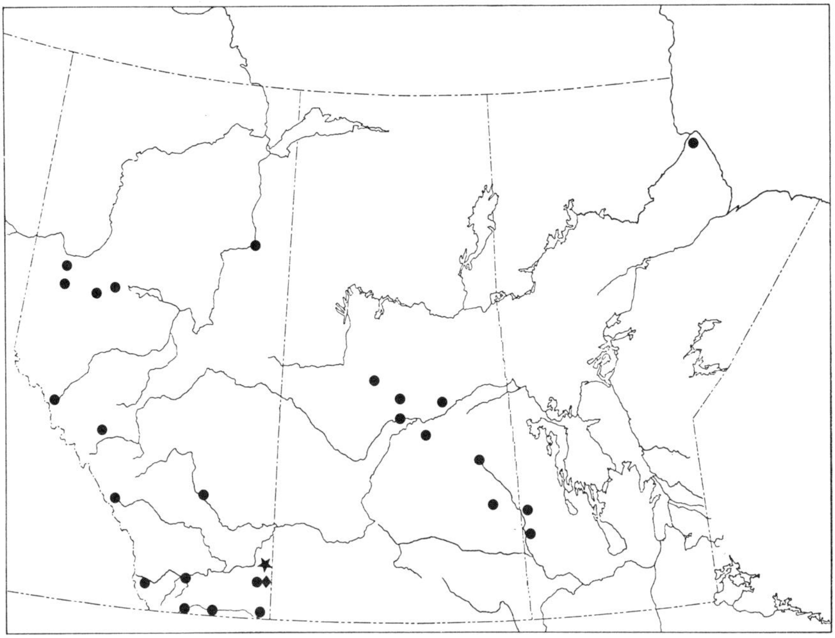

Map 99. Collection localities for *Chlamydatus artemisiae* (★), *C. pallidicornis* (◆), and *C. pullus* (●).

Length 2.50–2.80 mm; width 1.19–1.33 mm. Head black. First antennal segment black, second segment pallid. Rostrum 1.12–1.19 mm long. Pronotum, scutellum, and hemelytra black; pubescence black. First and second femora yellow, hind femora black; tibiae yellow without spots at bases of spines.

Remarks. This species is distinguished by the black pubescence on the hemelytra.

Habitat. Collected on *Potentilla plattensis*.

Distribution. Northwestern States; Alberta (Map 99).

Chlamydatus pullus (Reuter)

Map 99

Agalliastes pullus Reuter, 1871:324.
Chlamydatus pullus: Atkinson, 1890:174.
Chlamydatus pulicarius, Knight, 1925*b*:81; Kelton, 1965*b*:1132.

Length 2.17–2.80 mm; width 1.19–1.26 mm. Heads black. First and second antennal segments black, second segment often partly pale. Rostrum 1.05–1.15 mm long. Pronotum, scutellum, and hemelytra black, pruinose; pubescence black. Femora black; tibiae yellow with black spots at bases of spines.

Remarks. This species is similar to *pallidicornis* except for the black front femora and the spotting on the tibiae.

Former references to *pulicarius* from North America should probably refer to *pullus*. The presence of *pulicarius* in the Nearctic region is now doubtful. The femora of *pulicarius* are yellow with large, black spots; these conditions are often encountered in teneral specimens of *pullus*.

Habitat. Collected on a great variety of plants.

Distribution. Alaska, Colorado, New Mexico, northern Canada, Ontario, Quebec, Newfoundland, British Columbia; Prairie Provinces (Map 99).

Genus *Tytthus* Fieber

Small, elongate-oval species with simple pubescence. Head vertical, globose; black with two pale spots at vertex. Pronotum campanulate, pruinose; calli tumid. Hemelytra pruinose; pubescence golden, appressed. Legs pale greenish yellow. Species resemble those of *Mecomma* Fieber.

Carvalho and Southwood (1955) reviewed the genus. Eight species are known from North America; two are Holarctic in distribution.

Key to species of *Tytthus*

1. First antennal segment black with apex pale .... ***pygmaeus* (Zetterstedt)** (p. 302)
   First antennal segment mostly pale .... ***geminus* (Flor)** (p. 303)

Tytthus pygmaeus (Zetterstedt)

Fig. 237; Map 100

Capsus pygmaeus Zetterstedt, 1840:279.
Tytthus pygmaeus: Fieber, 1864:83.
Tytthus pygmaeus Carvalho & Southwood, 1955:23.
Tytthus pygmaeus Kelton, 1980*a*:285.

Length 2.45–2.80 mm; width 0.84–1.26 mm. First antennal segment black, apex pale; other segments black. Rostrum 0.84–0.98 mm long.

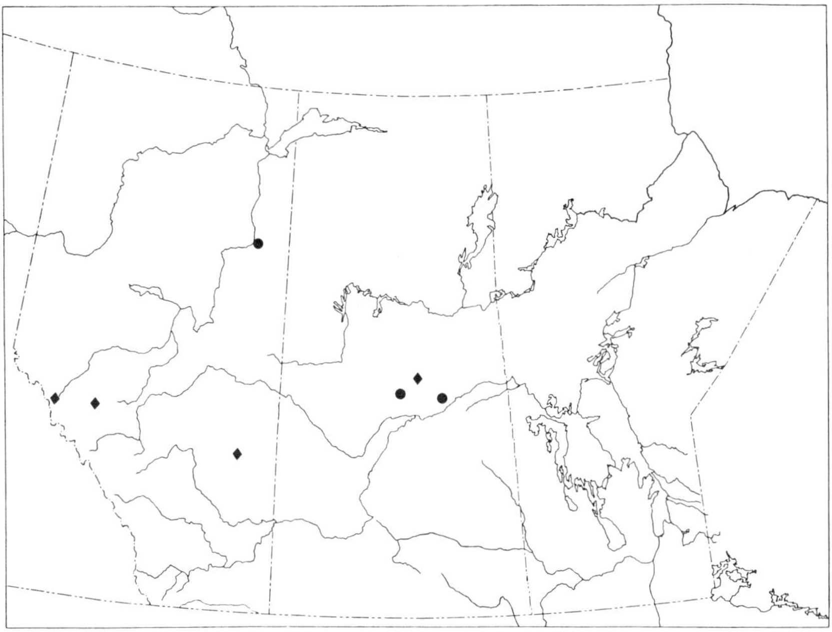

Map 100. Collection localities for *Tytthus pygmaeus* (●) and *T. geminus* (◆).

Pronotum and scutellum black. Hemelytra uniformly pale. Ventral surface black, abdomen in female mostly pale green. Male genital capsule without tuft of long pubescence.

Remarks. Kelton (1980*a*) reported this species from North America. It is distinguished by the black first antennal segment with pale apex (Fig. 237), and by the absence of long pubescence on the genital capsule.

Habitat. Collected on *Carex* spp.

Distribution. Holarctic; Yukon, Newfoundland, Ontario, British Columbia; Alberta, Saskatchewan (Map 100).

Tytthus geminus (Flor)

Map 100

Capsus geminus Flor, 1860:464.
Tytthus geminus Fieber, 1864:83.
Cyrtorhinus pubescens Knight, 1931*b*:172.
Tytthus geminus Carvalho & Southwood, 1955:28.

Length 2.58–3.08 mm; width 0.84–1.12 mm. Head similar to *pygmaeus*. First antennal segment mostly pale, other segments black. Rostrum

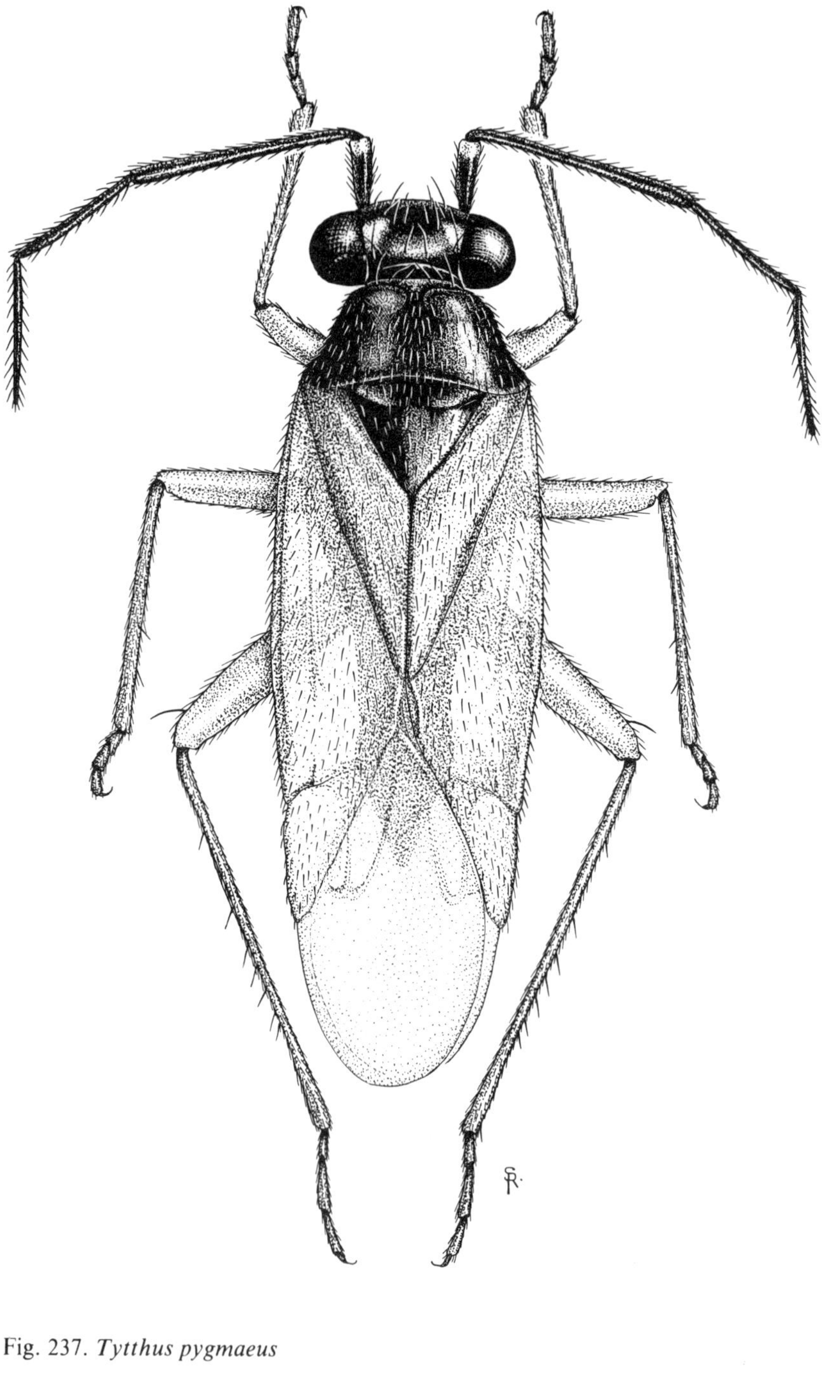

Fig. 237. *Tytthus pygmaeus*

0.87–1.12 mm long. Pronotum and scutellum black, basal angles or basal margin of pronotum often pale. Hemelytra uniformly pale. Ventral surface black, abdomen in female mostly pale; left side of genital capsule with tuft of long pale pubescence.

Remarks. This species is similar to *pygmaeus* in color and appearance, but the first antennal segment is mostly pale. The males are distinguished by the long pubescence on the genital capsule.

Habitat. Collected on *Carex* spp.

Distribution. Holarctic; Alaska, Colorado, now known to occur in Yukon, British Columbia; Alberta, Saskatchewan (Map 100).

Genus *Rhinocapsus* Uhler

Elongate-oval, reddish brown species with simple type of pubescence. Head oblique. Rostrum extending just beyond hind coxae. Pronotum trapeziform, lateral margins angulate, smooth, shiny; calli inconspicuous. Hemelytra uniformly reddish brown, shiny; pubescence brown to black, appressed. Ventral surface reddish brown, legs paler.

Three species are known from North America, two occur in Saskatchewan and Manitoba.

Key to species of *Rhinocapsus*

1. Clypeus, jugum, and lorum black; second antennal segment pale on basal half, black on apical half; femora spotted with black .... ***vanduzeei* Uhler** (p. 305)
   Clypeus partly red, jugum and lorum red; second antennal segment uniformly reddish brown or black; femora without spots (Fig. 238) .... ***rubricans* (Provancher)** (p. 306)

Rhinocapsus vanduzeei Uhler

Map 101

Rhinocapsus vanduzeei Uhler, 1890:82.

Length 3.64–3.85; width 1.47–1.61 mm. Head reddish brown, clypeus, jugum, and lorum black. First antennal segment pale, second segment pale on basal half, black on apical half. Rostrum 1.45–1.50 mm long. Femora light yellowish brown, spotted with black.

Remarks. This species is distinguish by the uniformly black clypeus, jugum, and lorum, by the bicolored second antennal segment, and by the spotted femora.

Map 101. Collection localities for *Rhinocapsus vanduzeei* (●) and *R. rubricans* (◆).

Habitat. Collected on *Kalmia polifolia.*

Distribution. Eastern States, Missouri, Quebec, Ontario; now known to occur in Manitoba (Map 101).

Rhinocapsus rubricans (Provancher)

Fig. 238; Map 101

Plagiognathus rubricans Provancher, 1887:154.
Rhinocapsus rubricans: Van Duzee, 1912:323.

Length 3.64–4.20 mm; width 1.47–1.82 mm. Head reddish brown with clypeus slightly darker. Second antennal segment uniformly reddish brown or black. Pronotum and hemelytra similar to those of *vanduzeei.* Femora without spots.

Remarks. This species is similar in appearance to *vanduzeei*, but the clypeus, jugum, and lorum are not uniformly black, the second antennal segment is uniform in color, and the femora are without spots (Fig. 238).

Habitat. Collected on *Kalmia polifolia.*

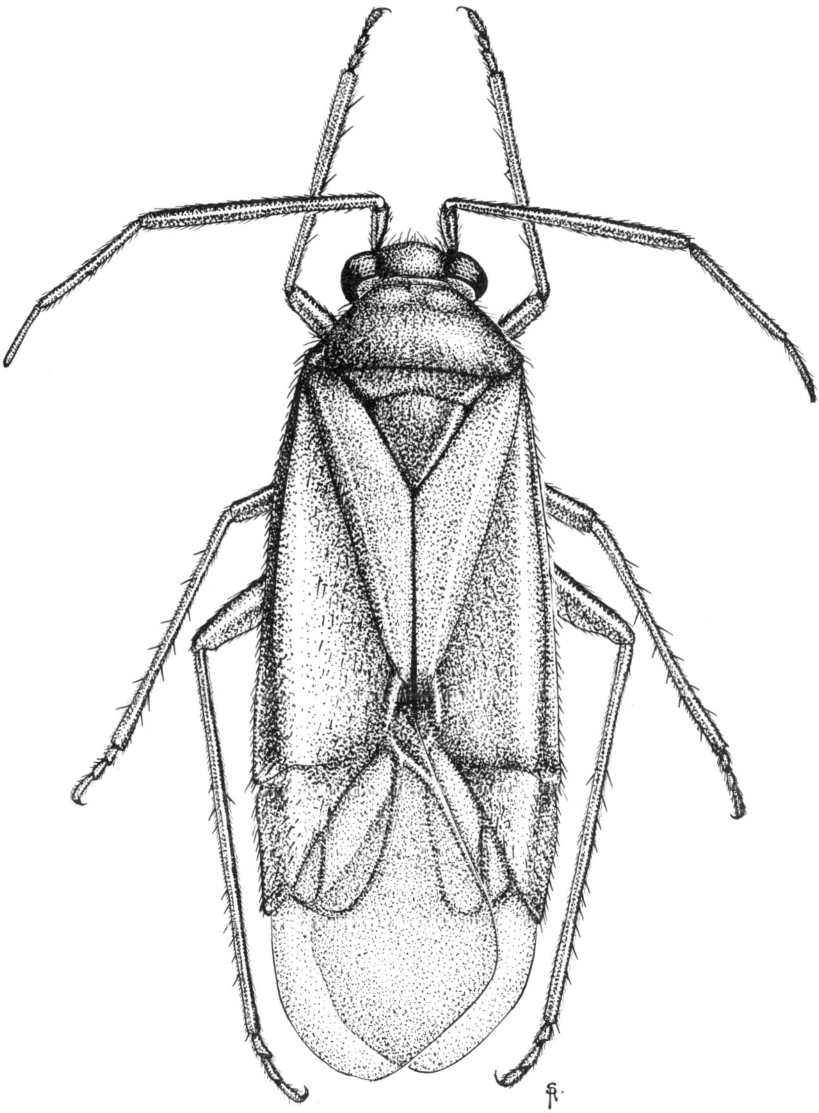

Fig. 238. *Rhinocapsus rubricans*

Distribution. Eastern USA, Quebec; now known to occur in Manitoba and Saskatchewan (Map 101).

Genus *Microphylellus* Reuter

Elongate-oblong, black species with simple type of pubescence, and without black spots on tibiae. Head oblique. Pronotum trapeziform, lateral margins angulate. Hemelytra shiny; pubescence long, dense. Ventral surface black, legs yellow.

Nineteen species are known from North America, five occur in the Prairie Provinces.

Key to species of *Microphylellus*

1. Pronotum and hemelytra black 2
   Pronotum and hemelytra with extensive pale markings 4
2. Rostrum 1.68 mm or longer extending to hind coxae or beyond .... ***longirostris* Knight** (p. 308)
   Rostrum less than 1.68 mm not extending to hind coxae 3
3. First antennal segment pale .... ***modestus* Reuter** (p. 308)
   First antennal segment black .... ***tumidifrons* Knight** (p. 309)
4. Hind femur without spots .... ***maculipennis* Knight** (p. 310)
   Hind femur with black spots .... ***fuscicornis* Knight** (p. 311)

Microphylellus longirostris Knight

Map 102

Microphylellus longirostris Knight, 1923*b*:458.

Length 3.50–3.92 mm; width 1.26–1.40 mm. Head black. Antennae yellow. Rostrum 1.68–1.82 mm long. Pronotum, scutellum, and hemelytra black; pubescence black, moderately long, appressed. Legs yellow.

Remarks. This species is distinguished by the uniformly black hemelytra and by the long rostrum.

Habitat. Collected on *Corylus americana*.

Distribution. Northeastern and North Central States, Quebec; now known to occur in Manitoba (Map 102).

Microphylellus modestus Reuter

Map 102

Microphylellus modestus Reuter, 1909:76.

Length 3.22–3.50 mm; width 1.19–1.40 mm. Similar to *longirostris* in color and appearance. Rostrum 1.13–1.22 mm long.

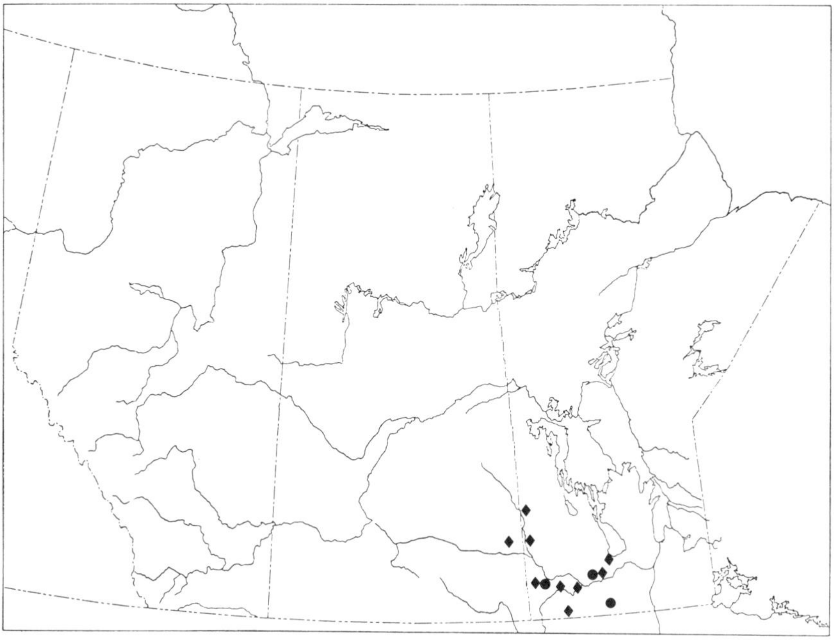

Map 102. Collection localities for *Microphylellus longirostris* (●) and *M. modestus* (◆).

Remarks. This species is separated from *longirostris* by the shorter rostrum.

Habitat. Collected on *Quercus macrocarpa.*

Distribution. Eastern States, Indiana, Missouri, Quebec, Ontario; now known to occur in Manitoba and Saskatchewan (Map 102).

Microphylellus tumidifrons Knight

Map 103

Microphylellus tumidifrons Knight, 1923*b*:455.

Length 3.08–3.36 mm; width 1.26–1.40 mm. Head black. First antennal segment black, other segments yellow. Rostrum 1.12–1.19 mm long. Pronotum, scutellum, and hemelytra black; pubescence black.

Remarks. This species resembles *modestus* except for the black first antennal segment.

Habitat. Collected on *Picea glauca.*

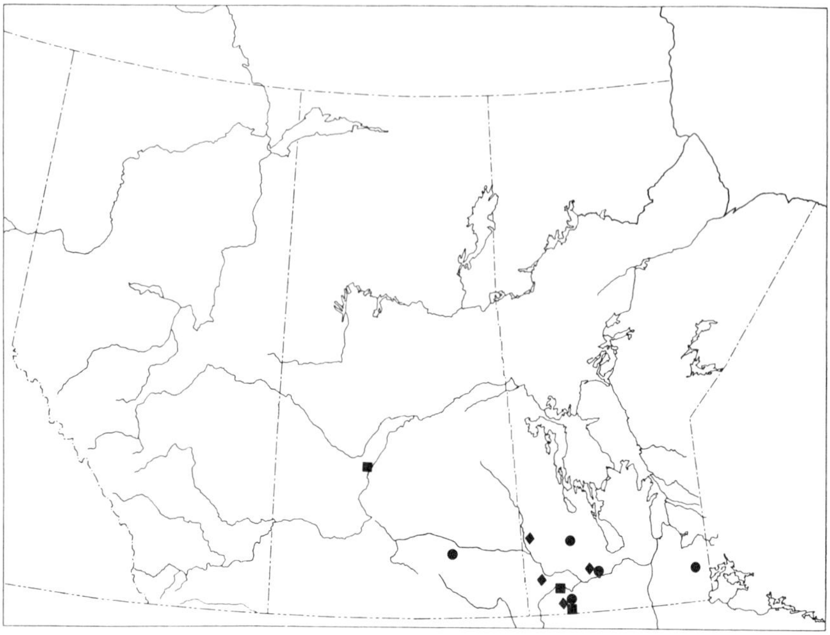

Map 103. Collection localities for *Microphylellus tumidifrons* (●), *M. maculipennis* (◆), and *M. fuscicornis* (■).

Distribution. Nova Scotia; now known to occur in Manitoba and Saskatchewan (Map 103).

Microphylellus maculipennis Knight

Map 103

Microphylellus maculipennis Knight, 1923*b*:456.

Length 3.08–3.36 mm; width 1.26–1.40 mm. Head black, vertex often yellow. Antennae yellow, base of first segment often black. Rostrum 1.19–1.26 mm long. Pronotum black with area behind calli and basal angles pale. Scutellum pale, median longitudinal line black. Hemelytra black, basal half of corium, embolium, and cuneus pale yellow. Ventral surface black, legs pale yellow.

Remarks. This species is distinguished by the pale markings on the hemelytra.

Habitat. Collected on *Quercus macrocarpa.*

Distribution. North Central States, Texas; now known to occur in Manitoba (Map 103).

Microphylellus fuscicornis Knight

Map 103

Microphylellus maculipennis var. *fuscicornis* Knight, 1923*b*:457.

Length 3.50–3.85 mm; width 1.40–1.54 mm. Head similar to *maculipennis*. Antennae yellow, base of first segment and base of second black. Rostrum 1.33–1.40 mm long. Pronotum, scutellum, and hemelytra similar to *maculipennis* in color. Ventral surface black, legs yellow, hind femur spotted with black.

Remarks. This species is slightly larger than *maculipennis* with longer rostrum and spotted hind femur.

Habitat. Collected on *Ulmus americana*.

Distribution. Maine; now known to occur in Manitoba and Saskatchewan (Map 103).

Genus *Phyllopidea* Knight

Elongate-oval, black and gray species with long, erect, black pubescence. Head vertical, yellow with extensive black markings. Eyes large, bulging. Rostrum extending to middle coxae. Pronotum trapeziform, lateral margins angulate, gray; calli black. Hemelytra gray, corium with longitudinal black line adjacent to radial vein; cuneus with large black central spot. Ventral surface black with extensive yellow markings; legs yellowish spotted with black.

Four species are known from North America, one occurs in Alberta.

Phyllopidea montana Knight

Fig. 239; Map 104

Phyllopidea montana Knight, 1968*a*:33.

Length 4.55–5.67 mm; width 1.68–1.96 mm. Head with distinctive black pattern on frons. Rostrum 1.12–1.26 mm long. Dorsal surface with black, long, erect hairs.

Remarks. This species is distinguished by the large size, by the long, erect, black pubescence on the dorsal surface, and by the contrasting black spot on the cuneus (Fig. 239).

Habitat. Collected on *Artemisia ludoviciana*.

Distribution. Colorado, Idaho, Wyoming; now known to occur in Alberta (Map 104).

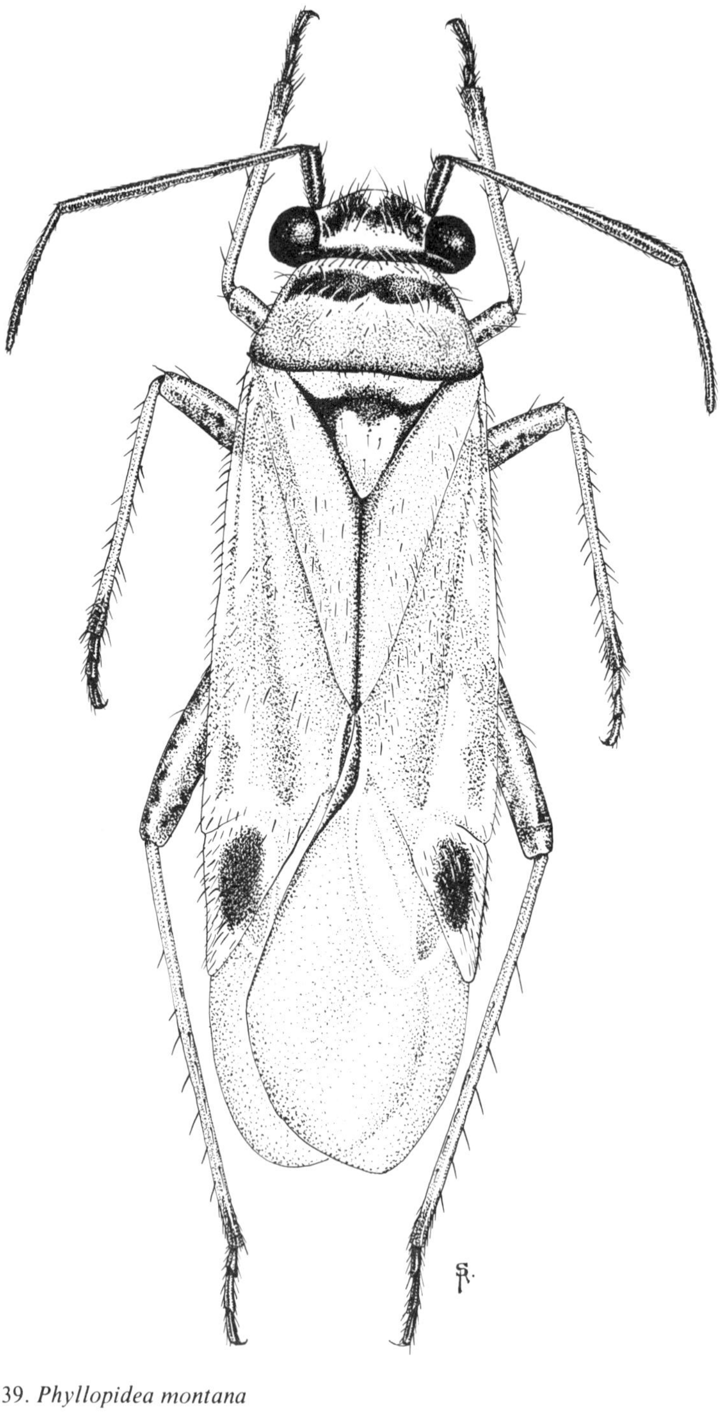

Fig. 239. *Phyllopidea montana*

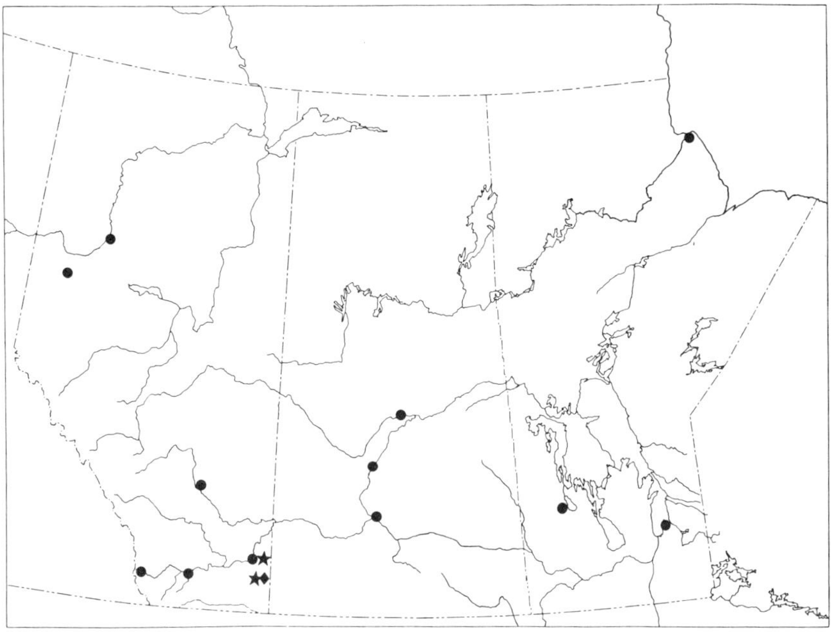

Map 104. Collection localities for *Phyllopidea montana* (★), *Hoplomachus affiguratus* (◆), and *Monosynamma bohemanni* (●).

Genus *Hoplomachus* Fieber

Elongate-oval, gray species with simple, black pubescence. Head oblique. Pronotum trapeziform, lateral margins angulate; pubescence black, erect with black spots at bases. Hemelytra shiny; pubescence as on pronotum. Tibiae spotted with black.

One species is known from North America.

Hoplomachus affiguratus (Uhler)

Fig. 240; Map 104

Macrotylus affiguratus Uhler, 1895:50.
Hoplomachus affiguratus: Carvalho, 1955*b*:226.

Length 4.34–4.90 mm; width 1.68–2.03 mm. Head mostly black, frons pale. Antennae black. Rostrum 1.40–1.54 mm long. Pronotum and hemelytra shiny gray; calli yellowish, hind margins black, depressed; pubescence black and erect with black spots at bases. Ventral surface mostly black, legs yellowish spotted with black.

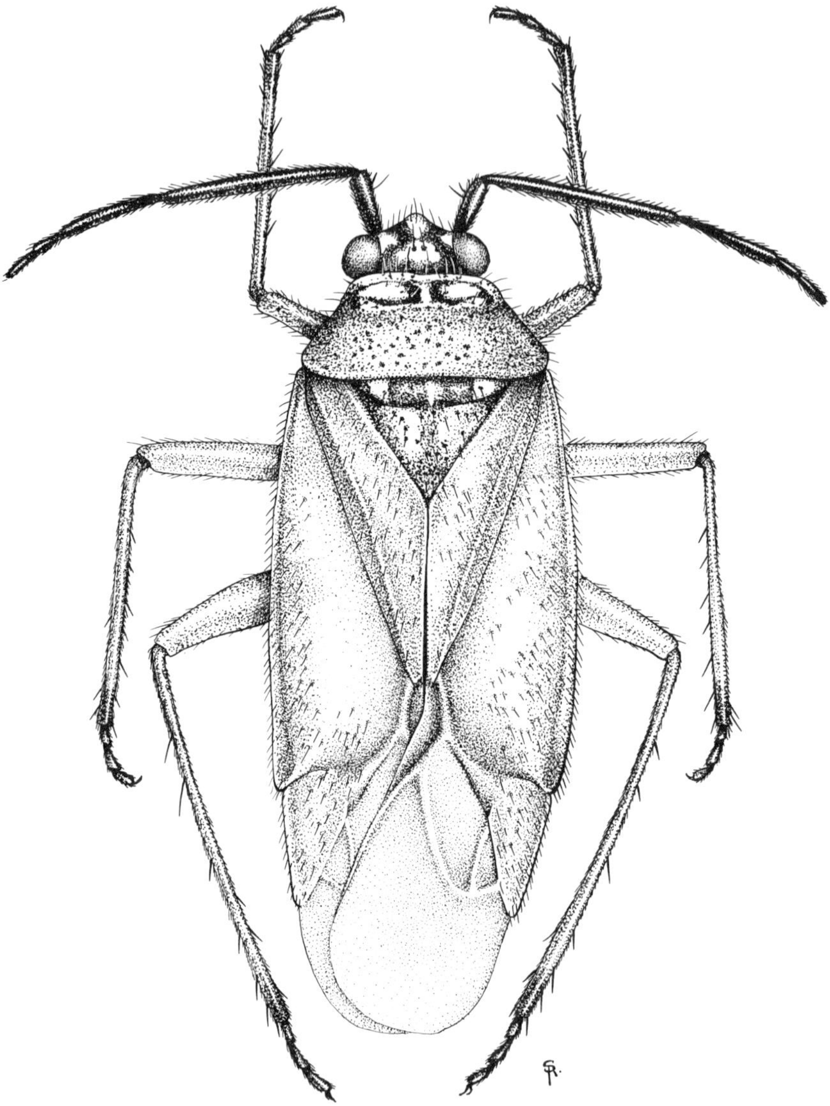

Fig. 240. *Hoplomachus affiguratus*

Remarks. This species is distinguished by the gray color and black spotting on the pronotum and hemelytra, and by the black and erect pubescence (Fig. 240).

Habitat. Collected on *Delphinium bicolor*.

Distribution. Western States; now known to occur in Alberta (Map 104).

Genus *Monosynamma* Scott

Elongate-oval species with distinctive color pattern. Head oblique. Pronotum trapeziform, lateral margins broadly angulate. Hemelytra shiny; pubescence pale and simple, dense and appressed. Tibiae spotted with black.

One species is known from North America.

Monosynamma bohemanni (Fallén)

Fig. 241; Map 104

Phytocoris bohemanni Fallén, 1829:106.
Monosynamma bohemanni: Oshanin, 1910:935.

Length 3.50–3.84 mm; width 1.47–1.68 mm. Head pale yellow, clypeus and lorum black, frons black or with transverse reddish bars. Second antennal segment black. Rostrum 1.40–1.54 mm long. Pronotum finely rugose, usually black or gray, calli slightly raised and marked with reddish brown. Scutellum black, median longitudinal line pale. Hemelytra pale marked with black. Legs pale, femora often black at middle; tibiae pale spotted with black, spines black.

Remarks. This species is distinguished by the simple pubescence on the hemelytra, by the spotted tibiae and the black spines, and by the pale median line on the scutellum (Fig. 241).

Habitat. Collected on *Salix interior*.

Distribution. Northern USA, Newfoundland, British Columbia; now known to occur in the Prairie Provinces (Map 104).

Genus *Plagiognathus* Fieber

Elongate-oval, black and green species with simple, pale pubescence and spotted legs. Head oblique. Pronotum trapeziform, lateral margins angulate. Hemelytra shiny, pubescence black or pale, long and dense. Tibial spines black with spots at bases.

Approximately 27 species are known from North America, 20 occur in the Prairie Provinces.

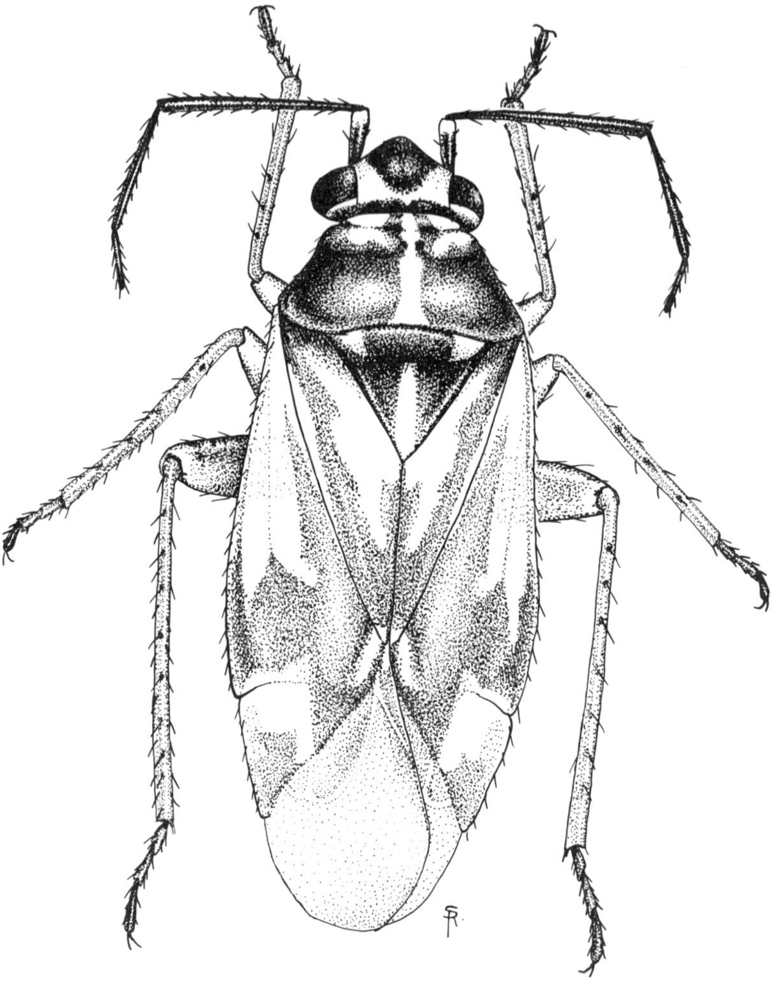

Fig. 241. *Monosynamma bohemanni*

Key to species of *Plagiognathus*

1. Head green; pronotum and hemelytra green without black markings 2
   Head black or green with black markings; pronotum and hemelytra black, sometimes with pale markings 3
2. Pubescence on hemelytra mostly black; rostrum 1.19–1.26 mm long .... ***guttatipes* (Uhler)** (p. 317)
   Pubescence on hemelytra pale (Fig. 249); rostrum 1.40–1.54 mm long .... ***flavidus* Knight** (p. 318)

3. Pronotum, scutellum, or hemelytra fuscous with pale or green areas 4
Pronotum and scutellum black; hemelytra black with cuneus sometimes pale, or uniformly pale 10
4. Head greenish yellow, clypeus partly black .... ***blatchleyi* Reuter** (p. 319)
Head black, only vertex pale 5
5. Large, robust species; rostrum 2.10 mm or longer .... ***shoshonea* Knight** (p. 320)
Smaller, not robust species; rostrum shorter than 2.10 mm 6
6. Front and middle femora black or partly fuscous; oval species (Fig. 250) .... ***albonotatus* Knight** (p. 320)
Front and middle femora yellow, spotted with black 7
7. Cuneus uniformly pale; rostrum 1.68 mm or longer; elongate species (Fig. 251) .... ***alboradialis* Knight** (p. 321)
Cuneus marked with black; rostrum shorter than 1.68 mm 8
8. Pronotum black, or pale at basal angles (Fig. 252) .. ***negundinis* Knight** (p. 322)
Pronotum black, or pale on median area 9
9. Pronotum and scutellum black; hemelytra arcuate; female abdomen black (Fig. 253) .... ***medicagus* Arrand** (p. 322)
Pronotum black, or pale on median area; sides of scutellum often pale; hemelytra almost parallel sided (Fig. 254); female abdomen with pale markings .... ***obscurus* Uhler** (p. 323)
10. Front femur mostly black, pale only at base and apex 11
Front femur pale, often fuscous and spotted 13
11. Second antennal segment pale, black at base .... ***fuscosus* (Provancher)** (p. 324)
Second antennal sgement black 12
12. Species 3.5 mm or longer; second antennal segment as long as or longer than width of pronotum at base .... ***nigritus* Knight** (p. 325)
Species shorter than 3.5 mm; second antennal segment much shorter than width of pronotum at base .... ***nigronitens* Knight** (p. 325)
13. First antennal segment pale, black at base .... ***davisi* Knight** (p. 326)
First antennal segment black, pale at apex 14
14. Rostrum 1.40 mm or shorter 15
Rostrum 1.41 mm or longer 17
15. Hemelytra lighter than pronotum and scutellum .. ***suffuscipennis* Knight** (p. 327)
Hemelytra concolorous with pronotum and scutellum 16
16. Second antennal segment pale; hind femur pale .... ***dispar* Knight** (p. 328)
Second antennal segment black; hind femur fuscous .. ***laricicola* Knight** (p. 328)
17. Rostrum 1.68 mm or longer, extending beyond hind coxae; second antennal segment black .... ***fuscipes* Knight** (p. 329)
Rostrum 1.54 mm or shorter, extending just beyond middle coxae; second antennal segment black, pale at middle .... ***annulatus* Uhler** (p. 329)

Plagiognathus guttatipes (Uhler)

Map 105

Lygus guttatipes Uhler, 1895:35.
Plagiognathus guttatipes: Knight, 1917*b*:639.

Length 3.78–4.06 mm; width 1.40–1.68 mm. Head pale green. Antennae yellowish green. Rostrum 1.19–1.26 mm long. Pronotum, scutellum, and

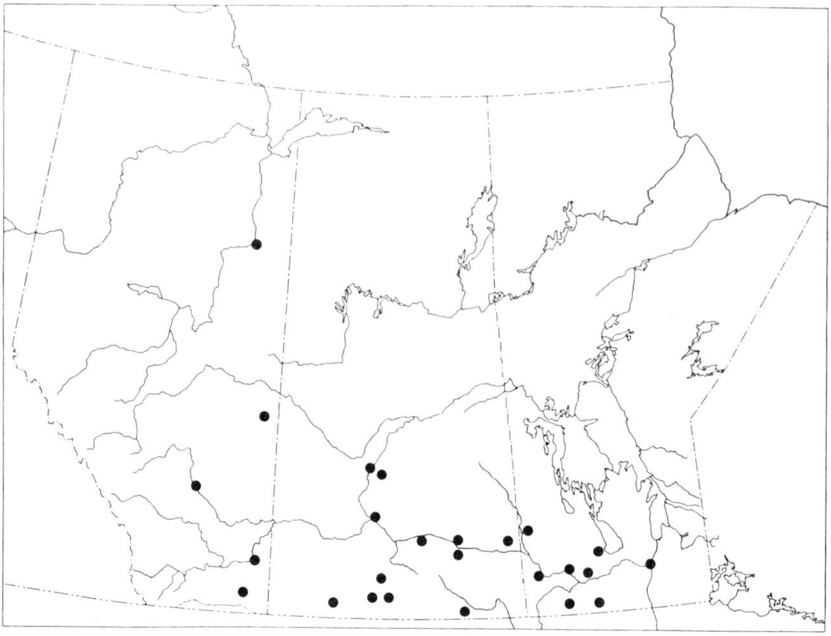

Map 105. Collection localities for *Plagiognathus guttatipes.*

hemelytra pale green; pubescence black and yellow. Ventral surface pale green; legs yellowish green spotted with black.

Remarks. This species is distinguished by the uniformly pale green color, by the yellowish green antennae without black markings, and by the black pubescence.

Habitat. Collected on *Glycyrrhiza lepidota.*

Distribution. North Central States; Alberta, now known to occur in Saskatchewan and Manitoba (Map 105).

Plagiognathus flavidus Knight

Fig. 249; Map 106

Plagiognathus shepherdiae var. *flavidus* Knight, 1929*a*:71.

Length 3.85–4.20 mm; width 1.54–1.68 mm. Head pale green, often with two to four brown spots on frons. Antennae pale yellow, spot on first segment and bases of first and second segments black. Rostrum 1.40–1.54 mm long. Pronotum, scutellum, and hemelytra yellowish green; pubescence pale. Ventral surface pale green; legs pale green spotted with black.

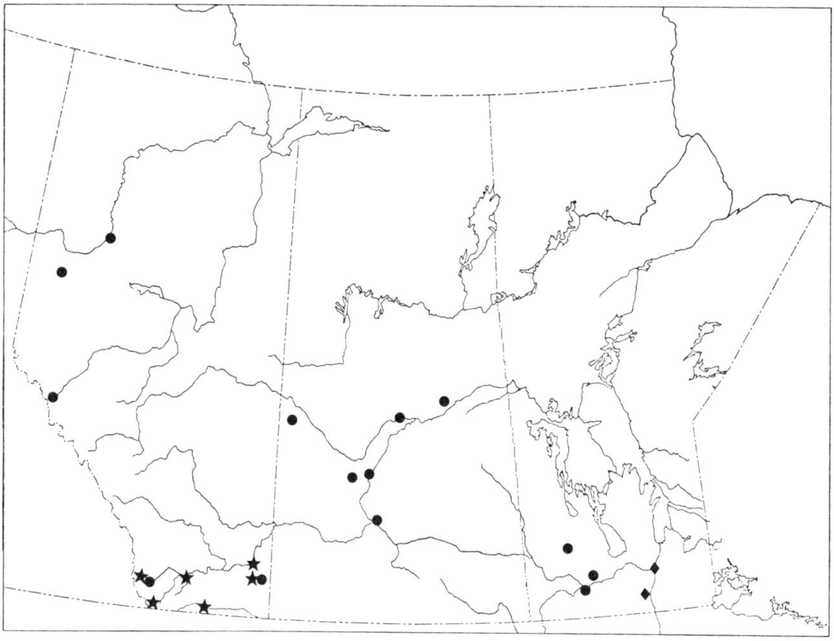

Map 106. Collection localities for *Plagiognathus flavidus* (●), *P. blatchleyi* (♦), and *P. shoshonea* (★).

Remarks. This species resembles *guttatipes* in size and color, but is separated from it by the black markings on the antennae and by the pale pubescence (Fig. 249).

Habitat. Collected on *Shepherdia canadensis* and *Elaeagnus commutata.*

Distribution. South Dakota; now known to occur in the Prairie Provinces (Map 106).

Plagiognathus blatchleyi Reuter

Map 106

Plagiognathus blatchleyi Reuter, 1912:61.

Length 4.20–4.55 mm; width 1.68–1.89 mm. Head greenish yellow, clypeus partly black. Antennae black. Rostrum 1.54–1.68 mm long. Pronotum green, basal half fuscous. Scutellum pale green. Hemelytra pale green, inner clavus and apical corium fuscous to black. Ventral surface and legs green; tibiae spotted with black.

Remarks. This species is distinguished by the greenish yellow head and the partly black clypeus, and by the fuscous basal half of the pronotum.

Habitat. Collected on *Ambrosia trifida*.

Distribution. Eastern USA, Quebec, Ontario; now known to occur in Manitoba (Map 106).

Plagiognathus shoshonea Knight

Map 106

Plagiognathus shoshonea Knight, 1964:142.

Length 5.04–5.46 mm; width 1.96–2.24 mm. Head black, vertex pale. Antennae black. Pronotum pale green to fuscous, calli black. Scutellum black. Hemelytra pale green, inner corium, embolium, and apical half of cuneus black. Ventral surface black; femora mostly yellowish spotted with black; tibiae pale green spotted with black, hind tibiae fuscous to black.

Remarks. This is the largest species in the Prairie Provinces, and is also distinguished by the color pattern on the hemelytra.

Habitat. Collected on *Geranium bicknellii*.

Distribution. Northwestern States, British Columbia; now known to occur in Alberta (Map 106).

Plagiognathus albonotatus Knight

Fig. 250; Map 107

Plagiognathus albonotatus Knight, 1923*b*:437.

Length 3.50–3.82 mm; width 1.47–1.75 mm. Head black, vertex often pale. First antennal segment black; second segment pale, base black. Rostrum 1.40–1.54 mm long. Pronotum black in male, central area pale green in female. Scutellum black, lateral margins often pale in female. Hemelytra black, base of corium and basal half of cuneus pale green. Ventral surface black; coxae mostly pale yellow; femora mostly black, bases and apices yellow; tibiae pale spotted with black.

Remarks. This species is distinguished by the mostly pale second antennal segment, by the color pattern on the hemelytra, and by the pale areas on the pronotum and scutellum in the female (Fig. 250).

Habitat. Collected on *Spiraea alba* and *Salix* spp.

Distribution. Northeastern and North Central States, Quebec; now known to occur in Manitoba and Saskatchewan (Map 107).

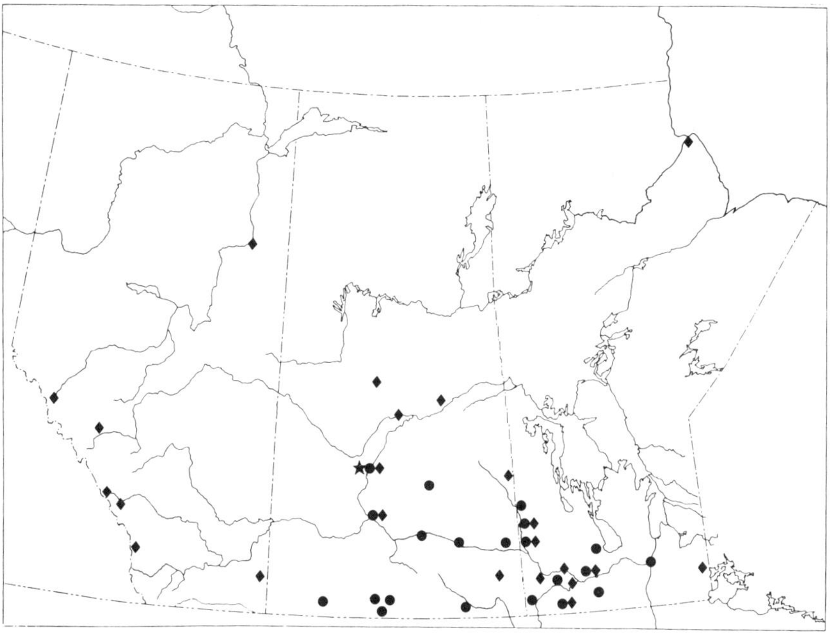

Map 107. Collection localities for *Plagiognathus albonotatus* (●), *P. alboradialis* (◆), and *P. negundinis* (★).

Plagiognathus alboradialis Knight

Fig. 251; Map 107

Plagiognathus alboradialis Knight, 1923*b*:439.

Length 4.55–4.90 mm; width 1.54–1.75 mm. Head black, vertex pale. First and second antennal segments black. Rostrum 1.68–1.82 mm long. Pronotum black, median area often pale. Scutellum black, side margins often pale. Hemelytra predominantly black in male with base of corium and cuneus pale; predominantly pale in female with few fuscous marks on clavus and corium. Ventral surface black in male, abdomen partly pale in female; legs pale and spotted with black; femora often fuscous.

Remarks. This species is distinguished by the uniformly pale cuneus (Fig. 251).

Habitat. Collected on *Salix* spp.

Distribution. Northeastern States, Quebec, Ontario; now known to occur in the Prairie Provinces (Map 107).

Plagiognathus negundinis Knight

Fig. 252; Map 107

Plagiognathus negundinis Knight, 1929*c*:263.

Length 3.78–4.06 mm; width 1.40–1.54 mm. Head brown, clypeus black, vertex pale. First and second antennal segments black, terminal segments fuscous. Rostrum 1.40–1.54 mm long. Pronotum black, basal angles often pale. Scutellum black. Hemelytra dark brown to black, basal half of corium and basal half of cuneus often pale. Ventral surface dark brown; legs yellow, femora and tibiae spotted with black.

Remarks. The pronotum and hemelytra are black in mature specimens and partly pale in teneral specimens (Fig. 252). This species resembles *obscurus*, but may be separated from it by the pale basal angles on the pronotum and by the yellow femora spotted with black.

Habitat. Collected on *Acer negundo*.

Distribution. North Central States; now known to occur in Saskatchewan (Map 107).

Plagiognathus medicagus Arrand

Fig. 253; Map 108

Plagiognathus medicagus Arrand, 1958:498.

Length 3.73–4.48 mm; width 1.34–1.89 mm. Head black, vertex yellow. First and second antennal segments black, terminal segments pale to fuscous. Rostrum 1.43–1.61 mm long. Pronotum and scutellum black. Hemelytra black, basal half and outer apex of corium, and base of cuneus white. Ventral surface black; legs generally pale; femora fuscous spotted with black, often with longitudinal lines on anterior and posterior surfaces; tibiae spotted with black.

Remarks. This species is distinguished by the black pronotum and scutellum, and by the arcuate margins of the hemelytra (Fig. 253). The abdomen in the female is always black.

Habitat. Collected on *Medicago sativa* and many other herbaceous plants.

Distribution. Northwestern States, British Columbia; Prairie Provinces (Map 108).

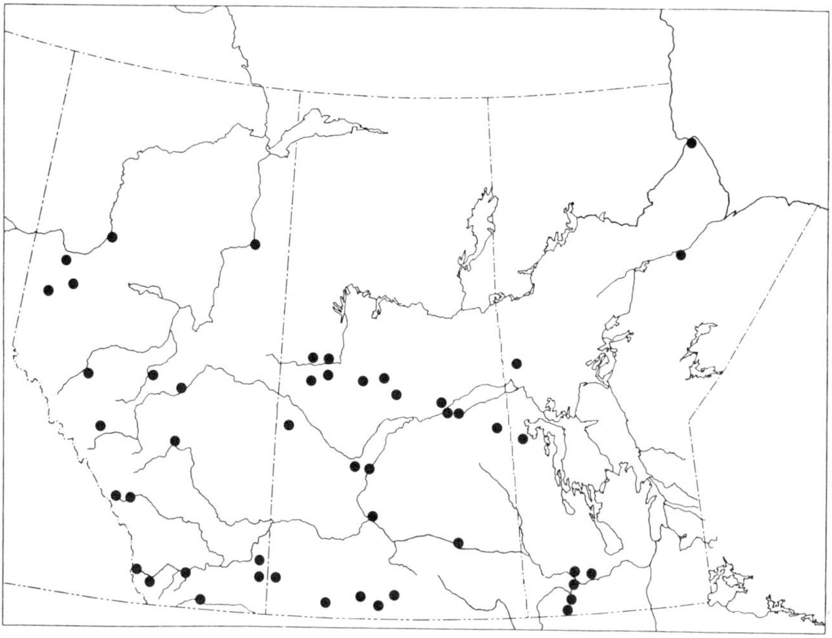

Map 108. Collection localities for *Plagiognathus medicagus.*

Plagiognathus obscurus Uhler

Fig. 254, Map 109

Plagiognathus obscurus Uhler, 1872:418.
Lygus brunneus, Provancher 1872:104.

Length 4.20–4.62 mm; width 1.54–1.68 mm. Head black, vertex yellowish. First and second antennal segments black, terminal segments pale. Rostrum 1.48–1.68 mm long. Pronotum black, often pale at middle. Scutellum black, side margins often pale. Hemelytra black, outer margins of clavus, basal half of corium, and basal half of cuneus white. Ventral surface black; legs pale, hind femora often fuscous; tibiae spotted with black.

Remarks. This species is similar to *medicagus* in size, color, and appearance, and the two are difficult to separate. The hemelytra in *obscurus* are almost parallel sided, and the pronotum and scutellum often have pale areas (Fig. 254). The female abdomen has pale markings.

Habitat. Collected on *Medicago sativa* and many other herbaceous plants.

Distribution. Widespread in USA, Eastern Canada, British Columbia; Prairie Provinces (Map 109).

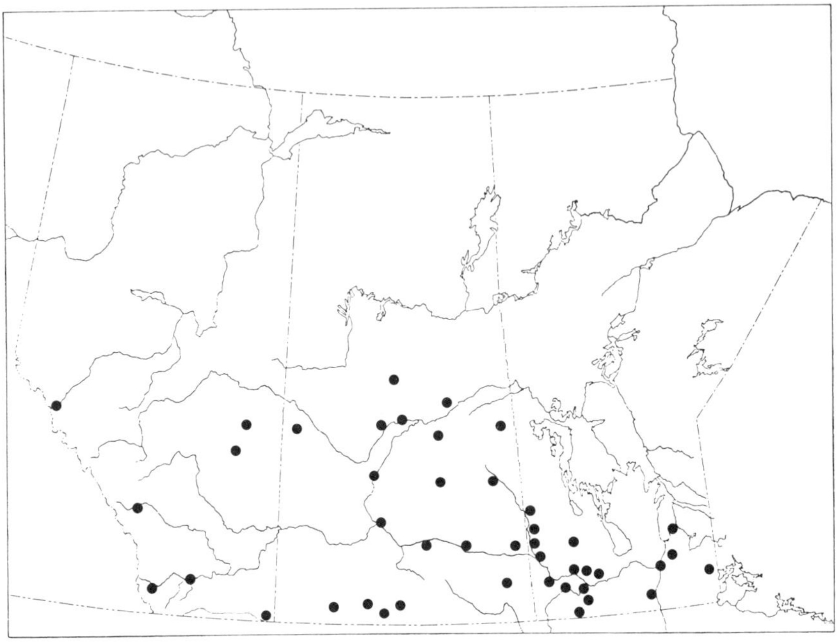

Map 109. Collection localities for *Plagiognathus obscurus*.

Plagiognathus fuscosus (Provancher)

Map 110

Lygus fuscosus Provancher, 1872:105.
Plagiognathus fuscosus: Provancher, 1877:153.
Plagiognathus politus var. *pallidicornis* Knight, 1923*b*:435.
Plagiognathus pallidicornis Knight, 1941*c*:28.
Plagiognathus fuscosus Kelton, 1968*a*:1074.

Length 3.50–3.78 mm; width 1.40–1.68 mm. Head black, vertex pale. First antennal segment black; second segment pale, base black; terminal segments pale. Rostrum 1.33–1.40 mm long. Pronotum and scutellum black. Hemelytra black, cuneus often pale at base. Ventral surface black; femora black; tibiae pale spotted with black.

Remarks. This species is distinguished by the partly pale second antennal segment and black femora, and by the partly pale cuneus.

Habitat. Collected on *Salix interior, Lathyrus venosus, Spiraea alba, Potentilla fruticosa*, and *Symphoricarpos occidentalis*.

Distribution. Northeastern States, Quebec, Ontario; now known to occur in the Prairie Provinces (Map 110).

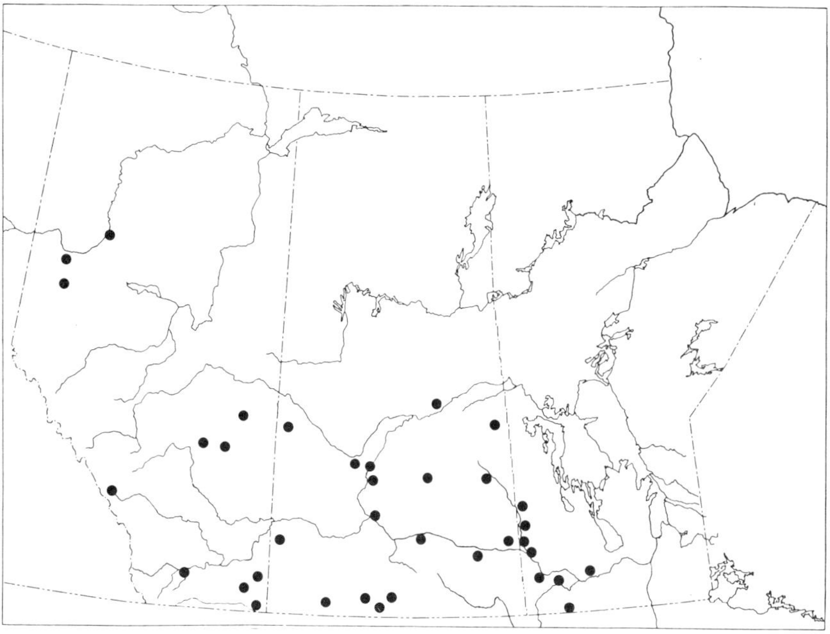

Map 110. Collection localities for *Plagiognathus fuscosus*.

Plagiognathus nigritus Knight

Map 111

Plagiognathus nigritus Knight, 1923*b*:441.

Length 4.06–4.76 mm; width 1.82–1.96 mm. Head black, vertex pale. First and second antennal segment black, terminal segments fuscous. Rostrum 1.33–1.40 mm long. Pronotum and hemelytra black. Ventral surface black; femora black, apices pale; tibiae pale spotted with black.

Remarks. This species is distinguished by the relatively large size, by the long second antennal segment, and by the mostly black femora.

Habitat. Collected on *Medicago sativa* and *Lathyrus palustris*.

Distribution. Northeastern States, Colorado, Quebec; now known to occur in the Prairie Provinces (Map 111).

Plagiognathus nigronitens Knight

Map 111

Plagiognathus nigronitens Knight, 1923*b*:435.

Length 2.94–3.08 mm; width 1.19–1.40 mm. Head black, vertex often pale. First and second antennal segments black, terminal segments fuscous.

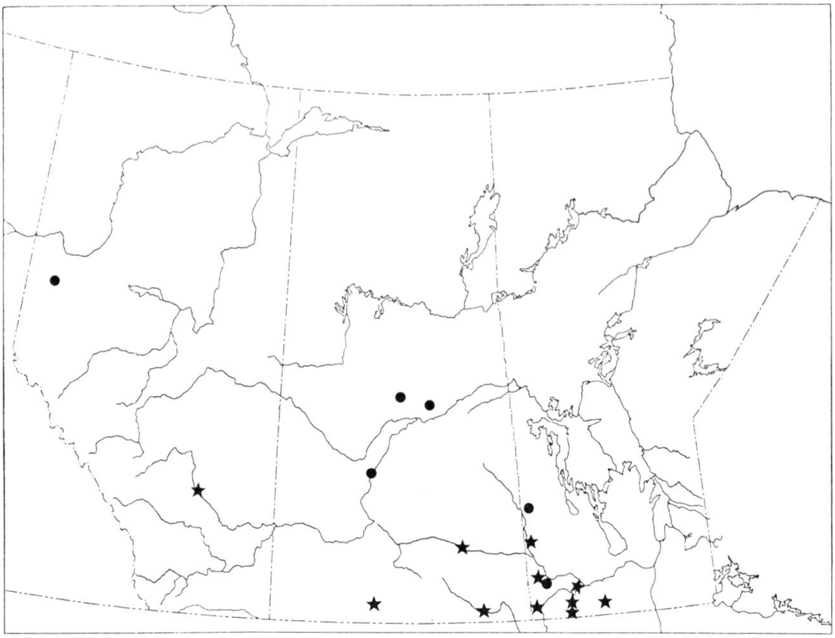

Map 111. Collection localities for *Plagiognathus nigritus* (●) and *P. nigronitens* (★).

Rostrum 0.84–0.98 mm long. Pronotum, scutellum, and hemelytra black. Ventral surface black; femora black, apices pale; tibiae pale spotted with black.

Remarks. This species is distinguished by the small size, by the black hemelytra, and by the short rostrum.

Habitat. Collected on *Helianthus* spp.

Distribution. Northeastern and North Central States, Mississippi, Quebec, Ontario; now known to occur in the Prairie Provinces (Map 111).

Plagiognathus davisi Knight

Map 112

Plagiognathus davisi Knight, 1923*b*:452.

Length 3.08–3.43 mm; width 1.19–1.40 mm. Head black, vertex pale. First antennal segment pale, base black; second segment pale, base black. Rostrum 1.40–1.54 mm long. Pronotum, scutellum, and hemelytra black. Ventral surface black; legs pale, tibiae spotted with black.

Remarks. This species is distinguished by the pale first antennal segment and by the pale legs.

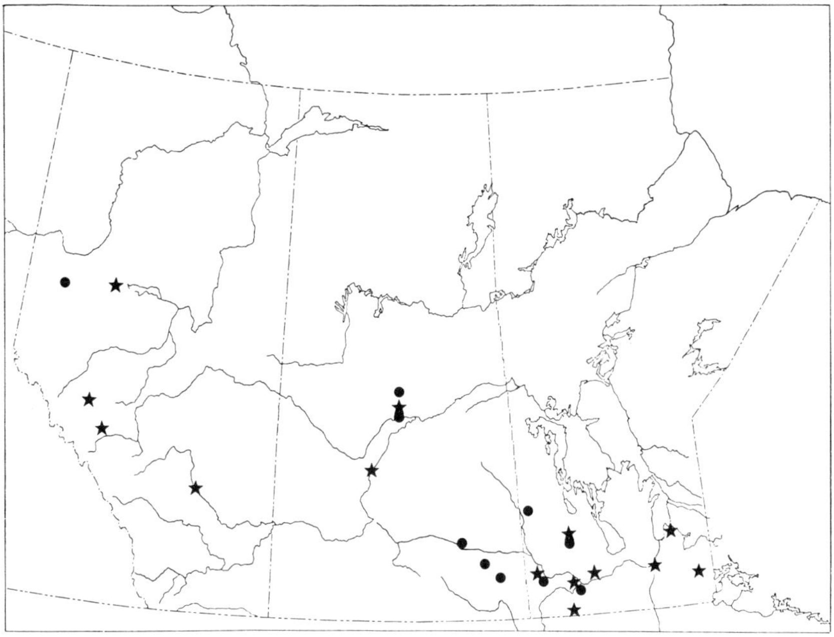

Map 112. Collection localities for *Plagiognathus davisi* (●) and *P. suffuscipennis* (★).

Habitat. Collected on *Lathyrus palustris.*

Distribution. New York, Iowa; now known to occur in the Prairie Provinces (Map 112).

Plagiognathus suffuscipennis Knight

Map 112

Plagiognathus suffuscipennis Knight, 1923*b*: 454.

Length 3.08–3.43 mm; width 1.26–1.40 mm. Head black, vertex often pale. First antennal segment black; second segment pale, base black. Rostrum 1.33–1.40 mm long. Pronotum and scutellum black. Hemelytra pale to light brown, contrasting with black pronotum and scutellum. Ventral surface black; legs generally pale spotted with black.

Remarks. This species is distinguished by the lighter color of the hemelytra which sharply contrasts with the black pronotum and scutellum.

Habitat. Collected on *Picea glauca.*

Distribution. Northeastern States, Quebec, Ontario; now known to occur in the Prairie Provinces (Map 112).

Plagiognathus dispar Knight

Map 113

Plagiognathus punctatipes var. *dispar* Knight, 1923*b*:451.
Plagiognathus crataegi Knight, 1929c:264.

Length 3.08–3.57 mm; width 1.26–1.40 mm. Head black, vertex often pale. First antennal segment black, second segment pale. Rostrum 1.19–1.26 mm long. Pronotum, scutellum, and hemelytra black. Ventral surface black; legs pale.

Remarks. This species is distinguished by the pale second antennal segment and by the pale hind femur.

Habitat. Collected on *Picea glauca.*

Distribution. Northeastern and North Central States, Quebec, Ontario; now known to occur in the Prairie Provinces (Map 113).

Plagiognathus laricicola Knight

Map 113

Plagiognathus laricicola Knight, 1923*b*:452.

Length 3.50–3.85 mm; width 1.40–1.54 mm. Head black, vertex pale. First and second antennal segments black. Rostrum 1.26–1.33 mm long.

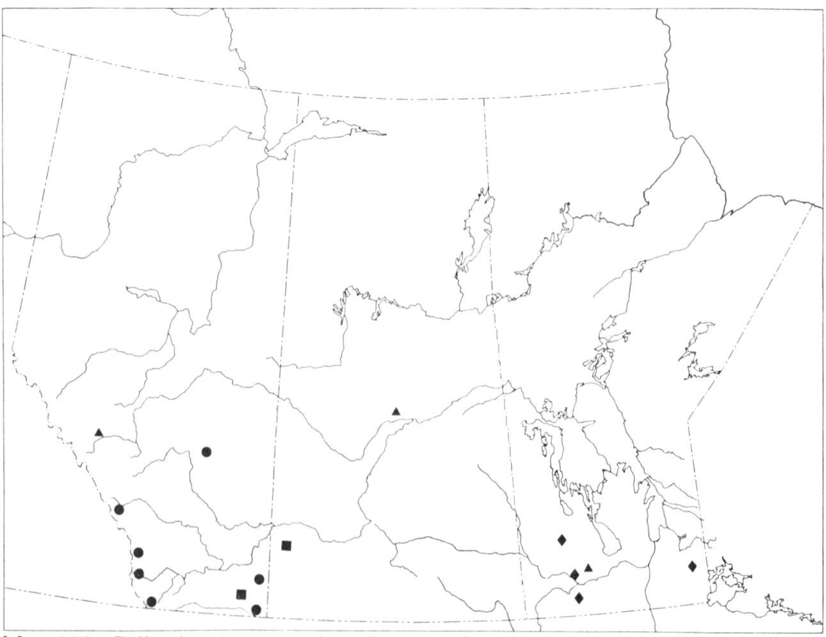

Map 113. Collection localities for *Plagiognathus dispar* (◆), *P. laricicola* (▲), *P. fuscipes* (●), and *P. annulatus* (■).

Pronotum, scutellum, and hemelytra black. Ventral surface black; legs pale, hind femur often fuscous.

Remarks. This species resembles *dispar*, but is slightly larger, with longer rostrum, and black second antennal segment.

Habitat. Collected on *Larix laricina.*

Distribution. Northeastern and North Central States; Alberta, now known to occur in Saskatchewan and Manitoba (Map 113).

Plagiognathus fuscipes Knight

Map 113

Plagiognathus fuscipes Knight, 1929*c*:268.

Length 3.64–3.85 mm; width 1.40–1.54 mm. Head black, vertex pale. First and second antennal segments black. Rostrum 1.68–1.75 mm long. Pronotum, scutellum, and hemelytra black. Ventral surface black; legs pale, femora often fuscous.

Remarks. This species is similar to *laricicola* in appearance. It is separated from it by the longer rostrum.

Habitat. Collected on *Potentilla fruticosa.*

Distribution. Colorado; now known to occur in the Prairie Provinces (Map 113).

Plagiognathus annulatus Uhler

Map 113

Plagiognathus annulatus Uhler, 1895:51.

Length 3.92–4.20 mm; width 1.40–1.54 mm. Head black, vertex pale. First antennal segment black; second segment black, middle area often pale. Rostrum 1.47–1.54 mm long. Pronotum, scutellum, and hemelytra black. Ventral surface black; femora pale, often with longitudinal black lines; tibiae spotted with black.

Remarks. This species is distinguished by the pale area on the second antennal segment.

Habitat. Unknown.

Distribution. Northeastern and North Central States, Quebec, Ontario; now known to occur in Alberta and Saskatchewan (Map 113).

Genus *Pseudatomoscelis* Poppius

Small, pale yellowish species, densely spotted with brown. Head oblique. Pronotum trapeziform, lateral margins sharply angulate. Hemelytra shiny; pubescence sericeous, often in clumps, intermixed with more erect, brownish hairs. Femora and tibiae spotted with brown.

One widely distributed species occurs in North America.

Pseudatomoscelis seriatus (Reuter)

Fig. 228; Map 114

Atomoscelis seriatus Reuter, 1875*c*:91.
Pseudatomoscelis seriatus: Knight, 1968*a*:55.

Length 2.80–3.01 mm; width 1.19–1.26 mm. Head pale yellowish green spotted with brown. First and second antennal segments spotted with black. Rostrum 1.40–1.47 mm long. Pronotum, scutellum, and hemelytra spotted with brown; spot at apex of corium and spot on membrane near tip of cuneus black. Ventral surface greenish yellow; legs green spotted with black.

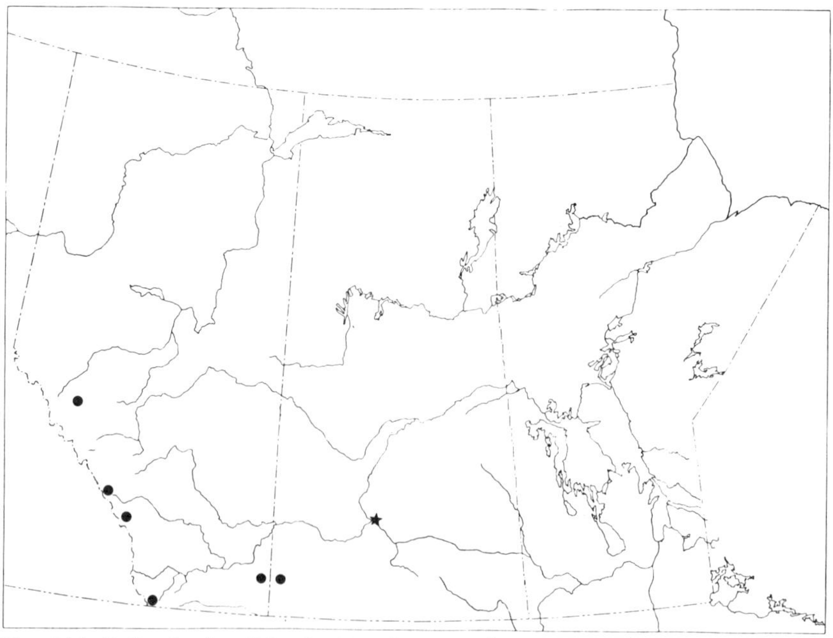

Map 114. Collection localities for *Pseudatomoscelis seriatus* (★) and *Lepidopsallus longirostris* (●).

Remarks. This species is distinguished by the black spots on the first and second antennal segments, and by numerous brown spots on the head and on the dorsal surface (Fig. 228).

Habitat. Unknown in the Prairie Provinces, elsewhere feeds on *Croton* spp. and cotton.

Distribution. Widespread in USA; now known to occur in Saskatchewan (Map 114).

Genus *Lepidopsallus* Knight

Ovate, black species with two types of pubescence. Head oblique, lora depressed. Second antennal segment shorter than head width. Pronotum trapeziform, lateral margins angulate. Hemelytra shiny; pubescence scaly, intermixed with slanting, black hairs. Femora without spots; tibiae with black spines and often with black spots at bases.

Approximately 17 species are known from North America, three occur in the Prairie Provinces.

Key to species of *Lepidopsallus*

1. Rostrum 1.68 mm or longer .... ***longirostris* Knight** (p. 331)
   Rostrum 1.54 mm or shorter 2
2. Hemelytra black, sometimes with pale or yellowish orange areas; rostrum 1.26 mm or shorter .... ***rubidus* (Uhler)** (p. 332)
   Hemelytra black; rostrum 1.36 mm or longer .... ***rostratus* Knight** (p. 333)

Lepidopsallus longirostris Knight

Fig. 242; Map 114

Lepidopsallus longirostris Knight, 1968*a*:54.

Length 3.36–3.50 mm; width 1.40–1.61 mm. Head black. First antennal segment black, other segments brown. Rostrum 1.68–1.82 mm long. Pronotum, scutellum, and hemelytra black (Fig. 242). Ventral surface black; femora black; tibiae pale.

Remarks. This species is distinguished by the long rostrum.

Habitat. Collected on *Pinus contorta.*

Distribution. West Central States; now known to occur in Alberta and Saskatchewan (Map 114).

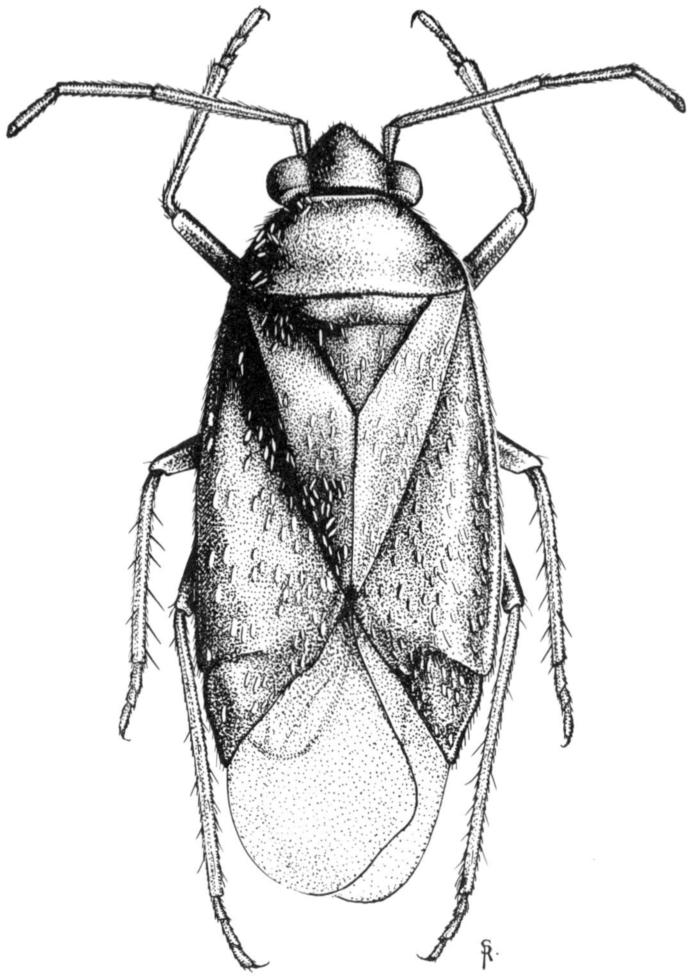

Fig. 242. *Lepidopsallus longirostris*

Lepidopsallus rubidus (Uhler)

Map 115

Sthenarus rubidus Uhler, 1895:41.
Lepidopsallus rubidus: Knight, 1923*b*:470.

Length 3.15–3.50 mm; width 1.47–1.61 mm. Head black, often light brown. Antennae black, second segment often orange. Rostrum 1.12–1.26 mm long. Pronotum, scutellum, and hemelytra black, yellowish orange in teneral specimens. Ventral surface black, often reddish orange; femora black or orange; tibiae yellow, often marked with black or orange.

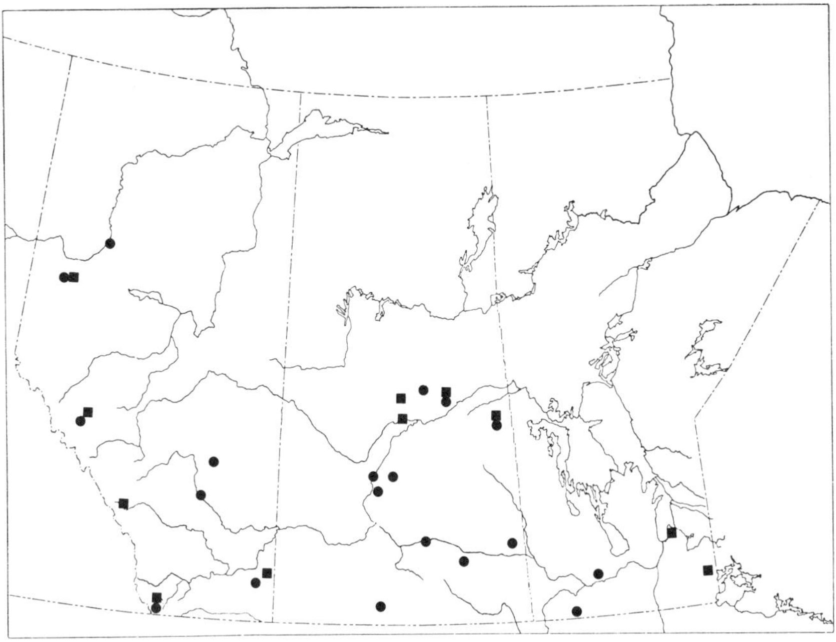

Map 115. Collection localities for *Lepidopsallus rubidus* (●) and *L. rostratus* (■).

Remarks. This species is strongly yellowish orange when teneral, gradually turning to black. It is separated from the other species by the short rostrum.

Habitat. Collected on *Salix candida.*

Distribution. Widespread in USA, Quebec, Ontario; now known to occur in the Prairie Provinces (Map 115).

Lepidopsallus rostratus Knight

Map 115

Lepidopsallus rostratus Knight, 1923*b*:470.

Length 2.80–3.08 mm; width 1.26–1.54 mm. Head black. First antennal segment black, second segment pale; terminal segments black. Rostrum 1.36–1.47 mm long. Pronotum, scutellum, and hemelytra black. Ventral surface black; femora black, tibiae pale.

Remarks. This species is smaller than *longirostris* and *rubidus*, and the rostrum is intermediate in length.

Habitat. Collected on *Pinus banksiana* and *P. contorta.*

Distribution. North Central States; now known to occur in the Prairie Provinces (Map 115).

Genus *Atractotomus* Fieber

Black, dimorphic species; males elongate, females oval. Head oblique. Pronotum trapeziform, shiny; lateral margins acutely angled. Hemelytra shiny; pubescence sericeous, appressed, intermixed with longer black, appressed hairs. Ventral surface black; legs brown.

There are 13 species of *Atractotomus* in North America, two of which are European introductions; one native species occurs in the Prairie Provinces.

Atractotomus hesperius (Uhler)

Fig. 4; Map 116

Dacota hesperia Uhler, 1872:413.
Atractotomus hesperius: Knight, 1931:36.

Length male 6.16–6.30 mm, female 4.69–4.90 mm; width 2.10–2.24 mm. Second antennal segment 1.68–1.96 mm long, slender in male, thickened in female. Rostrum 1.82–1.96 mm long. Pulvilli extending to tips of claws.

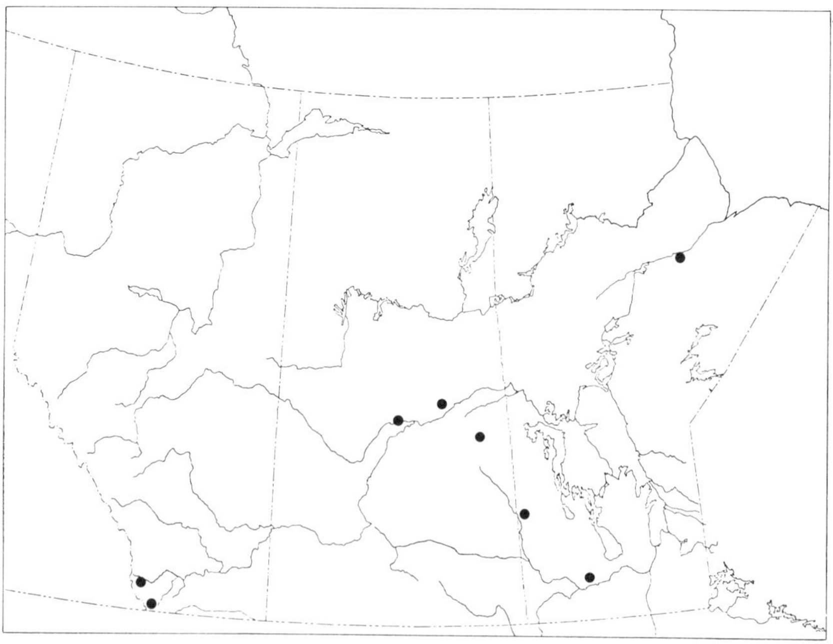

Map 116. Collection localities for *Atractotomus hesperius*.

Remarks. This species is distinguished by the black color, by the elongate males and the oval females, by the thickened second antennal segment in the female, and by the long pulvilli (Fig. 4).

Habitat. Collected on *Potentilla fruticosa.*

Distribution. Western States; now known to occur in the Prairie Provinces (Map 116).

Genus *Criocoris* Fieber

Oval, small, black species. Head oblique, clypeus prominent, jugum depressed. Pronotum trapeziform, lateral margins feebly angulate. Hemelytra shiny, faintly rugose; pubescence scaly, silvery, appressed, intermixed with slanting, black hairs.

One species is known from North America.

Criocoris saliens (Reuter)

Fig. 243; Map 117

Strongylotes saliens Reuter, 1875*c*:88.
Criocoris saliens: Van Duzee, 1914:29.

Length 2.59–2.80 mm; width 1.26–1.54 mm. Head tumid between eyes, first antennal segment enlarged in both sexes, second segment in male thickly fusiform, black; slender and pale at middle in female. Rostrum 0.98–1.05 mm long. Ventral surface and legs black, femora with scaly pubescence.

Remarks. This species is distinguished by the scaly pubescence on the femora, and by the fusiform second antennal segment in the male (Fig. 243).

Habitat. Collected on *Galium boreale.*

Distribution. Widespread in USA, Quebec, Ontario; now known to occur in the Prairie Provinces (Map 117).

Genus *Atomoscelis* Reuter

Small, oval, green species, spotted with yellowish brown. Head vertical, tumid on front. Pronotum trapeziform, calli inconspicuous, lateral margins angulate. Hemelytra pale green, shiny, with large light brown spots; pubescence sericeous, appressed, intermixed with longer slanting hairs. Ventral surface green, femora and tibiae spotted with black.

One species is known from North America.

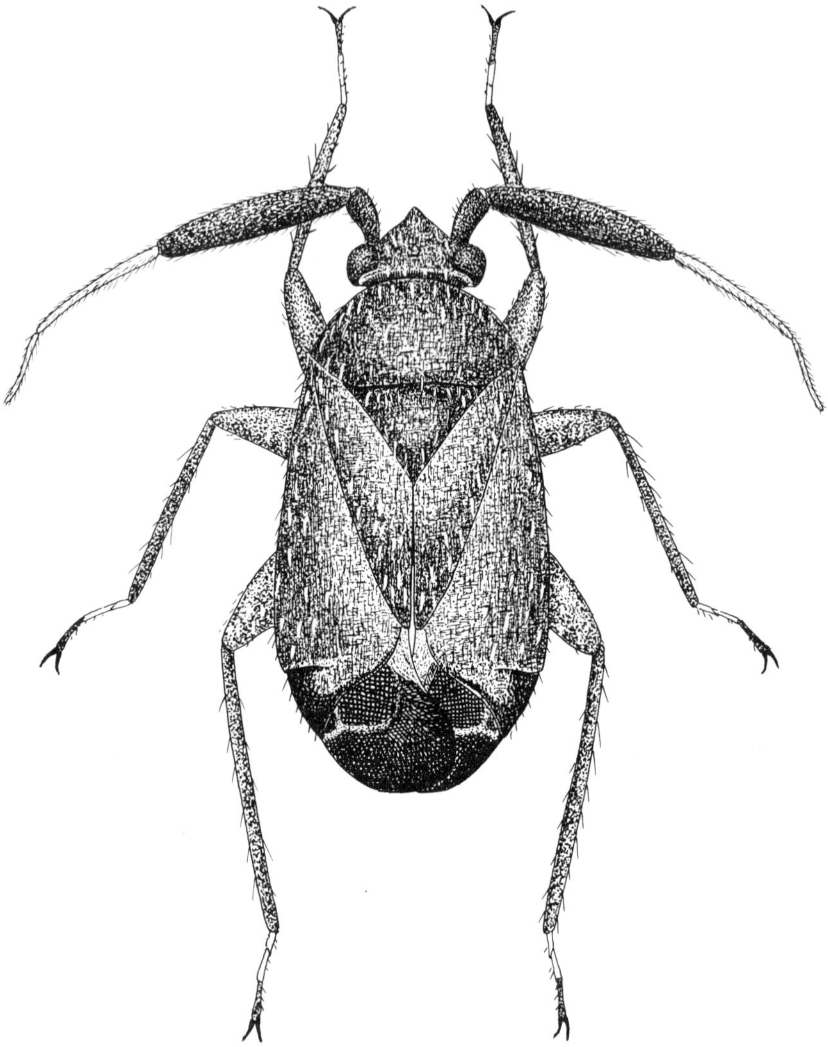

Fig. 243. *Criocoris saliens*

Atomoscelis modestus (Van Duzee)

Fig. 244; Map 117

Tuponia modesta Van Duzee, 1914:30.
Atomoscelis modestus: Van Duzee, 1917:414.

Length 2.38–2.66 mm; width 0.98–1.19 mm. Head, pronotum, and hemelytra pale green with large, faint yellowish brown spots. First antennal

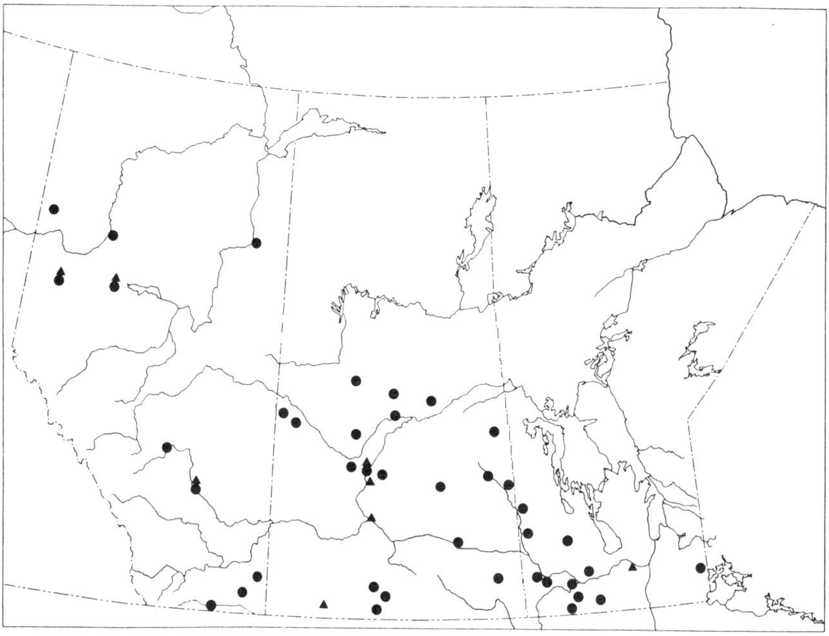

Map 117. Collection localities for *Criocoris saliens* (●) and *Atomoscelis modestus* (▲).

segment with black spot. Femora with black spots; tibiae with black spines and black spots.

Remarks. This species is distinguished by the small size and by the characters mentioned above (Fig. 244).

Habitat. Collected on *Chenopodium album, Suaeda depressa, Atriplex* spp., *Amaranthus* spp., and *Kochia scoparia.*

Distribution. Western States; now known to occur in the Prairie Provinces (Map 117).

Genus *Megalopsallus* Knight

Elongate-oval, green or brown species with two types of pubescence. Head almost vertical; lorum depressed; eyes large, protruding. Rostrum long, extending to hind coxae or beyond. Pronotum trapeziform, calli inconspicuous; lateral margins angulate. Hemelytra pruinose; pubescence sericeous, appressed, intermixed with longer, slanting, fuscous hairs. Femora and tibiae spotted with brown or red.

Nine species are known from North America, two occur in the Prairie Provinces.

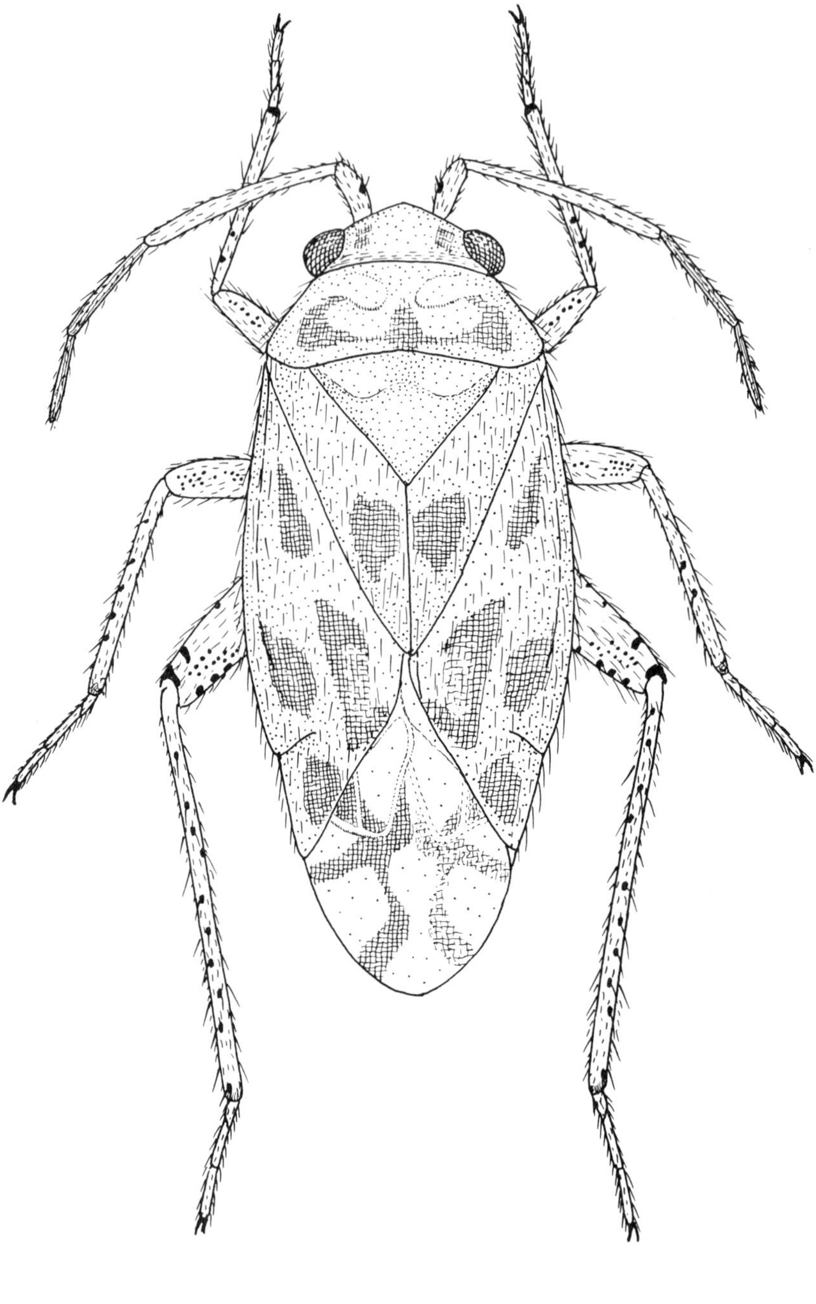

Fig. 244. *Atomoscelis modestus*

Key to species of *Megalopsallus*

1. Femora pale with numerous reddish dots; pronotum pale green (Fig. 245) .... ***rubropictipes* Knight** (p. 339)
   Femora reddish brown to black; pronotum brown to reddish brown (Fig. 246) .... ***femoralis* Kelton** (p. 341)

Megalopsallus rubropictipes Knight

Fig. 245; Map 118

Megalopsallus rubropictipes Knight, 1927:225.

Length 3.43–3.78 mm; width 1.24–1.40 mm. Head pale green, often marked with faint red spots. Antennae pale green. Rostrum 1.40–1.54 mm long. Pronotum pale green, calli yellowish, often with several reddish spots. Scutellum yellowish green. Hemelytra pale green, often with several fuscous spots. Wing membrane and cuneus often reduced in females. Femora pale with numerous reddish dots; tibiae pale green, spines pale with fuscous spots at bases.

Remarks. This species is distinguished by the pale green dorsal and ventral surfaces, and by the pale femora with reddish dots (Fig. 245).

Habitat. Collected on *Suaeda depressa* and *Atriplex* spp.

Distribution. Colorado; now known to occur in Alberta and Saskatchewan (Map 118).

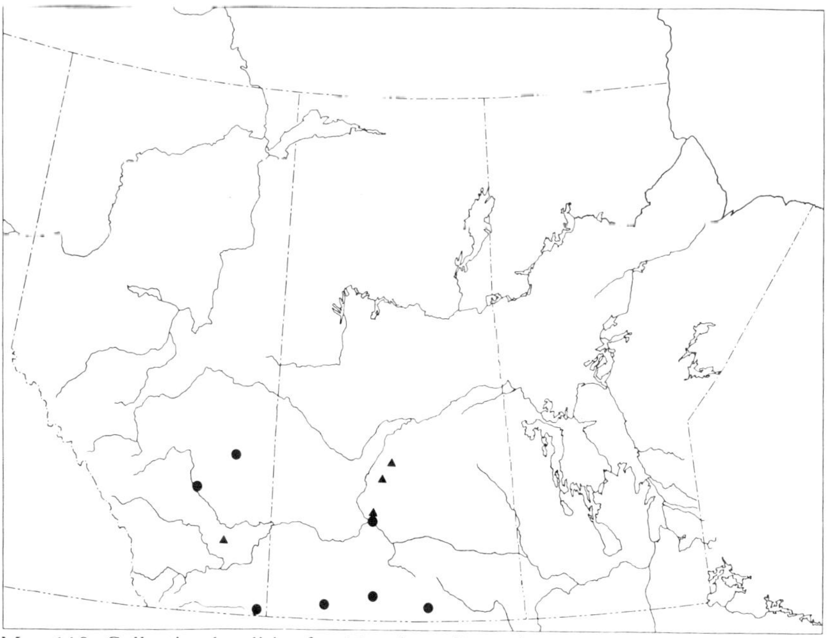

Map 118. Collection localities for *Megalopsallus rubropictipes* (●) and *M. femoralis* (▲).

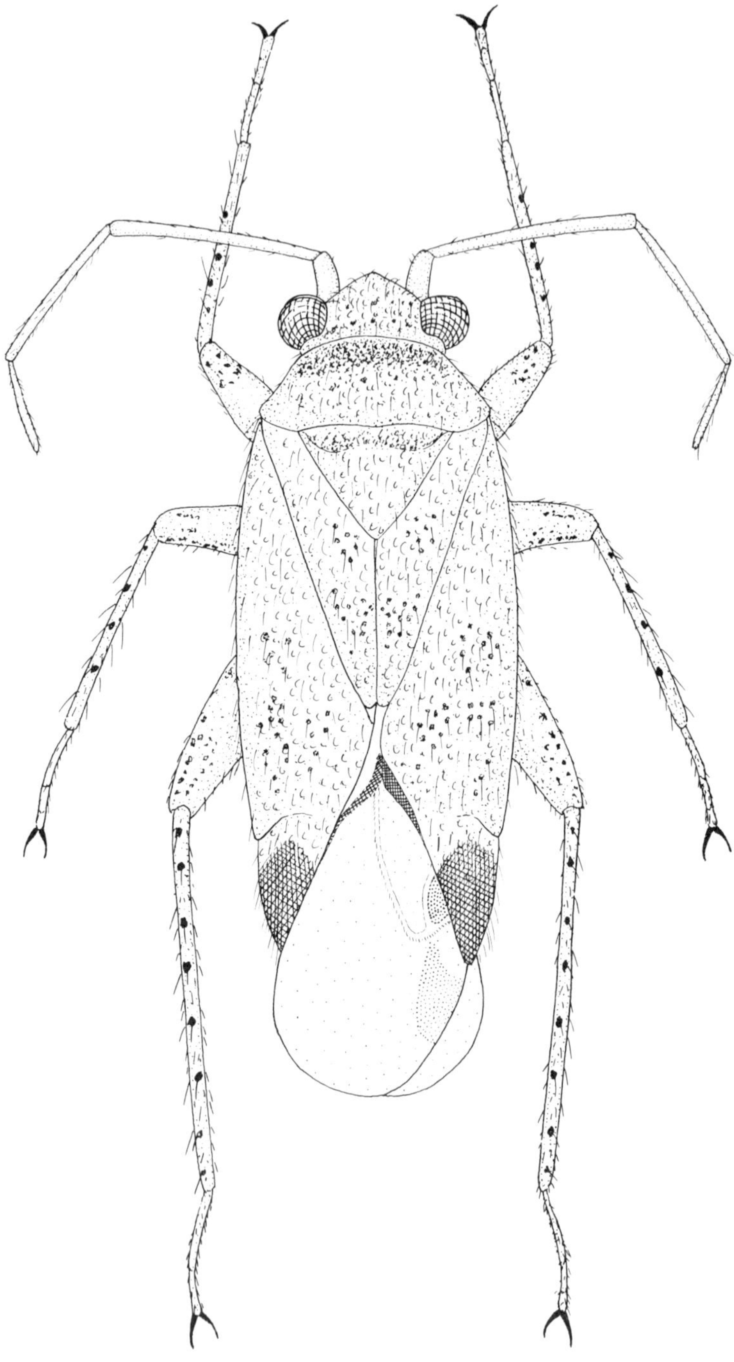

Fig. 245. *Megalopsallus rubropictipes*

Megalopsallus femoralis Kelton

Fig. 246; Map 118

Megalopsallus femoralis Kelton, 1980*a*:

Length 3.08–3.50 mm; width 1.26–1.47 mm. Head pale yellow or brownish marked with red; frons with transverse brownish bars. Antennae pale, first segment in darker specimens brown. Rostrum 1.36–1.54 mm long. Pronotum and scutellum fuscous to black. Hemelytra gray to dark fuscous. Ventral surface fuscous to black; femora reddish brown to black; tibiae pale, spines pale with black spots at bases.

Remarks. This species is darker than *rubropictipes*, especially on the head, pronotum, and scutellum. The femora are reddish brown (Fig. 246).

Habitat. Collected on *Salicornia rubra.*

Distribution. Alberta, Saskatchewan (Map 118).

Genus *Europiella* Reuter

Elongate-oval, shiny species with broad head, two types of pubescence, and spotted legs. Head vertical, short, broad; frons tumid, lora depressed; eyes protruding. Second antennal segment shorter than head width. Pronotum subrectangular, lateral margins angulate. Hemelytra black, gray, or fuscous; pubescence sericeous, intermixed with longer pale or fuscous hairs. Legs spotted with black.

Approximately 38 species are known from North America, three occur in the Prairie Provinces.

Key to species of *Europiella*

1. First antennal segment black (Fig. 247) .... ***yampae* Knight** (p. 341)
   First antennal segment pale or orange 2
2. Femora with prominent spots; spots on tibiae margined with reddish brown .... ***lycii* Knight** (p. 343)
   Femora with diffuse spots; spots on tibia black without reddish margins .... ***montanae* Knight** (p. 345)

Europiella yampae Knight

Fig 247; Map 119

Europiella yampae Knight, 1968*a*:43.

Length 3.08–3.64 mm; width 1.33–1.40 mm. Head black in male, pale in female. First antennal segment black, other segments fuscous to black in

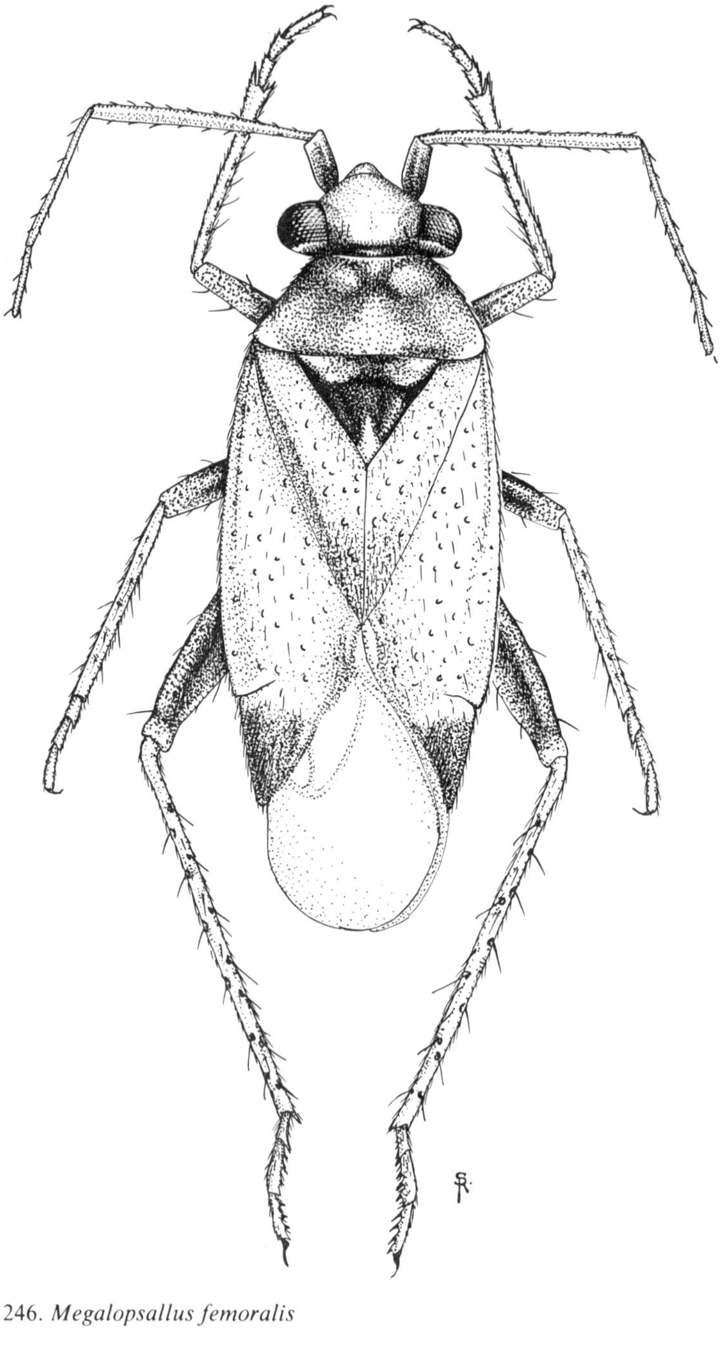

Fig. 246. *Megalopsallus femoralis*

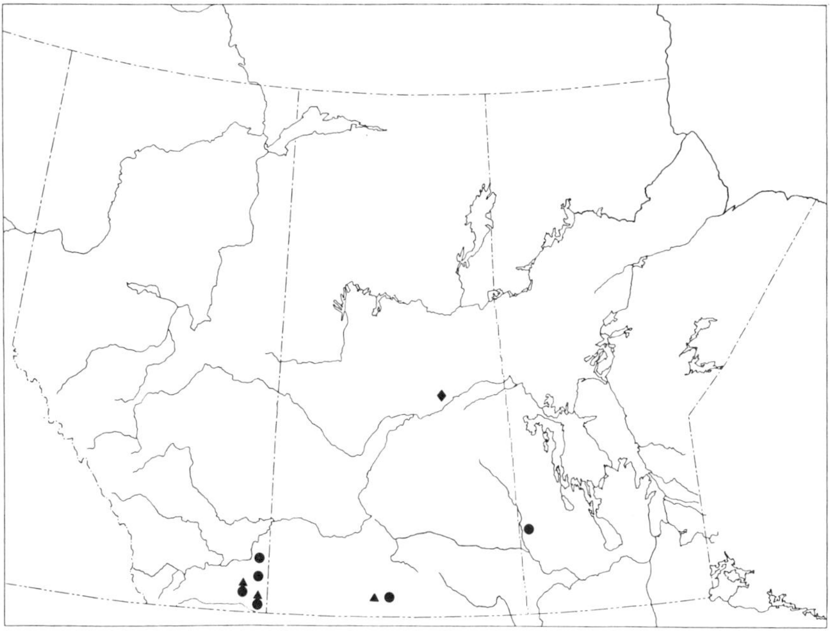

Map 119. Collection localities for *Europiella yampae* (●), *E. lycii* (▲), and *E. montanae* (◆).

male, pale in female. Pronotum pale green, calli black. Scutellum black in male, pale green in female. Hemelytra pale green, often grayish, cuneus darker. Legs pale spotted with black.

Remarks. This species is distinguished by the black first antennal segment (Fig. 247).

Habitat. Collected on *Artemisia cana* and *A. frigida.*

Distribution. Colorado; now known to occur in the Prairie Provinces (Map 119).

Europiella lycii Knight

Map 119

Europiella lycii Knight, 1968*a*:40.

Length 2.80–3.78 mm; width 1.26–1.40 mm. Head black. Antennae pale to yellowish orange. Pronotum and scutellum black. Hemelytra marked with fuscous. Femora brown to black spotted with red; tibiae pale, spots on tibiae margined with reddish brown.

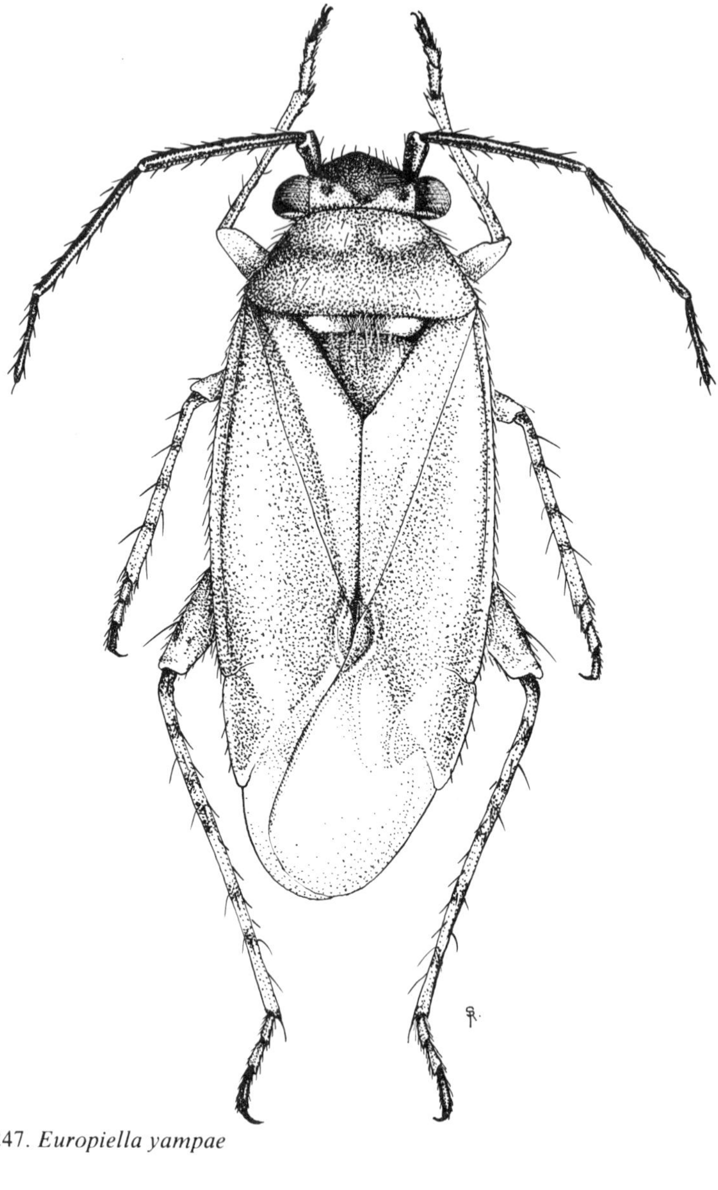

Fig. 247. *Europiella yampae*

Remarks. The males are elongate and the females are oval. Both have yellowish orange antennae and legs spotted with reddish brown.

Habitat. Collected on *Sarcobatus vermiculatus*.

Distribution. Nevada; now known to occur in Alberta and Saskatchewan (Map 119).

Europiella montanae Knight

Map 119

Europiella montanae Knight, 1968*a*:45.

Length 2.80–3.08 mm; width 1.33–1.40 mm. Head brown. Antennae pale yellow. Pronotum and scutellum brown. Hemelytra pallid to dusky. Ventral surface brown; femora brown without distinct spotting; tibia pale spotted with black.

Remarks. This species is distinguished by the pale yellow antennae, by the brown head and pronotum, and by the diffuse brown femora.

Habitat. Collected on *Pinus banksiana.*

Distribution. Montana; now known to occur in Saskatchewan (Map 119).

Genus *Psallus* Fieber

Elongate-oval species with two types of pubescence. Head oblique, frons tumid; eyes hairy. Second antennal segment longer than width of head. Pronotum trapeziform, wider than long; calli inconspicuous; lateral margins angulate. Hemelytra pruinose; pubescence sericeous, intermixed with longer simple hairs. Femora usually spotted with black; tibiae with black spines and black spots at bases.

Approximately 40 species are known from North America, five occur in the Prairie Provinces.

Key to species of *Psallus*

1. Head strongly marked with reddish tinge ...... ***alnicola* Douglas & Scott** (p. 345)
   Head not marked with reddish tinge 2
2. Species 4.10 mm or longer 3
   Species 4.09 mm or shorter 4
3. Femora black .... ***drakei* Knight** (p. 346)
   Femora pale to fuscous .... ***parshleyi* Knight** (p. 347)
4. Antennae black (Fig. 248) .... ***artemisicola* Knight** (p. 347)
   Antennae pale .... ***piceicola* Knight** (p. 348)

Psallus alnicola Douglas & Scott

Map 120

Psallus alnicola Douglas & Scott, 1865:414.

Length 3.36–3.50 mm; width 1.40–1.61 mm. Head yellowish brown, strongly marked with reddish tinge. Antennae pale brown. Rostrum

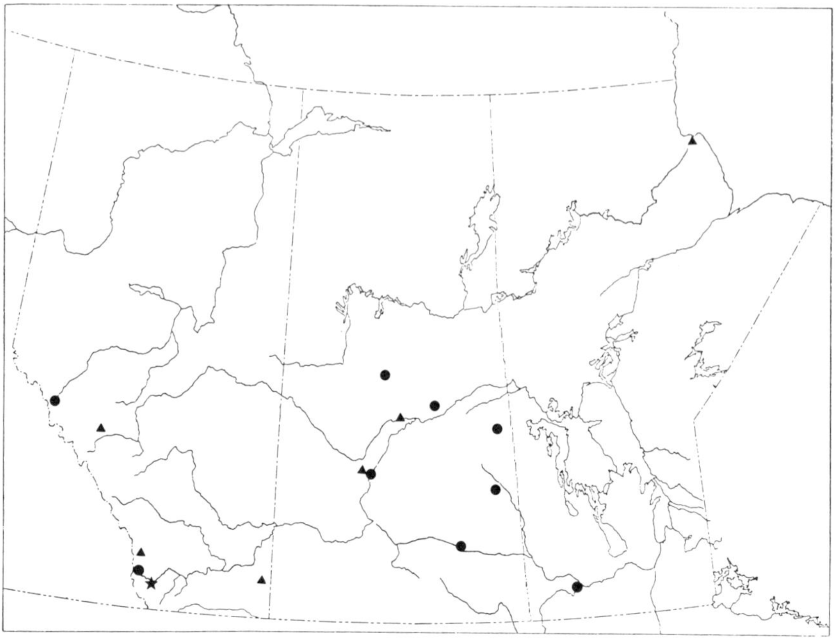

Map 120. Collection localities for *Psallus alnicola* (★), *P. drakei* (♦), and *P. parshleyi* (●).

1.33–1.40 mm long. Pronotum reddish brown, sometimes yellowish brown and irrorate with red. Scutellum and hemelytra reddish brown. Ventral surface reddish brown; legs yellowish brown spotted with reddish brown.

Remarks. This species is distinguished by the reddish brown color and by the reddish brown spots on the legs.

Habitat. Collected on *Alnus tenuifolia.*

Distribution. Holarctic; northern USA, Quebec, British Columbia; now known to occur in Alberta (Map 120).

Psallus drakei Knight

Map 120

Psallus drakei Knight, 1923*b*:464.

Length 4.20–4.74 mm; width 1.82–1.96 mm. Head black, area between eyes often yellowish. Antennae black. Rostrum 1.40–1.54 mm long. Pronotum, scutellum, and hemelytra black. Ventral surface black; femora black; tibiae pale spotted with black.

Remarks This species is distinguished by the large size and the black color.

Habitat. Collected on *Salix bebbiana* and *S. discolor.*

Distribution. New York, Colorado; Alberta, now known to occur in Saskatchewan and Manitoba (Map 120).

Psallus parshleyi Knight

Map 120

Psallus parshleyi Knight, 1923*b*:465.

Length 4.13–4.76 mm; width 1.68–1.82 mm. Head black, area between eyes often pale. Antennae black, terminal segments pale. Rostrum 1.40–1.54 mm long. Pronotum and scutellum black. Hemelytra mostly black, base of corium, embolium, and base of cuneus pale. Ventral surface black; legs mostly pale with femora often fuscous, and tibiae spotted with black.

Remarks. This species is distinguished by the pale areas on the hemelytra. It resembles certain species of *Plagiognathus.*

Habitat. Collected on *Betula glandulosa.*

Distribution. Northeastern States; now known to occur in the Prairie Provinces (Map 120).

Psallus artemisicola Knight

Fig. 248; Map 121

Psallus artemisicola Knight, 1964:149.

Length 3.01–3.36 mm; width 1.12–1.40 mm. Head black, spot next to eye on vertex pale. Antennae black. Rostrum 1.26–1.33 mm long. Pronotum and scutellum black. Hemelytra generally fuscous, costal margin and basal one half of cuneus pale. Ventral surface black; hind femora black, front and middle femora pale to fuscous.

Remarks. This species is distinguished by the small size and by the pale spots between the eyes (Fig. 248). It was previously reported as *Psallus bakeri* Bergroth.

Habitat. Collected on *Artemisia ludoviciana.*

Distribution. North Central States; now known to occur in the Prairie Provinces (Map 121).

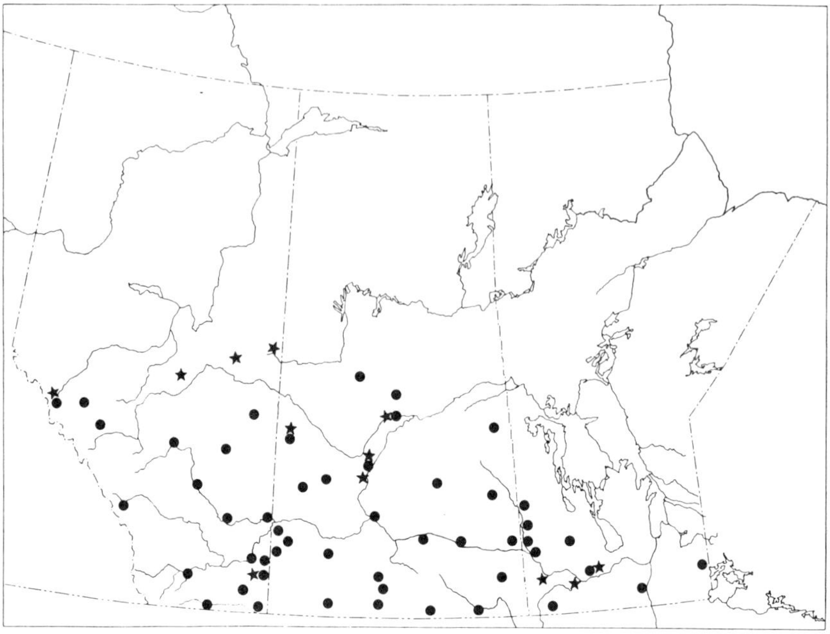

Map 121. Collection localities for *Psallus artemisicola* (●) and *P. piceicola* (★).

Psallus piceicola Knight

Map 121

Psallus piceicola Knight, 1923*b*:469.

Length 2.94–3.29 mm; width 1.26–1.40 mm. Head brown. Antennae pale. Rostrum 1.33–1.40 mm long. Pronotum and scutellum brownish black. Hemelytra brownish, base and tip of cuneus pale. Ventral surface and femora black; tibiae pale spotted with black.

Remarks. This species is distinguished by the small size and the overall blackish color, and by the long rostrum.

Habitat. Collected on *Picea glauca, Abies balsamea,* and *Pinus banksiana.*

Distribution. New York, North Central States, Quebec; now known to occur in the Prairie Provinces (Map 121).

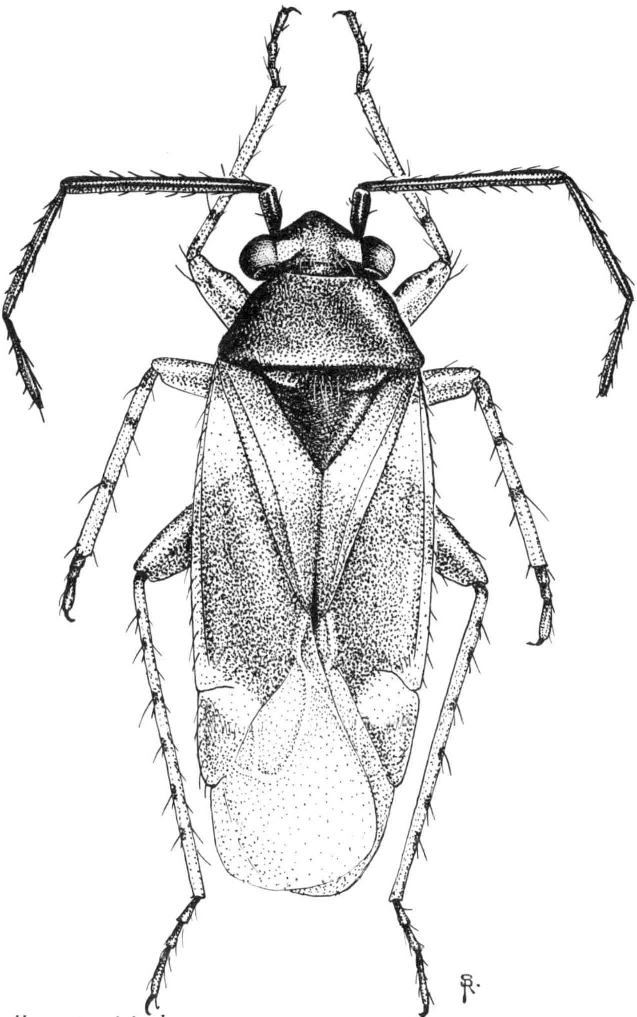

Fig. 248. *Psallus artemisicola*

Subfamily Deraeocorinae Douglas & Scott

This is a relatively small subfamily with only 21 genera and approximately 125 species reported from North America. In the Prairie Provinces the subfamily is represented by 3 tribes, 5 genera, and 19 species. Most of the species are predaceous.

The following are the subfamily characteristics: 1) straight, hairlike parempodia between the claws; 2) pulvilli absent; 3) pronotal collar distinct or modified into a hood; 4) hemelytra deeply punctate; and 5) male genitalia basically with membranous lobes, and flexible ductus semenis.

Key to tribes of Deraeocorinae

1. Pronotum with impressed line running from anterolateral angle to posterior margin of callus (Fig. 255) .... ***Clivinematini*** (p. 350)
   Pronotum without impressed line 2
2. Eyes situated forward from posterior margin of head (Fig. 256) .... ***Hyaliodini*** (p. 353)
   Eyes situated near posterior margin of head (Fig. 257) .... ***Deraeocorini*** (p. 355)

Tribe Clivinematini

In the Prairie Provinces the tribe is represented by two genera and two species.

Key to genera of Clivinematini

1. Pronotum and hemelytra pale; pronotum, scutellum, and pleura covered with white fluff .... ***Clivinema* Reuter** (p. 350)
   Pronotum and hemelytra dark reddish brown; white fluff absent .... ***Largidea* Van Duzee** (p. 351)

Genus *Clivinema* Reuter

Pronotum and scutellum covered in part with white fluff. Head vertical, frons convex projecting forward between eyes; eyes protruding, carina between them absent. Pronotum rugose, punctate, strongly pubescent, arched. Hemelytra shiny, finely pubescent. Ventral surface and legs brown.

Approximately 11 species are known from North America; one species occurs in the Prairie Provinces.

Clivinema villosum Reuter

Map 122

Clivinema villosa Reuter, 1875*c*:63.

Length 4.34 mm; width 1.61 mm. Head light brown marked with black. Rostrum 0.98 mm long. Second antennal segment almost as thick as first, terminal segments much thinner than second. Scutellum brown, hemelytra pale with cuneus slightly darker. Femora brown and spotted with black, tibia pale.

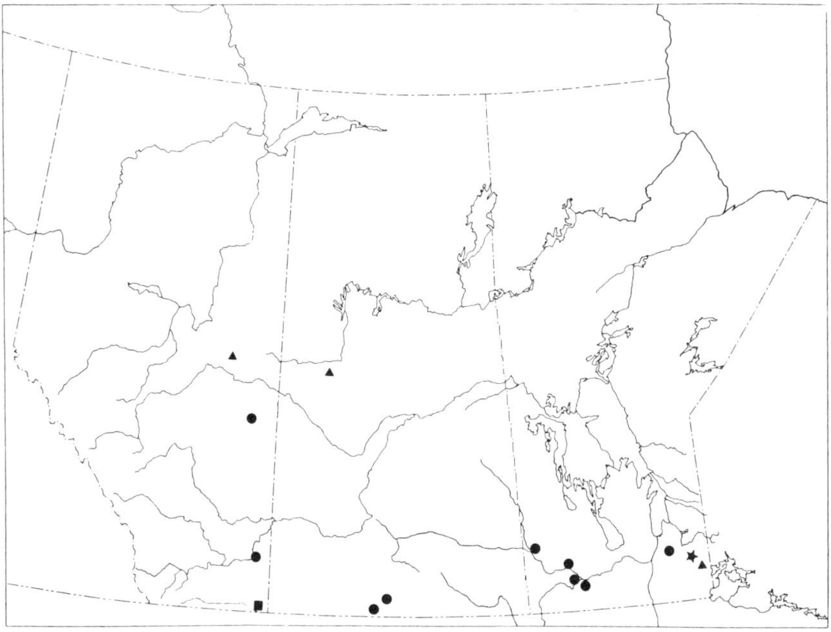

Map 122. Collection localities for *Clivinema villosum* (■), *Largidea shoshonea* (▲), *Hyaliodes harti* (●), and *Eustictus necopinus* (★).

Remarks. The mealy appearance would suggest that this species is predaceous, living in the moldy duff on trees infested with mealy bugs.

Habitat. Unknown, one specimen collected at light.

Distribution. Southwestern States; now known to occur in Alberta (Map 122).

Genus *Largidea* Van Duzee

Reddish brown. Head vertical, eyes protruding, carina between them absent. Pronotum coarsely punctuate. Hemelytra shiny, punctate, finely pubescent. First tarsal segment broad and scooped out below.

Ten species are known from North America, one occurs in the Prairie Provinces.

Largidea shoshonea Knight

Fig. 255; Map 122

Largidea shoshonea Knight, 1968*a*:87.

Length 4.76–5.25 mm; width 1.82–1.96 mm. Head reddish brown to black. Rostrum 1.33–1.49 mm long. Second antennal segment 1.26–1.40

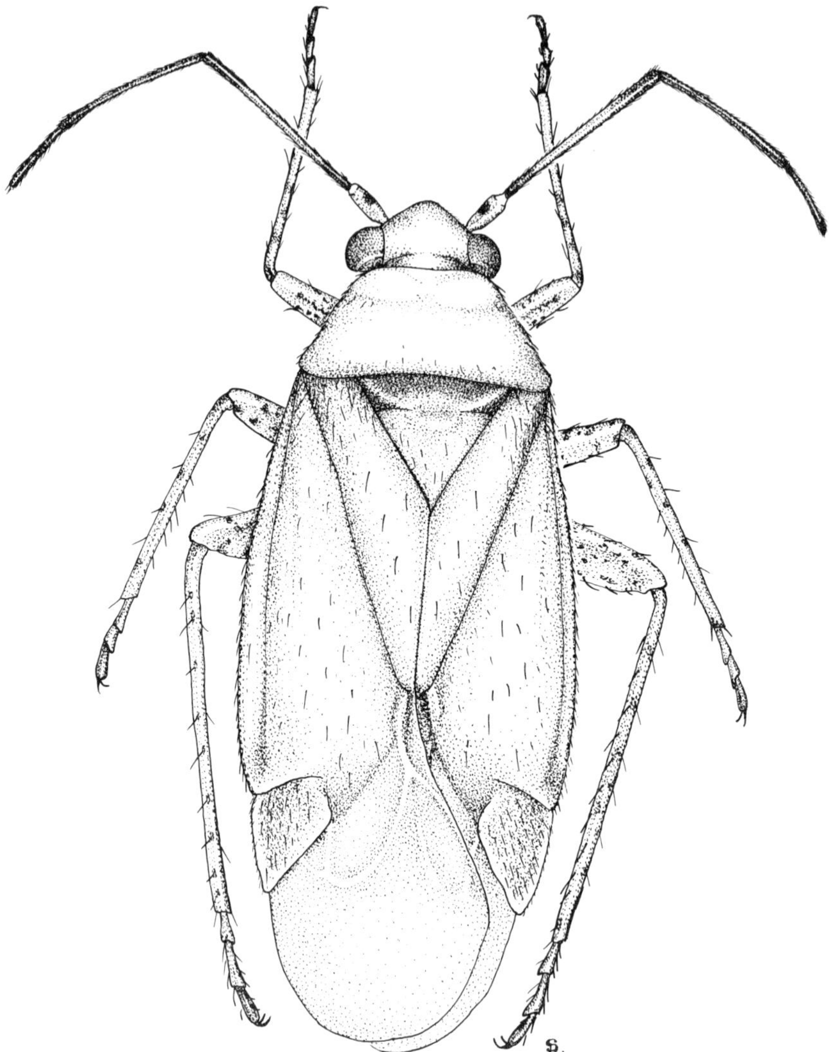

Fig. 249. *Plagiognathus flavidus*

mm long, as thick as first, terminal segments short and thin. Pronotum with lateral margins carinate. Scutellum not as punctate as pronotum. Hemelytra reddish brown. Legs reddish brown.

Remarks. This species is smaller than *L. grossa* Van Duzee and has shorter and thinner second antennal segment (Fig. 255).

Habitat. Collected on *Pinus banksiana*; probably preys on aphids.

Distribution. Northwestern States; now known to occur in the Prairie Provinces (Map 122).

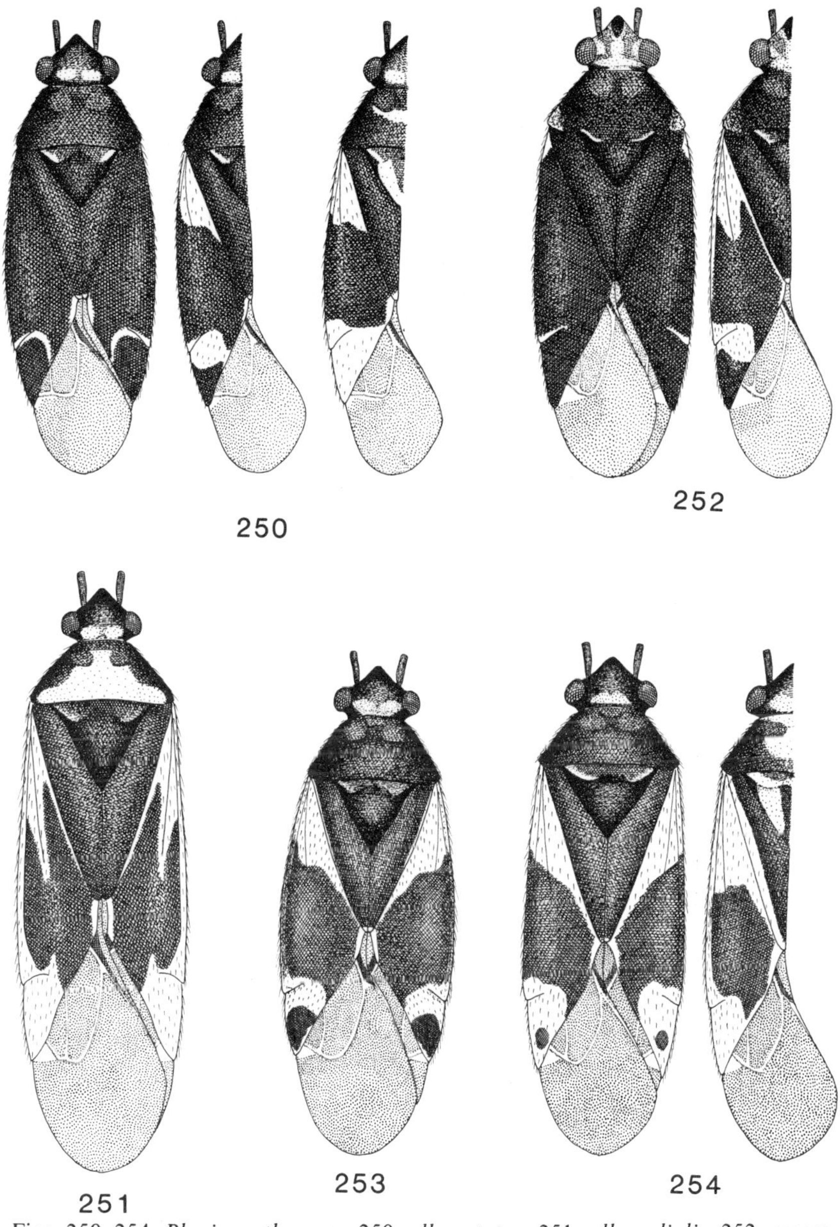

Figs. 250–254. *Plagiognathus* spp. 250, *albonotatus*; 251, *alboradialis*; 252, *negundinis*; 253, medicagus; 254, *obscurus*.

Tribe Hyaliodini

In the Prairie Provinces the tribe is represented by one genus and one species.

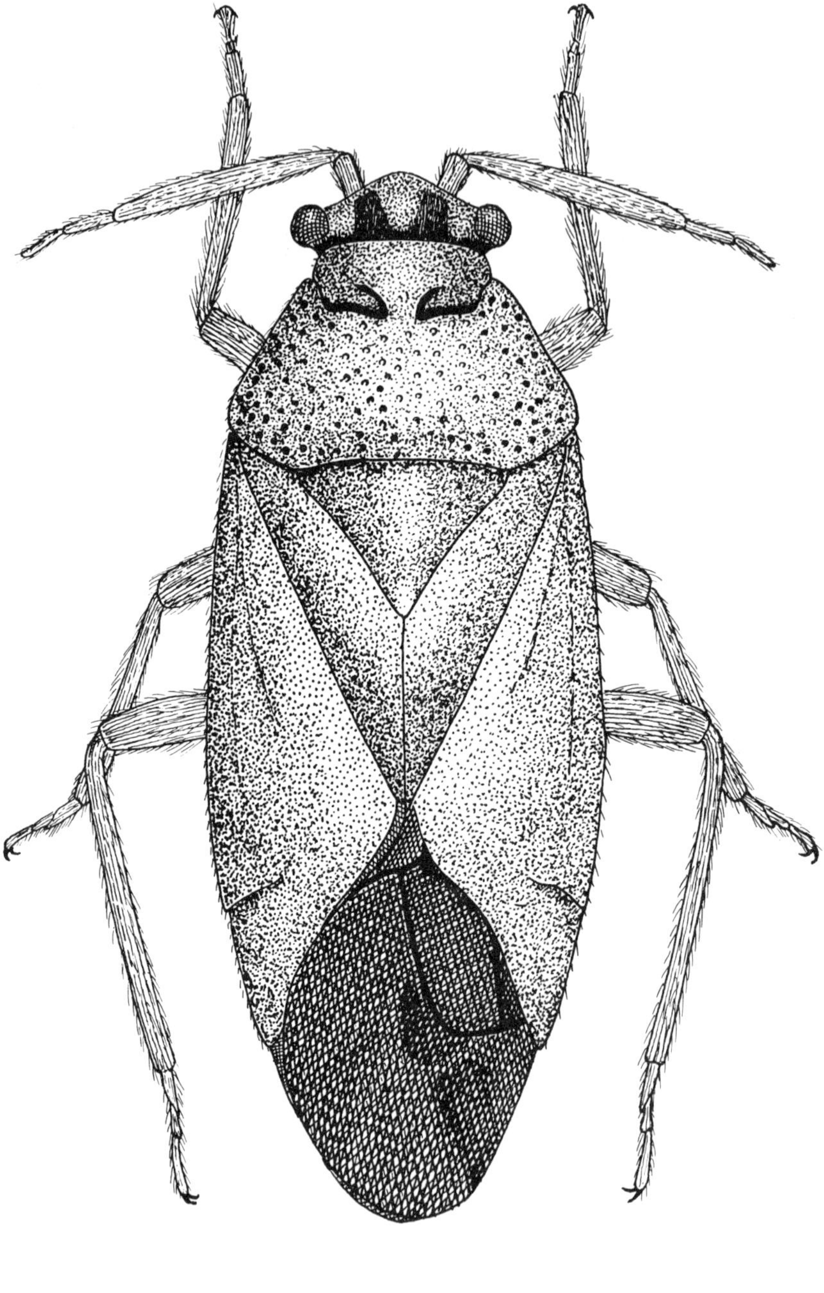

Fig. 255. *Largidea shoshonea*

Genus *Hyaliodes* Reuter

Elongate, shiny species. Head vertical, strongly narrowed behind to form a distinct neck, eyes large, almost covering side of head, carina between them absent. Pronotum subtriangular, strongly punctate, lateral margins rounded, collar present. Hemelytra smooth, shiny; wing membrane with one cell. Legs long, slender.

Only four species are known from North America, one is found in the Prairie Provinces.

Hyaliodes harti Knight

Fig. 256; Map 122

Hyaliodes harti Knight, 1941*c*:57.

Length 5.04–5.39 mm; width 1.54–1.82 mm. Head pale, area between eyes dark brown. Rostrum 1.26–1.40 mm long. First antennal segment mostly red, almost as long as or longer than rostrum. Pronotum mostly pale, collar and calli area black. Scutellum white. Hemelytra mostly clear, inner clavus black, apical corium red. Legs pale.

Remarks. This species is distinguished by its shape and color (Fig. 256).

Habitat. Collected on *Corylus americana* and *Quercus macrocarpa*; predaceous on plant lice.

Distribution. Eastern and North Central States; now known to occur in the Prairie Provinces (Map 122).

Tribe Deraeocorini

In the Prairie Provinces the tribe is represented by two genera and 16 species.

Key to genera of Deraeocorini

1. Frons transversely striate or grooved .... ***Eustictus* Reuter** (p. 357)
   Frons smooth, not striate or grooved .... ***Deraeocoris* Kirschbaum** (p. 357)

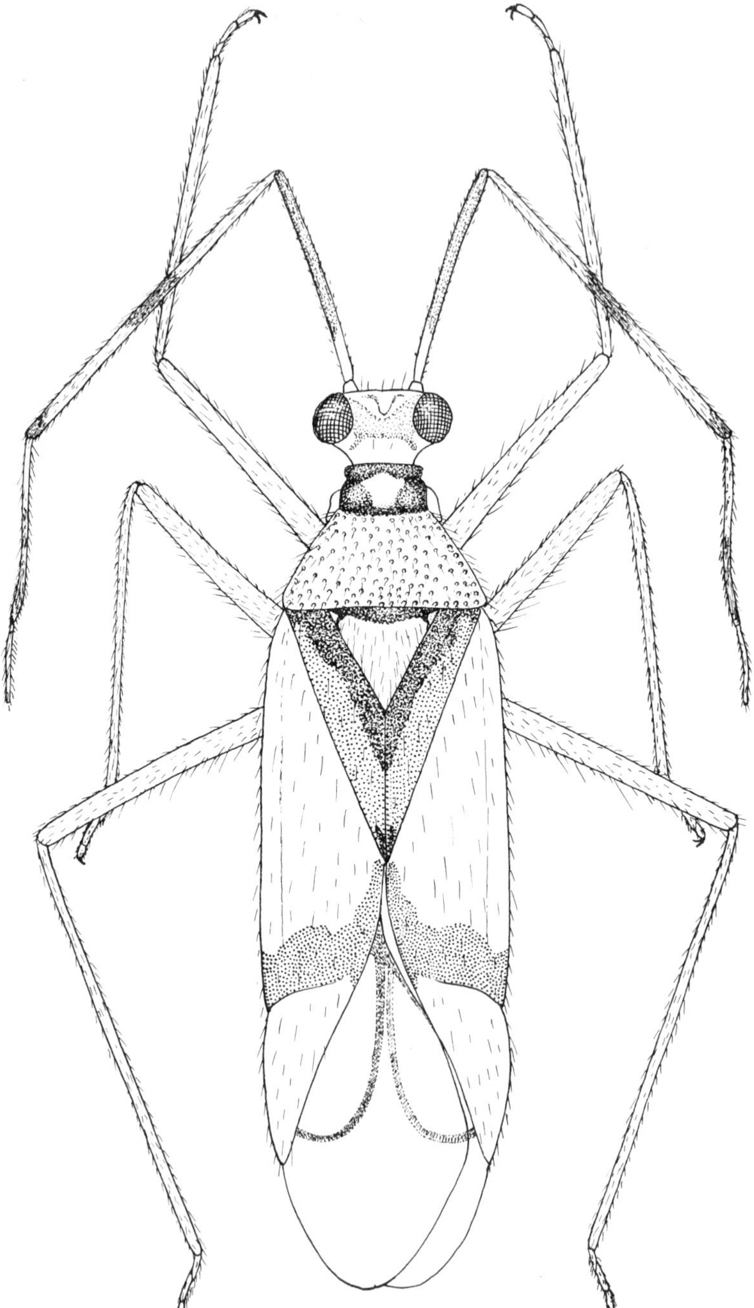

Fig. 256. *Hyaliodes harti*

Genus *Eustictus* Reuter

Shiny, glabrous species. Head vertical, eyes prominent, frons longitudinally sulcate. Antennae of nearly equal thickness throughout. Pronotum punctate, lateral margins carinate, collar present. Hemelytra shiny.

Approximately 22 species are known from North America, one occurs in the Prairie Provinces.

Eustictus necopinus Knight

Map 122

Eustictus necopinus Knight, 1923*b*:481.

Length 5.60–6.30 mm; width 2.10–2.66 mm. Head light yellowish brown, striations brown. Rostrum 2.80–3.01 mm long. Second antennal segment 2.38–2.66 mm long. Pronotum dark brown, often with small, pale area at middle, basal and lateral margins white. Scutellum black, apex pale, basal angles and median line often pale. Hemelytra pale, irregularly marked with dark brown. Legs pale, femora spotted apically, tibiae with three brown bands.

Remarks. This species is distinguished by the shiny and glabrous appearance, by the vertical head and prominent eyes, and by the striate frons.

Habitat. Unknown in the Prairie Provinces, collected at light, elsewhere collected on *Populus* spp.

Distribution. Northeastern States, Missouri, British Columbia, Ontario; now known to occur in Manitoba (Map 122).

Genus *Deraeocoris* Kirschbaum

Robust, shiny, strongly punctate species. Head oblique, frons smooth, eyes prominent, carina between them distinct or absent. Pronotum punctate, lateral margins carinate, collar present. Scutellum smooth or punctate. Hemelytra punctate, glabrous, or pubescent.

Approximately 56 species are known from the Nearctic region, 15 occur in the Prairie Provinces.

Key to species of *Deraeocoris*

1. Dorsum bright red with large, black markings on pronotum, clavus, and corium .... ***histrio* (Reuter)** (p. 358)
   Dorsum uniformly reddish brown or black without sharply defined black markings 2

2. Scutellum punctate 3
Scutellum smooth, impunctate 4
3. Cuneus and scutellum marked with red .... ***ornatus* Knight** (p. 359)
Cuneus and scutellum not marked with red .... ***brevis* (Uhler)** (p. 360)
4. Claws deeply cleft (Fig. 6) 5
Claws not deeply cleft (Fig. 7) 10
5. Hemelytra densely pubescent; pubescence long, erect; larger species; left clasper (Fig. 258) .... ***sayi* (Reuter)** (p. 361)
Hemelytra glabrous, pubescence short, sparse 6
6. Tibiae uniformly pale yellow; wing membrane clear 7
Tibiae banded with brown; wing membrane fuscous 8
7. Calli on pronotum margined with black; pronotum without pale median line; left clasper (Fig. 259) .... ***nitenatus* Knight** (p. 361)
Calli on pronotum black; pronotum with pale median line; left clasper (Fig. 260) .... ***quercicola* Knight** (p. 363)
8. Larger, dark brown species; punctures on pronotum obscured by black color between them; left clasper (Fig. 261) .... ***triannulipes* Knight** (p. 364)
Smaller, lighter species; black punctures on pronotum visible 9
9. Lateral margin of pronotum as seen from above pubescent; calli not completely black; left clasper (Fig. 262) .... ***fasciolus* Knight** (p. 364)
Lateral margin of pronotum not pubescent; calli black; left clasper (Fig. 263) .... ***aphidiphagus* Knight** (p. 366)
10. Hemelytra glabrous, pubescence short, sparse 11
Hemelytra strongly pubescent, pubescence long, dense 13
11. Head of male black, spot between eyes on top pale; left clasper (Fig. 264); female cuneus reddish, wing membrane reduced .... ***piceicola* Knight** (p. 367)
Head of male pale with black markings; female cuneus brown, wing membrane not reduced 12
12. Tibiae banded, strongly pilose; second antennal segment pilose; pronotum light brown; left clasper (Fig. 265) .... ***laricicola* Knight** (p. 368)
Tibiae not banded, not pilose; second antennal segment not pilose; pronotum pale; left clasper (Fig. 266) .... ***rubripes* Kelton** (p. 368)
13. Hind tibiae reddish brown; left clasper (Fig. 267) .... ***albigulus* Knight** (p. 369)
Hind tibiae brown with two pale bands 14
14. Pronotum black; rostrum 2.10 mm or less; left clasper (Fig. 268) .... ***diveni* Knight** (p. 371)
Pronotum pale; rostrum 2.10 mm or more; left clasper (Fig. 269) .... ***kennicotti* Knight** (p. 372)

Deraeocoris histrio (Reuter)

Map 123

Callicapsus histrio Reuter, 1875*c*:75.
Deracocoris histrio: Knight, 1921*b*:100.

Length 4.55–5.25 mm; width 2.38–2.52 mm. Head black, carina between eyes yellow. Rostrum 1.96–2.10 mm long. Pronotum red, two large spots behind calli black. Scutellum punctate, red or orange. Hemelytra red, basal clavus and large spot on apical half of corium black. Ventral surface black; legs black, tibiae with two bands of white.

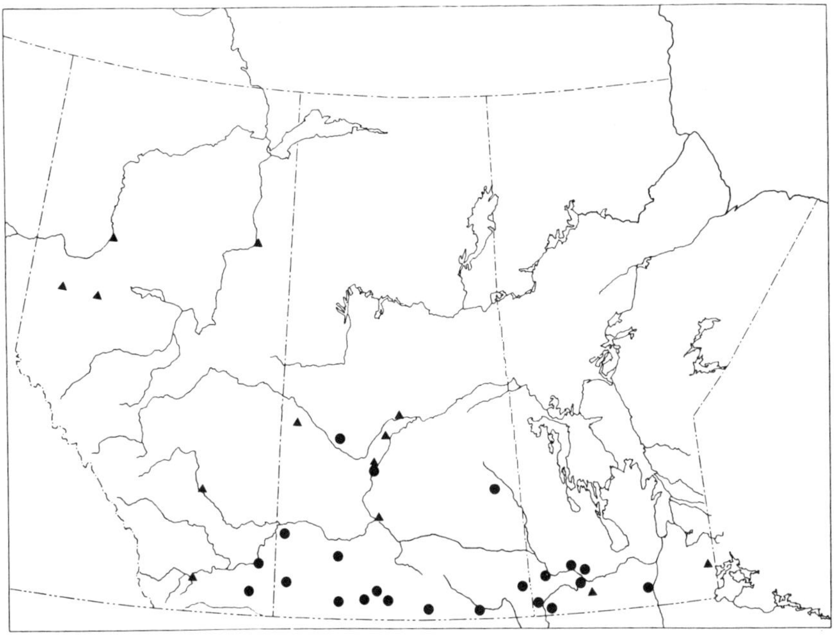

Map 123. Collection localities for *Deraeocoris histrio* (●) and *D. ornatus* (▲).

Remarks. The red and black color of this species is distinctive. The pronotum and hemelytra are glabrous.

Habitat. Collected on *Polygonum persicaria*; preys on plant lice. Often collected at light.

Distribution. Widespread in USA, Eastern Canada; Alberta, now known to occur in Saskatchewan and Manitoba (Map 123).

Deraeocoris ornatus Knight

Map 123

Deraeocoris (Camptobrochis) ornatus Knight, 1921*b*:99.
Deraeocoris ornatus: Blatchley, 1926:890.

Length 4.55–4.90 mm; width 2.10–2.38 mm. Head mostly pale, pattern on frons black. Pronotum mostly pale or light brown, calli and punctures black. Scutellum marked with red, side margins and apex pale. Hemelytra pale brown, cuneus marked with red, punctures black. Ventral surface mostly reddish brown or black; tibiae banded with white.

Remarks. This species is easily distinguished by the red cuneus.

Habitat. Collected on *Salix interior* and *Quercus macrocarpa*; probably predaceous on aphids.

Distribution. North Central States; now known to occur in the Prairie Provinces (Map 123).

Deraeocoris brevis (Uhler)

Map 124

Camptobrochis brevis Uhler, 1904:359.
Deraeocoris (Camptobrochis) brevis: Knight, 1921*b*:103.

Length 4.50–4.90 mm; width 1.96–2.24 mm. Head mostly pale, pattern on frons black. Rostrum 1.40–1.61 mm long. Pronotum light brown with calli and area behind black, or all black. Scutellum black, side margins and apex pale. Hemelytra brown to black, cuneus pale on basal half. Ventral surface dark brown to black; tibiae brown with pale bands.

Remarks. This species is variable in color, but the absence of red on the cuneus readily separates it from *ornatus*.

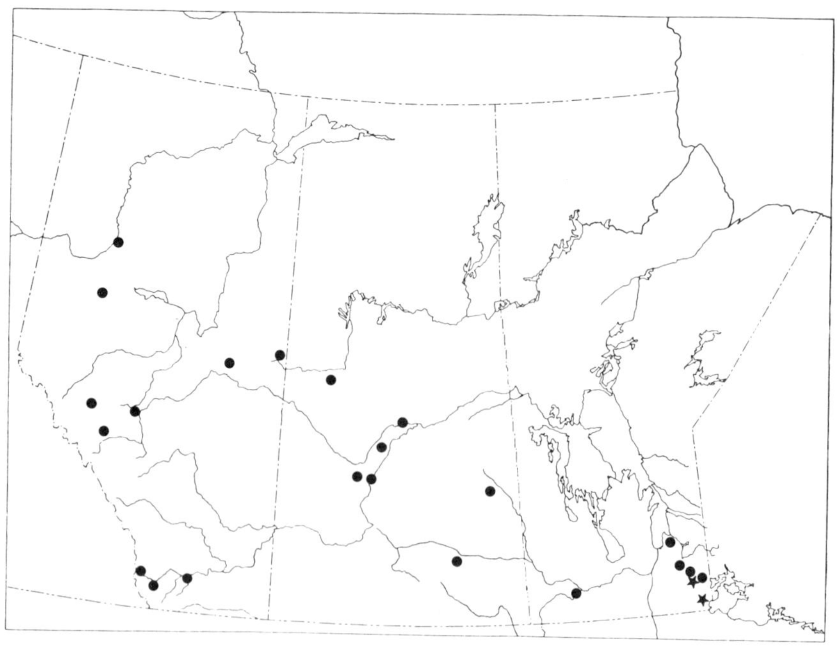

Map 124. Collection localities for *Deraeocoris brevis* (●) and *D. sayi* (★).

Habitat. Collected on *Acer negundo, Alnus* spp., *Juniperus communis, Pinus banksiana, P. contorta*, and *Abies balsamea*; preys on plant lice.

Distribution. Western USA, British Columbia; now known to occur in the Prairie Provinces (Map 124).

Deraeocoris sayi (Reuter)

Fig. 258; Map 124

Euarmosus sayi Reuter, 1875*c*:76.
Deraeocoris sayi: Knight, 1921*b*:175.

Length 6.44–7.70 mm; width 2.94–3.50 mm. Head black, basal half of clypeus, sides of frons, and along carina between eyes pale. Antennae strongly pilose. Rostrum 2.38–2.52 mm long. Pronotum black, median line at base and lateral margins often pale; strongly pilose. Scutellum black, median line near apex pale, in teneral specimens completely pale. Hemelytra black, costal margin often pale in immature specimens; strongly pilose. Ventral surface black; legs mostly black, tibiae with pale bands, strongly pilose.

Remarks. This is the largest species encountered, and is the only pubescent species in the Prairie Provinces with deeply cleft claws. The antennae and tibiae are strongly pilose. The genital claspers (Fig. 258) are distinctive.

Habitat. Collected on *Quercus macrocarpa*; probably preys on aphids.

Distribution. Widespread in USA; now known to occur in Manitoba (Map 124).

Deraeocoris nitenatus Knight

Fig. 259; Map 125

Deraeocoris nitenatus Knight, 1921*b*:141.

Length 5.60–6.30 mm; width 2.66–2.94 mm. Head mostly pale, frons with several transverse black bars. Rostrum 2.10–2.24 mm long. Pronotum light brown, calli margined with black, punctures black. Scutellum pale to light brown. Hemelytra light brown, corium often reddish, punctures black; glabrous. Wing membrane clear. Ventral surface light brown; legs pale.

Remarks. This species is distinguished by the pale tibiae and the genital claspers (Fig. 259).

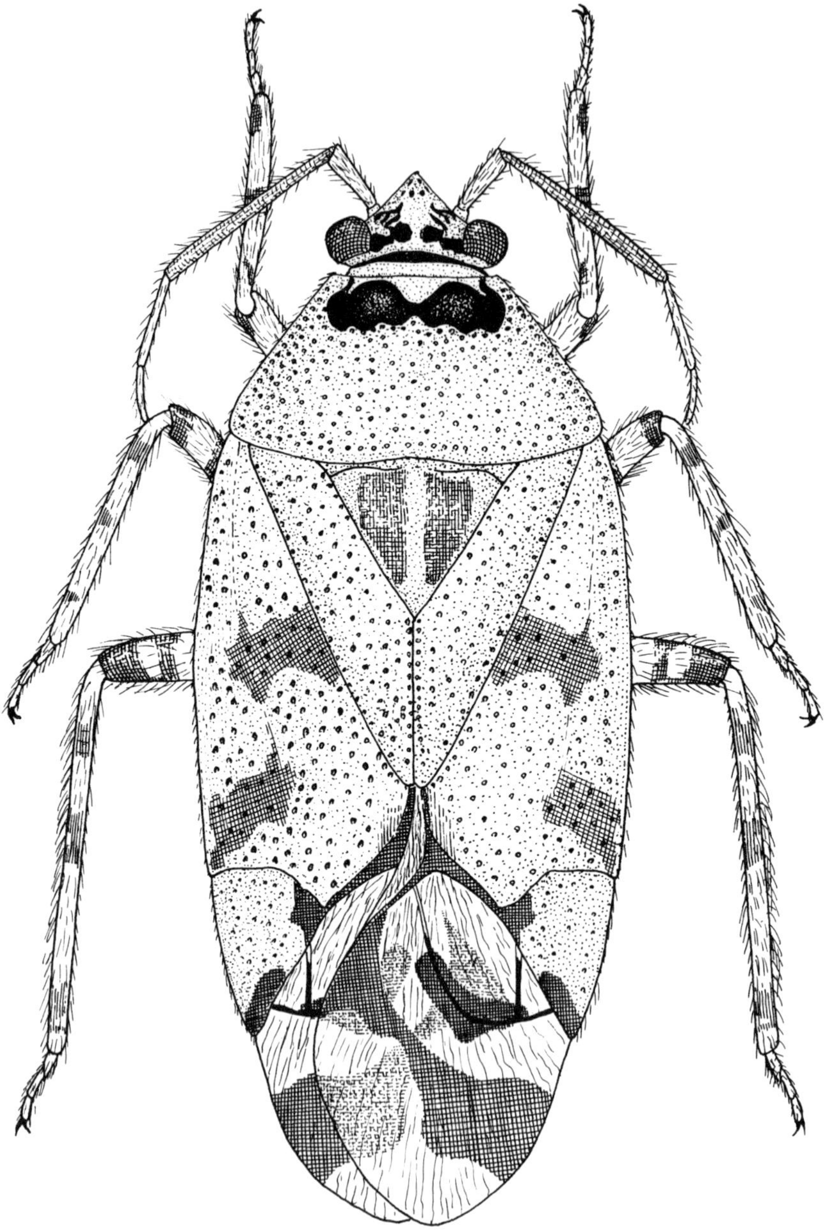

Fig. 257. *Deraeocoris aphidiphagus*

Habitat. Collected on *Ulmus americana* and *Quercus macrocarpa*; preys on plant lice.

Distribution. Eastern and North Central States, Quebec, Ontario; now known to occur in Saskatchewan (Map 125).

Deraeocoris quercicola Knight

Fig. 260; Map 125

Deraeocoris quercicola Knight, 1921*b*:138.

Length 5.88–6.30 mm; width 2.58–2.94 mm. Head mostly pale with few black markings. Rostrum 2.24–2.38 mm long. Pronotum light brown, calli and area behind usually black. Scutellum pale, longitudinal bar each side of middle brown. Hemelytra brown, corium on apical half often dark brown, cuneus with brown tip. Wing membrane clear. Ventral surface light brown; legs pale.

Remarks. This species has pale tibiae like *nitenatus*, but is readily separated from it by the longer rostrum and by the genital claspers (Fig. 260).

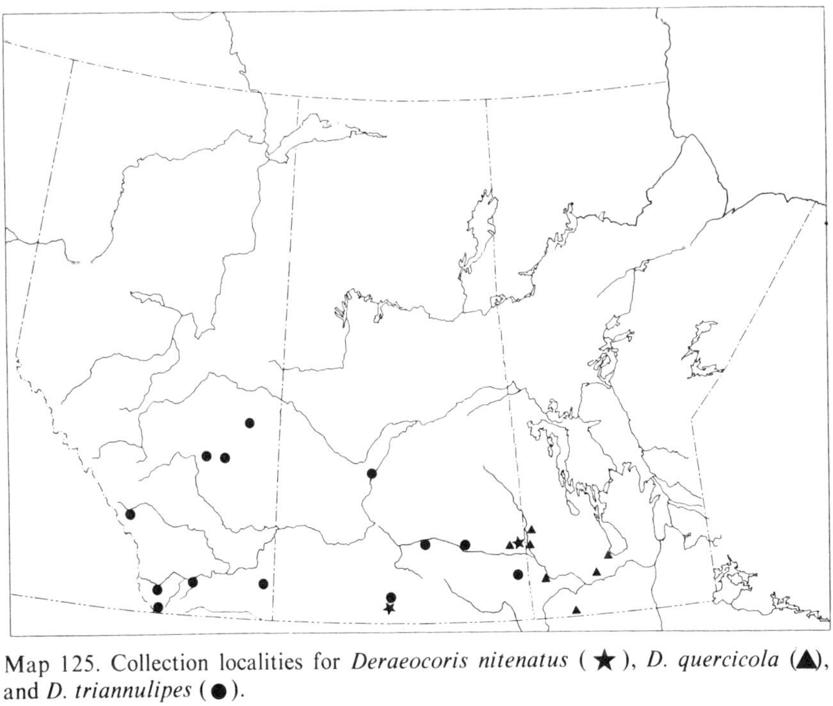

Map 125. Collection localities for *Deraeocoris nitenatus* (★), *D. quercicola* (▲), and *D. triannulipes* (●).

Habitat. Collected on *Quercus macrocarpa*; preys on aphids.

Distribution. Widespread in USA, Quebec, Ontario; now known to occur in Manitoba and Saskatchewan (Map 125).

Deraeocoris triannulipes Knight

Fig. 261; Map 125

Deraeocoris triannulipes Knight, 1921*b*:137.

Length 6.65–7.07 mm; width 3.08–3.22 mm. Head pale, clypeus and frons marked with black. Rostrum 2.10–2.24 mm long. Pronotum black, punctures obscured, basal and lateral margins pale. Scutellum black, basal angles and apex pale. Hemelytra dark brown, base of corium and middle of cuneus paler; wing membrane fuscous. Ventral surface dark brown; legs pale; femora biannulate with brown near apex, and tibia triannulate with brown.

Remarks. This species is darker than *nitenatus* and *quercicola* and the wing membrane is fuscous. The genital claspers (Fig. 261) are distinctive.

Habitat. Collected on *Populus tremuloides* and *Alnus* spp.; preys on aphids associated with the host plants.

Distribution. North Central States, Manitoba; now known to occur in Saskatchewan and Alberta (Map 125).

Deraeocoris fasciolus Knight

Fig. 262; Map 126

Deraeocoris fasciolus Knight, 1921*b*:123.

Length 6.30–7.00 mm; width 2.94–3.22 mm. Head pale with several black bars. Rostrum 2.17–2.31 mm long. Pronotum light brown, calli marked with black bars. Scutellum pale, wide bar each side of middle black. Hemelytra brown, corium with black spot at middle, cuneus pale, tip brown; wing membrane with rounded fuscous spot near apex. Ventral surface reddish brown, legs mostly pale banded with brown.

Remarks. This species is similar to *quericiola* in appearance, but is distinguished by the banded tibiae, by the fuscous spot on wing membrane, and by the genital claspers (Fig. 262).

Habitat. Collected on *Betula* spp., *Salix* spp., *Alnus* spp., *Quercus macrocarpa, Prunus pensylvanica, Rubus* spp., and *Acer negundo*; preys on aphids associated with the host plants.

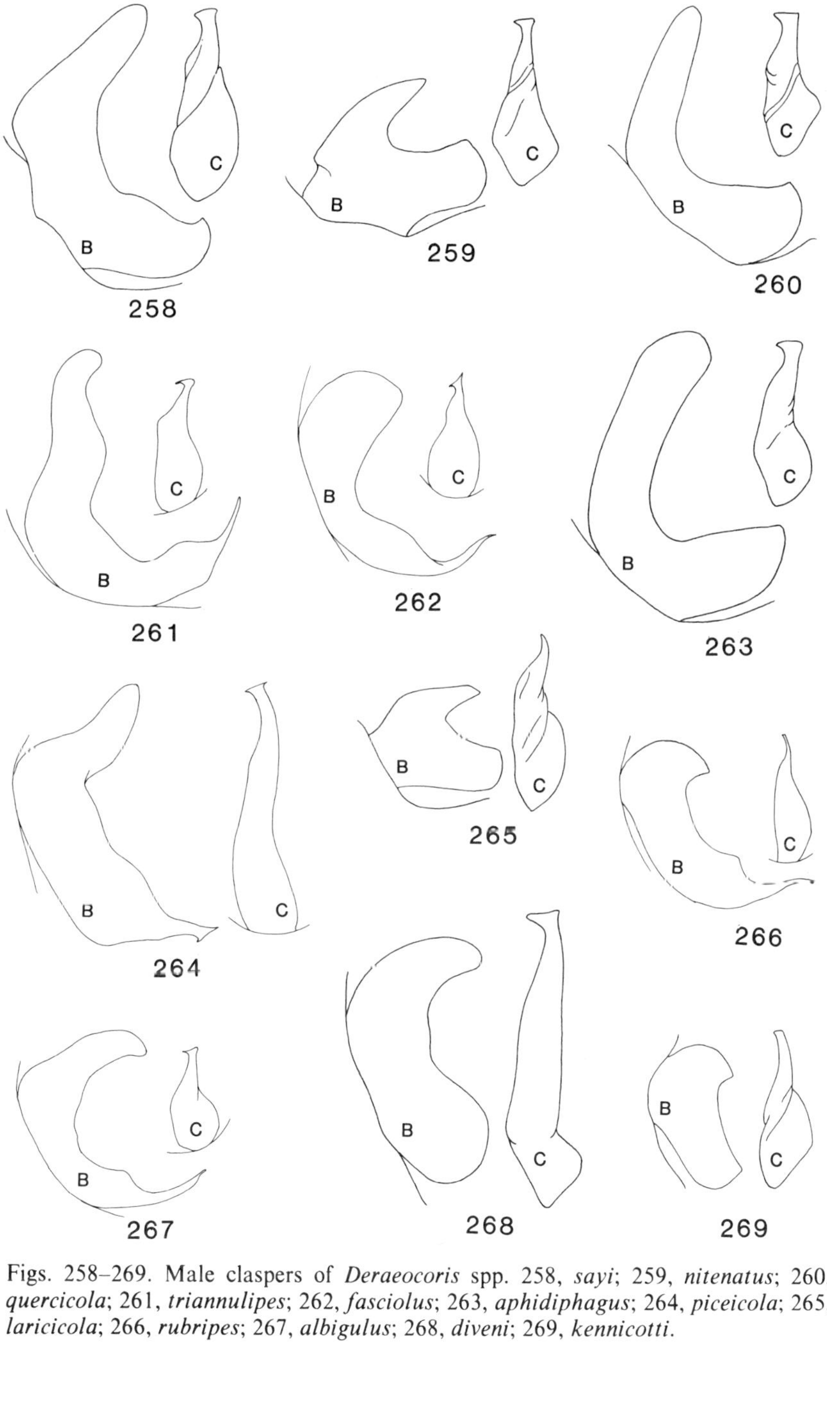

Figs. 258–269. Male claspers of *Deraeocoris* spp. 258, *sayi*; 259, *nitenatus*; 260, *quercicola*; 261, *triannulipes*; 262, *fasciolus*; 263, *aphidiphagus*; 264, *piceicola*; 265, *laricicola*; 266, *rubripes*; 267, *albigulus*; 268, *diveni*; 269, *kennicotti*.

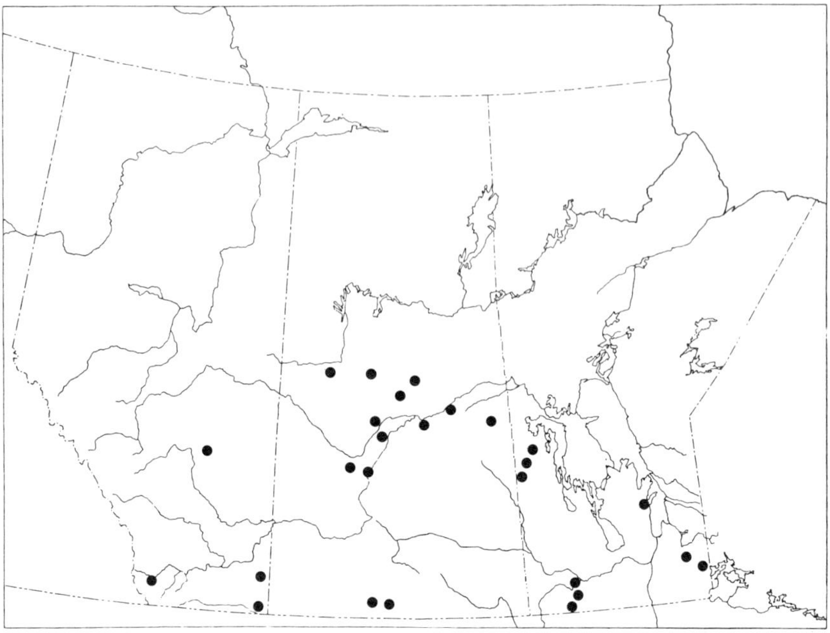

Map 126. Collection localities for *Deraeocoris fasciolus*.

Distribution. Northeastern and North Central States, Oregon, British Columbia, Eastern Canada; now known to occur in the Prairie Provinces (Map 126).

Deraeocoris aphidiphagus Knight

Figs. 257, 263; Map 127

Deraeocoris aphidiphagus Knight, 1921*b*:134.

Length 5.70–6.62 mm; width 2.84–3.22 mm. Head pale marked with black on clypeus and frons. Rostrum 2.24–2.38 mm long. Pronotum gray, calli black. Scutellum pale, wide longitudinal bar each side of middle black. Hemelytra grayish marked with black, cuneus pale, apex brown (Fig. 257). Ventral surface dark brown, legs mostly pale, hind femur brown on apical half, tibiae triannulate with brown.

Remarks. This species is distinguished by the genital claspers (Fig. 263).

Habitat. Collected on *Ulmus americana*; preys on aphids found in the leaf curls.

Distribution. Eastern half of USA, Eastern Canada; Alberta; now known to occur in Saskatchewan and Manitoba (Map 127).

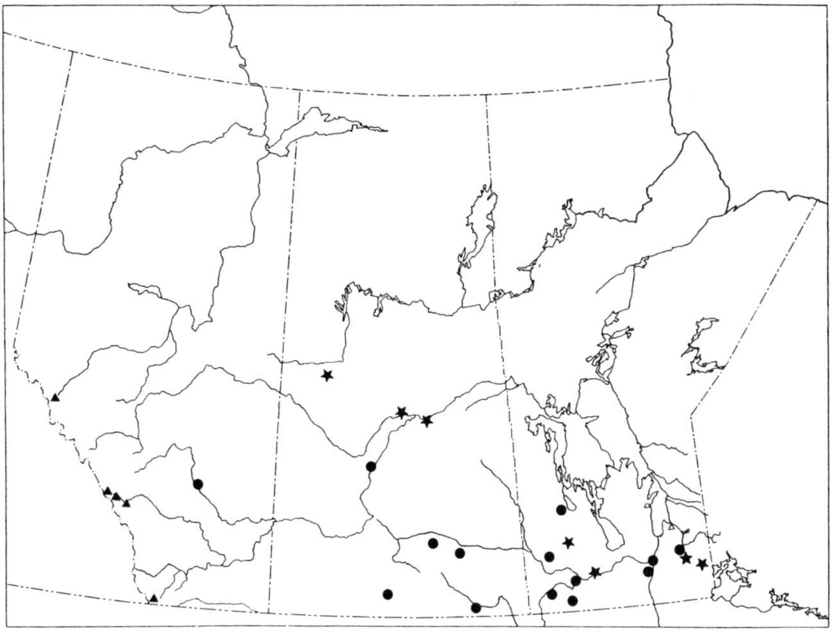

Map 127. Collection localities for *Deraeocoris aphidiphagus* (●), *D. piceicola* (▲), and *D. laricicola* (★).

Deraeocoris piceicola Knight

Fig. 264; Map 127

Deraeocoris piceicola Knight, 1927*c*:136.

Length 6.86–7.70 mm; width 2.94–3.36 mm. Head black in males, partly pale in female. Rostrum 2.80–3.08 mm long. Pronotum black in male, light brown with black calli in female. Scutellum black, basal angles and apex pale. Hemelytra mostly black in male, light brown in female with reddish cuneus; glabrous; wing membrane shorter in female. Ventral surface black; legs mostly black, tibiae with pale bands.

Remarks. The males of this species are long and narrow, and almost all black. The black head has a pale spot on top between the eyes. The females are shorter and broader, and have a reddish cuneus. The genital claspers (Fig. 264) are distinctive.

Habitat. Collected on *Picea glauca* infested with aphid galls; probably preys on these aphids.

Distribution. Colorado, Yukon, British Columbia; now known to occur in Alberta (Map 127).

Deraeocoris laricicola Knight

Fig. 265; Map 127

Deraeocoris laricicola Knight, 1921*b*:164.

Length 6.58–7.42 mm; width 2.80–3.08 mm. Head pale marked with black; second antennal segment pilose. Rostrum 2.31–2.38 mm long. Pronotum light brown, calli and punctures black; sparsely pubescent. Scutellum pale, wide bar each side of median line black; sparsely pubescent. Hemelytra pale marked with brown; sparsely pubescent. Ventral surface brown to black; femur and tibia black or reddish, bands on tibiae often obscure.

Remarks. This species appears glabrous, but has short pubescence on the hemelytra and longer hairs on the pronotum. The antennae and tibiae have prominent hairs. The genital claspers (Fig. 265) are distinctive.

Habitat. Collected on *Larix laricina*; probably preys on aphids normally found on the branches.

Distribution. Northeastern and North Central States; now known to occur in Saskatchewan and Manitoba (Map 127).

Deraeocoris rubripes Kelton

Figs. 266, 270; Map 128

Deraeocoris rubripes Kelton, 1980*a*:

Length 6.44–6.72 mm; width 2.59–2.94 mm. Head mostly pale, longitudinal lines on clypeus and two triangular spots between eyes brown. Rostrum 2.38–2.52 mm long. Pronotum pale, calli and punctures black; sparsely pubescent. Scutellum pale, bar each side of median line black. Hemelytra mostly pale, punctures black; apex of embolium and cuneus black. Ventral surface brown, genital segment yellowish brown; coxa and basal half of femur pale, apical half of femur and tibia red.

Remarks. This species is distinguished by the red hind tibia (Fig. 270). The genital claspers (Fig. 266) are similar to those of *D. barberi* Knight.

Habitat. Collected on *Picea glauca*; probably preys on aphids associated with the host plant.

Distribution. Manitoba (Map 128).

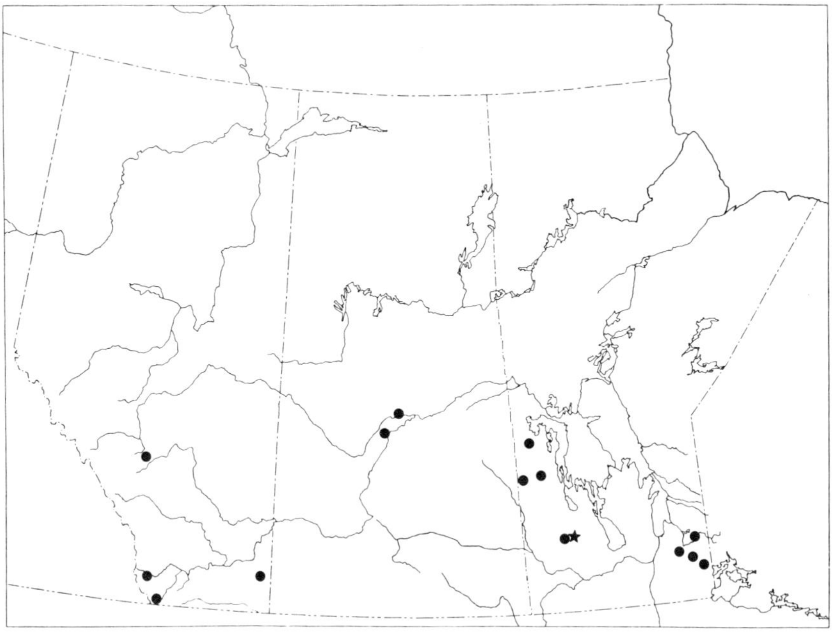

Map 128. Collection localities for *Deraeocoris rubripes* (★) and *D. albigulus* (●).

Deraeocoris albigulus Knight

Fig. 267; Map 128

Deraeocoris albigulus Knight, 1921*b*:171.

Length 6.09–6.65 mm; width 2.80–3.15 mm. Head generally black, parts of clypeus, jugum, and narrow streak each side of frons pale. Rostrum 2.38–2.52 mm long. Pronotum black, basal margin pale; densely pubescent. Scutellum black, basal angles and apex pale yellow; densely pubescent. Hemelytra light brown, spot at middle and at apex of corium, tip of embolium and apex of clavus, and tip of cuneus black; punctures black; densely pubescent. Ventral surface black; legs reddish, tibiae with paler bands near apices.

Remarks. This species is densely pubescent. The reddish legs and curved fingerlike projection on the left clasper (Fig. 267) are distinctive.

Habitat. Collected on *Pinus banksiana, P. contorta, Picea glauca,* and *Pseudotsuga menziesii*; probably preys on aphids associated with the host trees.

Distribution. Northeastern and North Central States; now known to occur in the Prairie Provinces (Map 128).

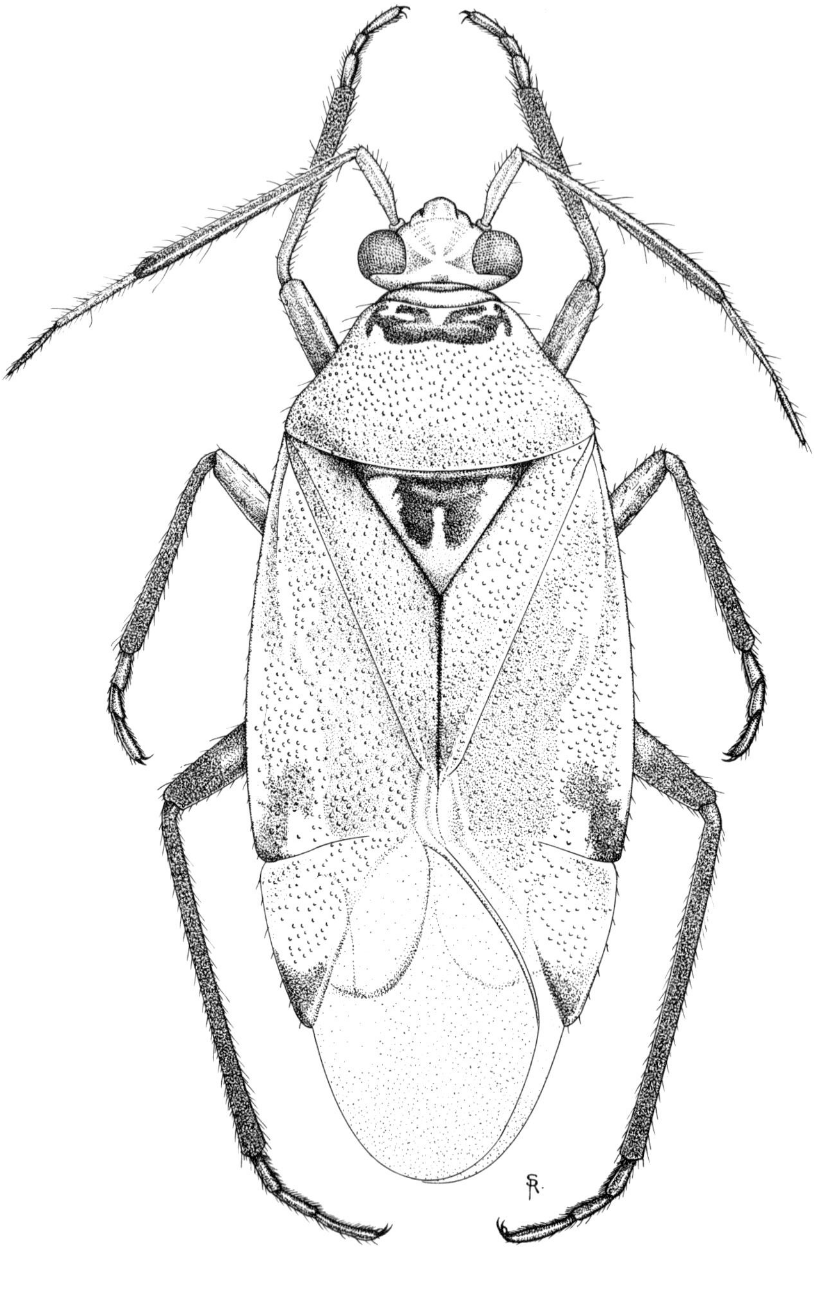

Fig. 270. *Deraeocoris rubripes*

Deraeocoris diveni Knight

Fig. 268; Map 129

Deraeocoris diveni Knight, 1921*b*:117.

Length 5.60–6.30 mm; width 2.38–2.52 mm. Head black, median line on clypeus, frons, and large spot between eyes pale yellow. Rostrum 1.96–2.10 mm long. Pronotum black, basal margin pale; densely pubescent. Scutellum black, basal angles and apex pale; densely pubescent. Hemelytra black in male, light brown with black patches in female; densely pubescent. Ventral surface black; legs mostly black, tibiae brown with two pale bands.

Remarks. This species is densely pubescent like *albigulus*, with exserted hairs on the antennae and tibiae. The genital claspers (Fig. 268) are distinctive.

Habitat. Collected on *Pinus contorta*; probably preys on aphids.

Distribution. Wyoming; now known to occur in Alberta (Map 129).

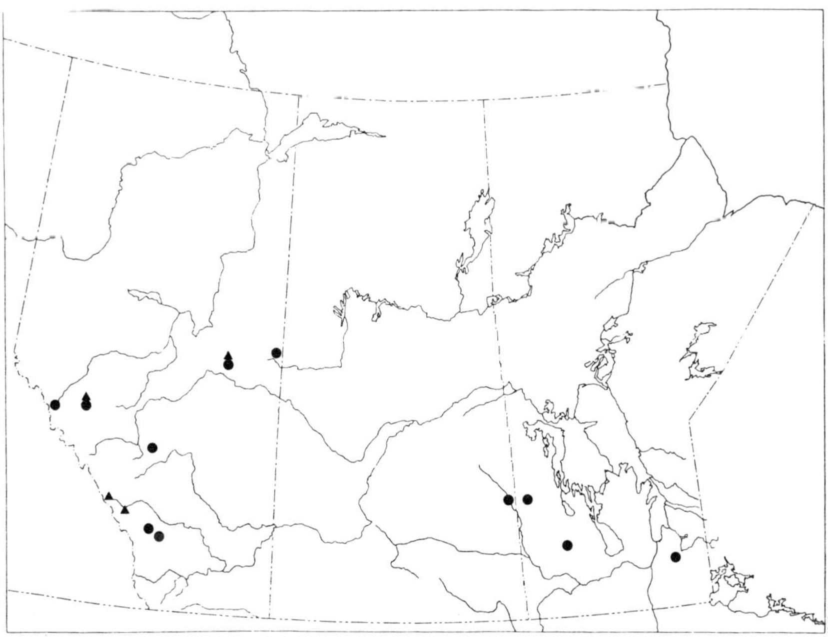

Map 129. Collection localities for *Deraeocoris diveni* (▲) and *D. kennicotti* (●).

Deraeocoris kennicotti Knight

Fig. 269; Map 129

Deraeocoris kennicotti Knight, 1921*b*:166.

Length 6.02–6.72 mm; width 2.52–2.80 mm. Head pale, jugum and sides of frons marked with black. Rostrum 2.24–2.38 mm long. Pronotum pale, calli and punctures black; densely pubescent. Scutellum black, basal angles and apex pale; strongly pubescent. Hemelytra basically pale, middle and apex of corium, tip of embolium, apex of clavus, and most of cuneus black; punctures black; densely pubescent. Ventral surface reddish brown to black; legs generally pale, femur often biannulate with brown at apex, tibia triannulate.

Remarks. This species resembles *laricicola*, but is densely pubescent, and the exserted hairs on the antennae and tibiae are shorter. The genital claspers (Fig. 269) are distinctive.

Habitat. Collected on *Picea glauca, P. mariana,* and *Pinus banksiana*; probably preys on aphids associated with the host trees.

Distribution. Maine, Northwest Territories; now known to occur in the Prairie Provinces (Map 129).

Subfamily Dicyphinae Reuter

In North America this subfamily contains only four genera and approximately 34 species.

The following are the subfamily characteristics: 1) large pronotal collar; 2) slender, delicate form; 3) second segment of hind tarsus longer than first; 4) tarsal claws sharply angled at their bases; 5) hairlike parempodia; and 6) pulvilli present.

In the Prairie Provinces the subfamily is represented by one genus and six species.

Genus *Dicyphus* Fieber

Elongate, slender species. Head nearly vertical, eyes large, considerably removed from pronotum, carina between them absent. Pronotum subcampanulate, lateral margins rounded, smooth; calli prominent, separated from basal half of pronotum by transverse impressed line; basal margin deeply concave. Hemelytra pale, opaque, marked with reddish tinge or fuscous, pubescence simple. Osteolar peritreme large or inconspicuous. Legs long, slender.

Approximately 23 species are known from North America, five species occur in the Prairie Provinces.

Key to species of *Dicyphus*

1. Osteolar peritreme elevated, prominent; pronotum mostly yellowish or reddish 2
   Osteolar peritreme depressed, inconspicuous; pronotum black or with yellow median line (Fig. 271) 3
2. First antennal segment mostly yellow marked with red; second segment longer than width of pronotum at base; left clasper (Fig. 272) .... ***discrepans*** **Knight** (p. 373)
   First antennal segment black; second segment shorter than width of pronotum at base; left clasper (Fig. 273) .... ***hesperus*** **Knight** (p. 374)
3. Hind tibia black; left clasper (Fig. 274) .... ***tibialis*** **Kelton** (p. 377)
   Hind tibia pale 4
4. Scutellum black; left clasper (Fig. 275) .... ***rubi*** **Knight** (p. 377)
   Scutellum with basal angles green or yellow 5
5. Species 4.90–5.25 mm; rostrum 1.48–1.61 mm; left clasper (Fig. 276) .... ***similis*** **Kelton** (p. 378)
   Species 4.48–4.90 mm; rostrum 1.31–1.48 mm; left clasper (Fig. 277) .... ***confusus*** **Kelton** (p. 379)

Dicyphus discrepans Knight

Fig. 272; Map 130

Dicyphus discrepans Knight, 1923*b*:476.

Length 3.22–3.92 mm; width 0.98–1.12 mm. Head brown, shiny, area extending from antennal fossa to collar yellow. Rostrum 1.48–1.61 mm long. First antennal segment 0.35 mm long, yellow marked with reddish tinge; second segment 0.94–0.98 mm long, base and apex reddish brown, median portion yellow. Hemelytra pale with several fuscous areas, tip of embolium and tip of cuneus reddish brown; pubescence long, slanting; wing membrane in female often abbreviated. Legs pale.

Remarks. This species is distinguished by the characters mentioned in the key. The left genital clasper (Fig. 272) is distinctive.

Habitat. Collected on *Aster* spp.

Distribution. Northeastern and northwestern States, Quebec, British Columbia; now known to occur in Alberta and Saskatchewan (Map 130).

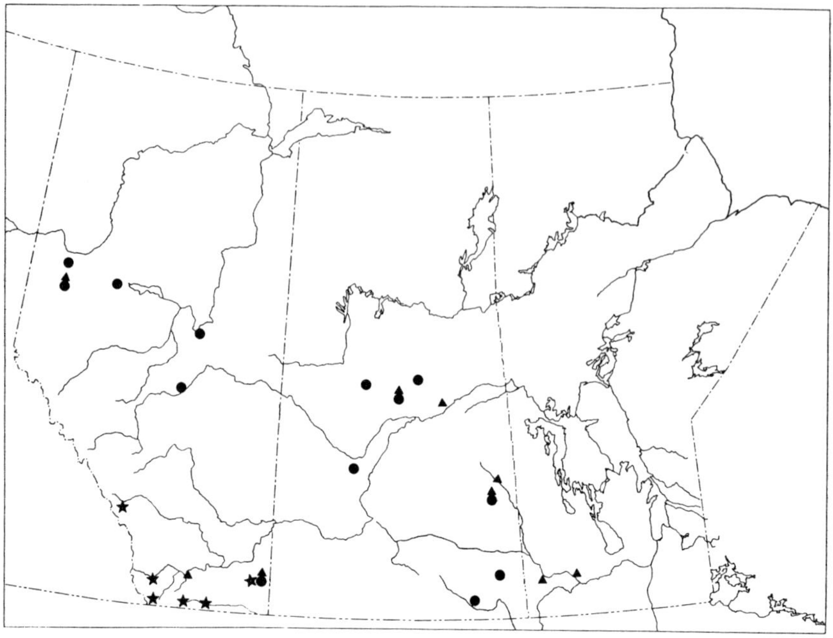

Map 130. Collection localities for *Dicyphus discrepans* (●), *D. hesperus* (▲), and *D. tibialis* (★).

Dicyphus hesperus Knight

Fig. 273; Map 130

Dicyphus hesperus Knight, 1943*a*:56.

Length 3.22–3.92 mm; width 0.98–1.12 mm. Head black, shiny, narrow area bordering eyes yellow. Rostrum 1.26–1.47 mm long. First antennal segment 0.24 mm long, black, apex white; second segment 0.66–0.87 mm long, yellow, apical one fourth reddish brown. Hemelytra similar to *discrepans* in color and pubescence, wing membrane in female fully developed.

Remarks. This species is similar to *discrepans* in color and appearance except for the shorter antennal segments and shorter rostrum. The first antennal segment in *hesperus* is mostly black, while in *discrepans* it is marked with red. The genital clasper (Fig. 272) is distinctive.

Habitat. Collected on *Mentha arvensis*.

Distribution. Western States; now known to occur in the Prairie Provinces (Map 130).

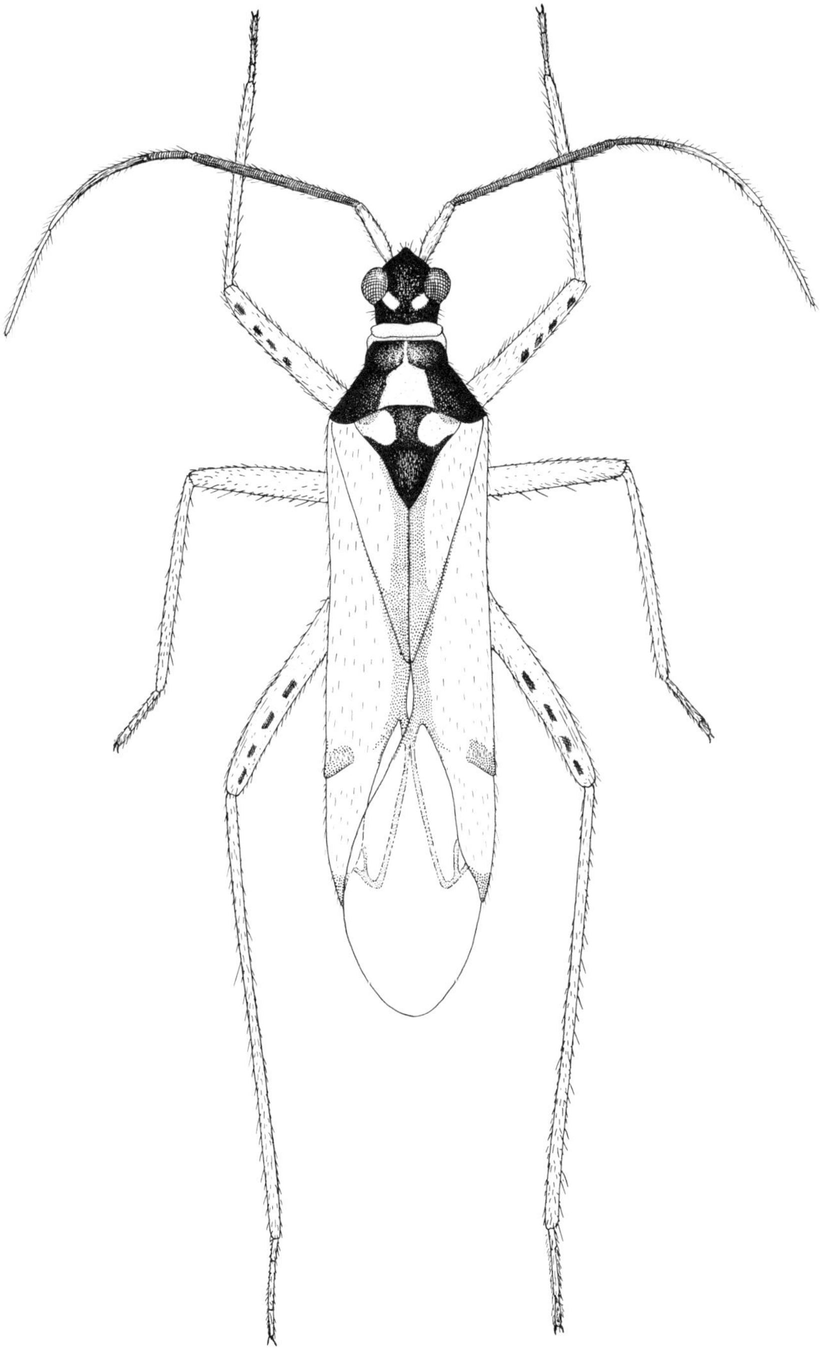

Fig. 271. *Dicyphus rubi*

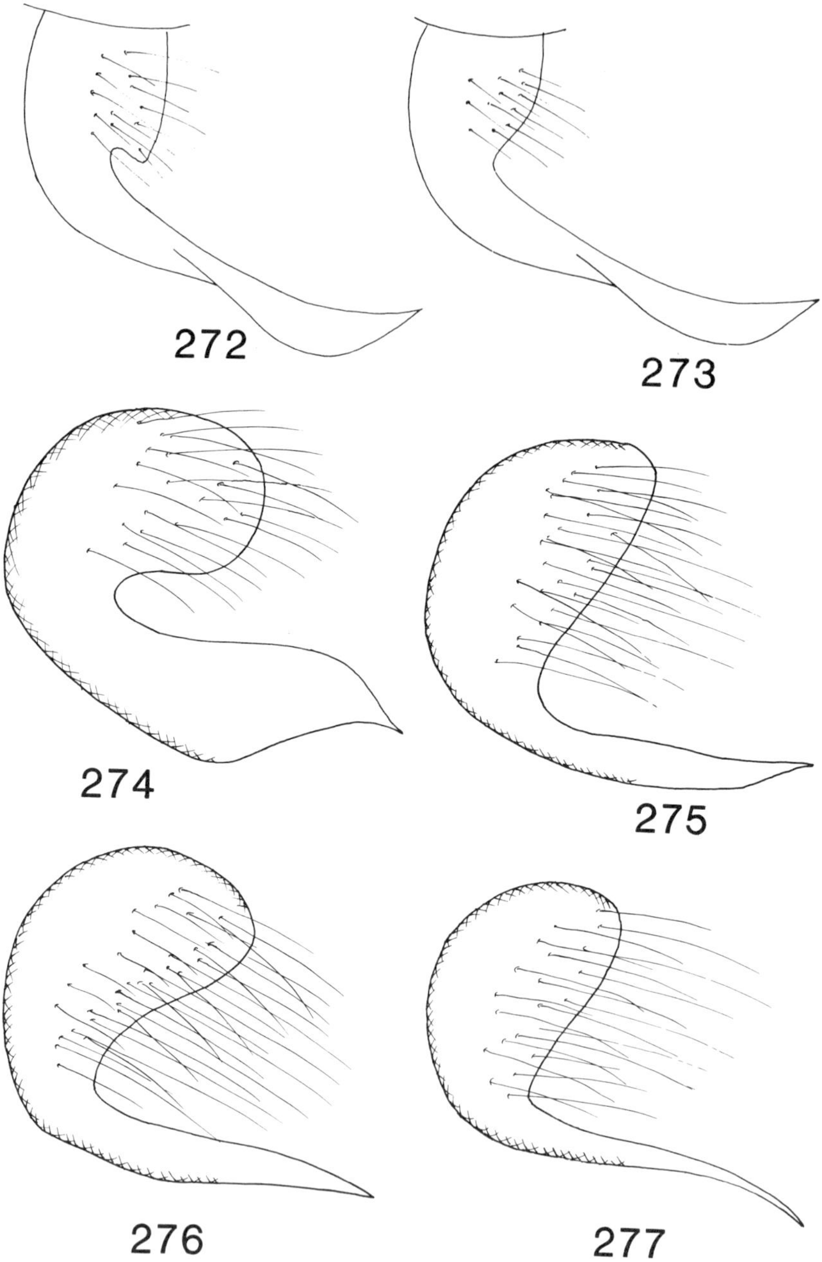

Figs. 272–277. Left claspers of *Dicyphus* spp. 272, *discrepans*; 273, *hesperus*; 274, *tibialis*; 275, *rubi*; 276, *similis*; 277, *confusus*.

Dicyphus tibialis Kelton

Fig. 274; Map 130

Dicyphus tibialis Kelton, 1980*d*:

Length 3.50–4.55 mm; width 0.98–1.40 mm. Head black, shiny; spot on top next to eye pale. Rostrum 1.40–1.61 mm long. First antennal segment black; second segment 0.84–0.98 mm long, black; third segment black, curved, as long as second; fourth segment light brown. Pronotum black, shiny, collar and median wedge-shaped area pale. Scutellum black, basal angles pale. Hemelytra pale, apical half of clavus reddish, apical margin of corium fuscous or reddish, and tip of cuneus fuscous; pubescence pale, long, slanting. Hind tibia black.

Remarks. The pale markings on the pronotum and scutellum are similar to those of *Dicyphus agilis*, but the black hind tibia readily distinguish the species. The genital clasper (Fig. 274) is distinctive.

Habitat. Collected on *Geranium viscosissimum.*

Distribution. Colorado, Montana, British Columbia; Alberta (Map 130).

Dicyphus rubi Knight

Fig. 275; Map 131

Dicyphus rubi Knight, 1968*a*:72.

Length 4.20–4.41 mm; width 0.98–1.12 mm. Head black, shiny, spot on top next to eye pale yellow. Rostrum 1.33–1.47 mm long. First antennal segment pale; second segment 1.05–1.12 mm long, black; third segment brown on basal half, pale on apical half, curved, as long as second segment; fourth segment pale. Pronotum black, often pale at middle, shiny, collar pale. Scutellum black. Hemelytra pale fuscous, apical margin of corium and apex of cuneus darker; pubescence pale, long, slanting. Legs pale.

Remarks. This species is distinguished by the black scutellum. The genital clasper (Fig. 275) is distinctive.

Habitat. Collected on *Rubus strigosus.*

Distribution. New York, now known to occur in Colorado, Michigan, Utah, Eastern Canada, British Columbia; Prairie Provinces (Map 131).

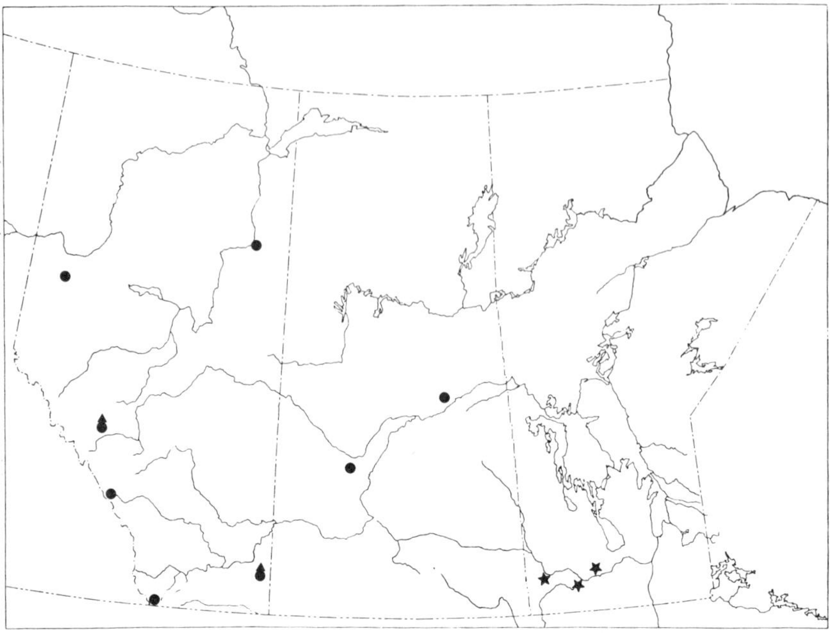

Map 131. Collection localities for *Dicyphus rubi* (★), *D. similis* (●), and *D. confusus* (▲).

Dicyphus similis Kelton

Fig. 276; Map 131

Dicyphus similis Kelton, 1980*d*:387.

Length 4.90–5.25 mm; width 1.12–1.29 mm. Head black, shiny, spot on top next to eye pale green. Rostrum 1.48–1.54 mm long. Second antennal segment black, 1.31–1.40 mm (male), 1.15–1.19 mm (female); third segment pale to black, curved, longer than second segment. Pronotum black, shiny, collar white, median wedge-shaped area greenish yellow. Scutellum black, basal angles green or yellow. Hemelytra pale green, apical half of clavus and adjacent corium fuscous, apex of corium bordering cuneus fuscous, cuneus margined with yellow, apex fuscous; pubescence simple, pale, long, slanting. Legs yellowish green.

Remarks. This species is similar to *confusus* in color and appearance, but is slightly larger and with a longer rostrum. The prong on the left genital clasper is thicker and more acutely angled (Fig. 276). References to *agilis* in Canada by Provancher (1887), Van Duzee (1889, 1908, 1912), Moore (1907, 1950), and Walley (1928) are probably for *similis*.

Habitat. Collected on *Mentha arvensis*.

Distribution. New York, New Hampshire, Nova Scotia, New Brunswick, Quebec, Ontario, British Columbia; Alberta (Map 131).

Dicyphus confusus Kelton

Fig. 277; Map 31

Dicyphus confusus Kelton, 1980*d*:387.

Length 4.48–4.90 mm; width 1.12–1.40 mm. Head black, shiny, spot on tip next to eye pale green. Rostrum 1.31–1.40 mm long. Second antennal segment black, apical half often pale yellow, 1.15–1.22 mm (male), 0.91–1.13 mm (female); third segment mostly yellow, base black, curved, longer than second segment. Pronotum black, median wedge-shaped area yellowish green; shiny, collar white. Scutellum black, basal angles yellow. Hemelytra pale green, apical half of clavus, apical margin of corium, and tip of cuneus fuscous; pubescence simple, pale, long, slanting. Legs pale green.

Remarks. This species is similar to *similis* in color and appearance, but is slightly smaller and with a shorter rostrum. The prong on the left genital clasper is thin and almost at right angle (Fig. 277). References to *agilis* in western Canada by Parshley (1919), Knight (1925*b*, 1941*c*), Blatchely (1926), Downes (1927), and Strickland (1953) are probably for *confusus, similis,* or some other species.

Habitat. Collected on *Geranium viscosissimum.*

Distribution. Colorado, Idaho, British Columbia; Alberta (Map 131).

Subfamily Bryocorinae Baerensprung

In North America this subfamily contains 24 genera and approximately 87 species. Most of the species are tropical.

The following are the subfamily characteristics: 1) wing membrane with one cell; 2) tibiae without spines; 3) last tarsal segment swollen, thicker than preceding; 4) hairlike parempodia, and 5) large pulvilli.

In the Prairie Provinces the subfamily is represented by two genera and two species.

Key to genera of Bryocorinae

1. Pronotum with distinct collar; pronotum finely punctate; head smooth .... ***Monalocoris* Dahlbom** (p. 380)
   Pronotum without collar; pronotum coarsely punctate; head punctate .... ***Sixeonotus* Reuter** (p. 383)

Genus *Monalocoris* Dahlbom

Small, oval species. Head vertical, smooth. Pronotum trapeziform, finely punctate, lateral margins angulate, anterior angle carinate; collar and calli distinct. Hemelytra shiny; pubescence pale, long, dense; wing membrane uniformly fuscous. Legs pale.

Two species are known from North America, one occurs in the Prairie Provinces.

Monalocoris americanus Wagner & Slater

Fig. 278; Map 132

Monalocoris americanus Wagner & Slater, 1952:279.

Length 2.24–2.66 mm; width 1.40–1.54 mm. Head smooth, yellow, clypeus black. Antennae pale, apex of second segment black. Rostrum 0.70–0.84 mm long. Pronotum dark brown, basal angles yellow. Hemelytra brown, embolium and cuneus pale yellow. Ventral surface brown; legs pale.

Map 132. Collection localities for *Monalocoris americanus* (●) and *Sixeonotus rostratus* (▲).

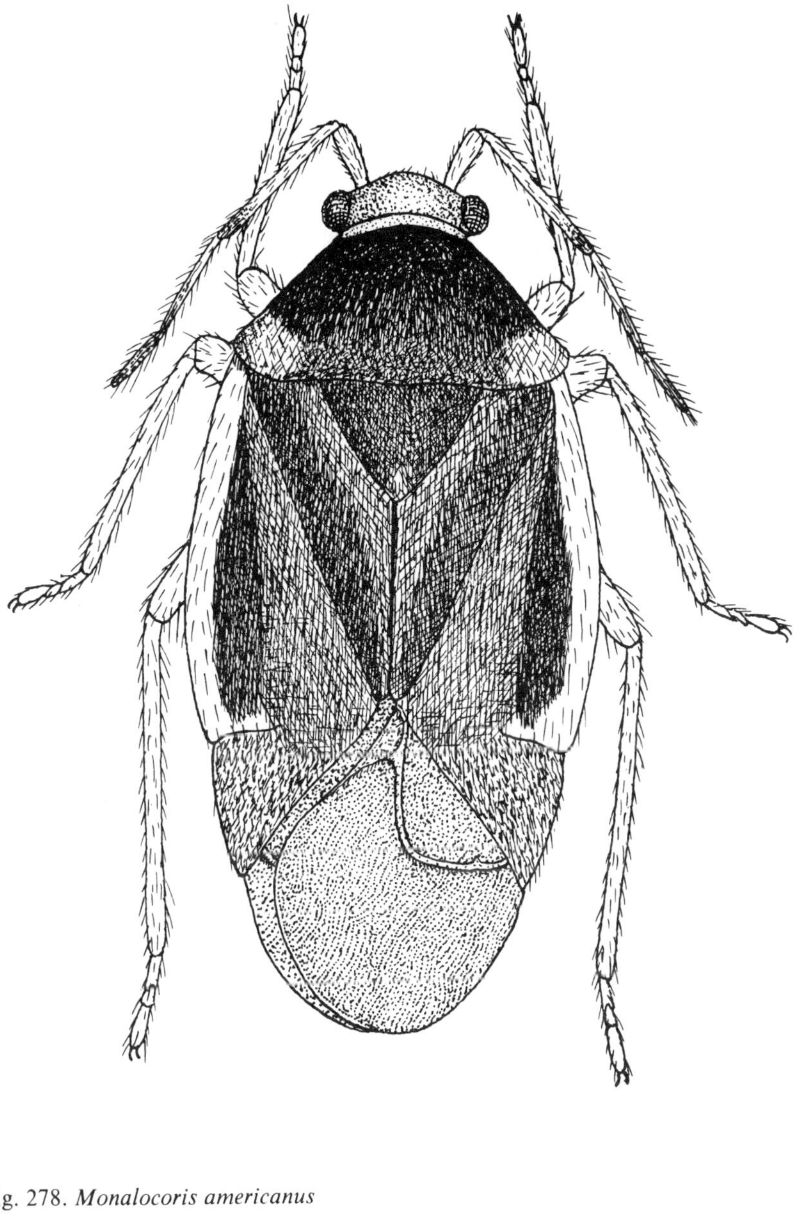

Fig. 278. *Monalocoris americanus*

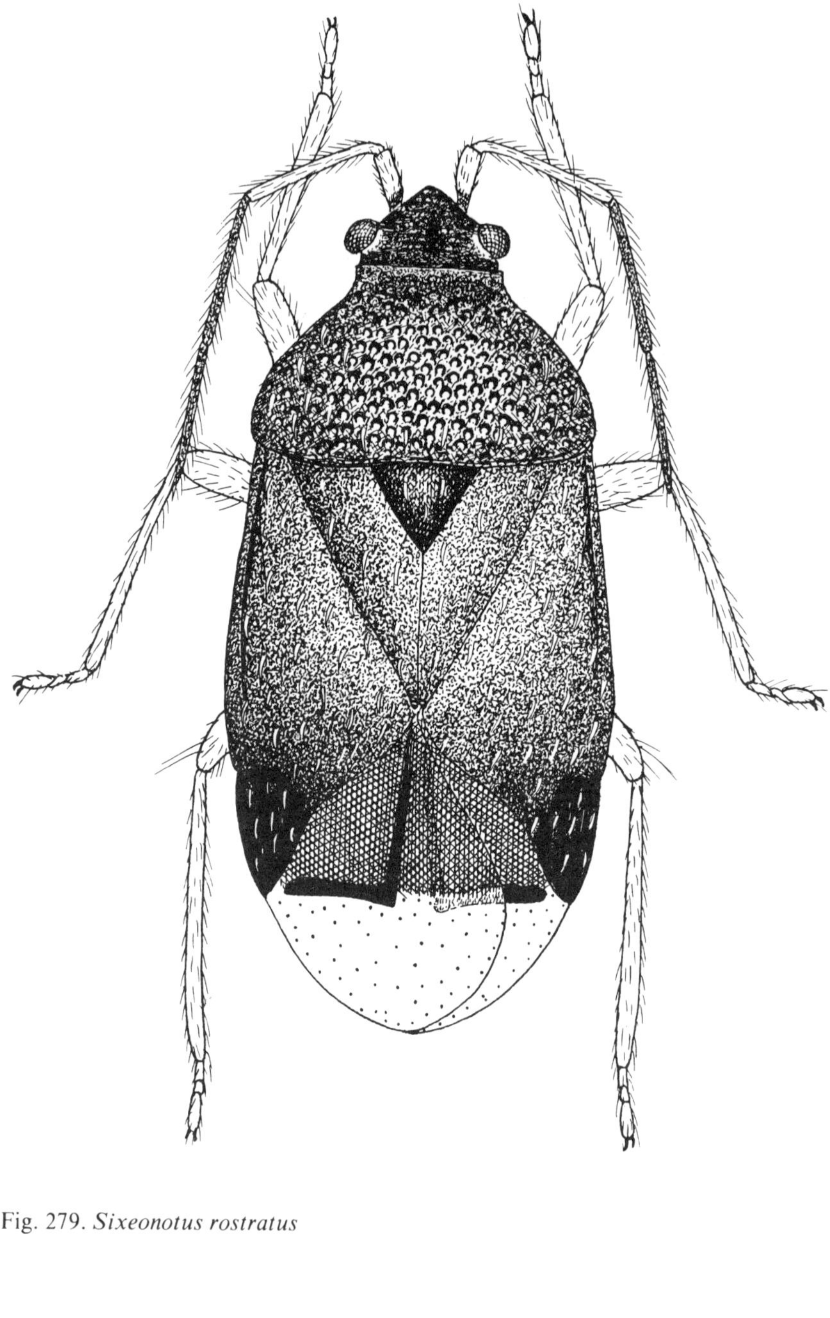

Fig. 279. *Sixeonotus rostratus*

Remarks. This species is distinguished by the distinct collar, by the smooth head, and by the pale basal angles of the pronotum (Fig. 278).

Habitat. Collected on *Dryopteris* spp.

Distribution. Eastern and North Central States, Alaska, Ontario, Quebec, British Columbia; Alberta, now known to occur in Saskatchewan and Manitoba (Map 132).

Genus *Sixeonotus* Reuter

Short, oval, black species. Head vertical, punctate; frons tumid, transversely striate; vertex with wide longitudinal impression; eyes globose, protruding. Pronotum strongly convex, coarsely punctate, side margins rounded; collar absent. Hemelytra partly pruinose, cuneus declivent. Ventral surface black, legs pale.

Twenty-two species are known from North America, one occurs in Alberta.

Sixeonotus rostratus Knight

Fig. 279; Map 132

Sixeonotus rostratus Knight, 1928*d*:245.

Length 3.01–3.50 mm; width 1.61–1.75 mm. Head black, spot next to eye on vertex yellow. First and second antennal segment pale, base of first segment black; terminal segments black. Rostrum 0.82–0.87 mm long. Pronotum black, shiny. Hemelytra almost smooth; pubescence pale, suberect; wing membrane fuscous on basal half, clear on apical half. Legs pale, hind femur swollen and fuscous near apex with several long and pale hairs.

Remarks. This species differs from *Monalocoris americanus* by the punctate head and pronotum, and by the absence of the collar (Fig. 279).

Habitat. Collected on *Helianthus* spp.

Distribution. Colorado; now known to occur in Alberta and Saskatchewan (Map 132).

Scientific and common names of plants

Abies balsamea	balsam fir
lasiocarpa	alpine fir
Acer negundo	Manitoba maple
Agropyron cristatum	crested wheat grass
repens	quack grass
Allium stellatum	pink-flowered onion
textile	prairie onion
Alnus spp.	alder
rugosa	speckled alder
tenuifolia	mountain alder
Amaranthus spp.	amaranth
Ambrosia trifida	giant ragweed
Anemone canadensis	Canada anemone
Antennaria campestris	prairie everlasting
Artemisia spp.	wormwood
cana	silver sagebrush
frigida	pasture sage
ludoviciana	prairie sage
tridentata	big sagebrush
Aster spp.	aster
Astragalus caryocarpus	large-fruited ground plum
Atriplex spp.	atriplex
Beckmannia syzigachne	slough grass
Betula spp.	birch
glandulosa	scrub birch
occidentalis	water birch
Bromus spp.	brome grass
Calamagrostis spp.	reed grass
Caragana arborescens	caragana
Carex spp.	sedge
Ceanothus spp.	redroot
Chenopodium album	lamb's-quarters
Chrysopsis villosa	hairy golden-aster
Cornus stolonifera	red-osier dogwood

Corylus spp.	hazelnut
americana	American hazelnut
Crataegus spp.	hawthorn
Croton spp.	croton
Delphinium bicolor	low larkspur
Descurainia sophia	flixweed
Dryopteris spp.	wood-fern
Elaeagnus commutata	silverberry
Erigeron spp.	fleabane
Fraxinus nigra	black ash
pennsylvanica	red ash
Galium boreale	northern bedstraw
Geranium bicknellii	Bicknell's geranium
viscosissimum	sticky geranium
Glycyrrhiza lepidota	wild licorice
Grindelia perennis	gumweed
Hedysarum spp.	sweet broom
Helianthus spp.	sunflower
Juniperus spp.	juniper
communis	common juniper
horizontalis	creeping juniper
Kalmia polifolia	bog-laurel
Kochia scoparia	summer cypress
Larix laricina	tamarack
Lathyrus spp.	vetchling
palustris	marsh vetchling
venosus	purple vetchling
Lupinus spp.	lupine
argenteus	silvery lupine
Malvastrum coccineum	scarlet mallow
Matricaria matricarioides	pineappleweed
Medicago sativa	alfalfa
Melilotus spp.	sweet-clover
Mentha arvensis	field mint
Monarda fistulosa	wild bergamot
Oenothera biennis var. *canescens*	western yellow evening-primrose
caespitosa	gumbo evening-primrose
serrulata	shrubby evening-primrose
Picea glauca	white spruce
mariana	black spruce

Pinus banksiana	jack pine
contorta	lodgepole pine
Polygonum persicaria	lady's-thumb
Populus tremuloides	aspen poplar
Potentilla fruticosa	shrubby cinquefoil
plattensis	low cinquefoil
Prunus pensylvanica	pin cherry
Pseudotsuga menziesii	douglas fir
Psoralea spp.	scurf-pea
lanceolata	lance-leaved psoralea
Quercus macrocarpa	bur oak
Rhus typhina	staghorn sumac
Rosa arkansana	prairie rose
Rubus spp.	bramble
strigosus	wild red raspberry
Salicornia rubra	red samphire
Salix spp.	willow
amygdaloides	peach-leaved willow
bebbiana	beaked willow
candida	hoary willow
discolor	pussy willow
interior	sandbar willow
Sambucus spp.	elder
Sarcobatus vermiculatus	greasewood
Scirpus spp.	bulrush
Senecio canus	silvery groundsel
eremophilus	cut-leaved ragwort
palustris	marsh ragwort
Shepherdia canadensis	russet buffaloberry
Solidago spp.	goldenrod
Spiraea alba	narrow-leaved meadowsweet
Suaeda depressa	western sea-blite
Symphoricarpos occidentalis	western snowberry
Thermopsis rhombifolia	golden-bean
Thuja occidentalis	eastern white cedar
Trifolium spp.	clover
repens	white clover
Ulmus americana	American elm
Vicia spp.	vetch

Glossary

Structures labeled in Figure 1 are not repeated here.

angulate Formed with corners.
annulate Banded.
apex That part of any joint or structure opposite the base by which it is attached.
appressed Pressed close to or lying flat.
apterous Lacking wings.
arcuate Arched.

bifurcate Divided into two branches.
brachypterous With short or abbreviated wings.

campanulate Bell-shaped.
carinate Ridged.
claspers The left and right structures on the male genital segment used in copulation.
clavate Club-shaped.
concave Hollowed inwards like the inside of a bowl.
contiguous Touching when in normal position.
convergent Coming together.
convex Rounded outwards like the outside of a bowl.
costal The lateral margin of the hemelytron.
cylindrical Having the form of a cylinder.

declivent Sloping gradually downwards.
diffuse Spread thinly or lightly, but concentrated.
dimorphic Occurring in two distinct forms.
divergent Drawing apart.
dorsum Dorsal surface.
ductus seminis The seminal duct of the male genitalia.
dusky Somewhat dark in color.

elliptical Shaped like an ellipse.

filiform Thin throughout.
fuscous Dusky; brownish gray, approaching black.

glabrous Hairs present, but too short to be seen readily.
globose Having the shape of a globe.

horizontal Relating to the horizontal plane.

impunctate Without punctures.
incrassate Thickened towards the apex.

macropterous Wings fully developed.
maculate Marked with spots; blotched.

nymph An immature mirid.

oblique Inclined, sloping.

pilose Covered with fine, long hairs.
pleuron The side of the thorax.
pollinose Covered with silvery bloom.
predaceous Preying on other insects.
pruinose Covered with whitish dust.
pubescence Fine, soft hair covering the surface.
punctate Marked with small depressions.

recumbent Reclining.
rugose Wrinkled.

saltatorial Adapted for leaping; e.g., saltatorial femur.
sclerite A hard sclerotized piece of integument, or covering.
sericeous Silky.
spiculum A slender sclerite.
striate Grooved.
stricture Depressed boundary marking the junction of the pronotum and collar.
submedian Below the median; e.g., a line on the frons between the median line and the eye.
subquadrate Nearly square.

trapeziform Having the shape of a trapezium; a quadrilateral having no two sides parallel.
trifurcate Divided into three branches.
truncate Squared rather than rounded or pointed at the tip.
tubercle A small or moderate projection.
tumid Swollen.

vesica The male phallus, or intromittent organ.

References

Arrand, J. C. 1958. A new species of *Plagiognathus* from alfalfa in western Canada (Hemiptera: Miridae). Can. Ent. 90:497–500.

Atkinson, E. T. 1890. Catalogue of the Insecta. II. Order Rhynchota, suborder Hemiptera-Heteroptera, Family Capsidae. J. Asiat. Soc. Beng. 58:25–199.

Blatchley, W. S. 1926. Heteroptera of eastern North America. Nature Publishing Co., Indianapolis, Ind. 1116 pp.

Carayon, J. 1958. Études sur les Hémiptères Cimicoidea. Mém. Mus. natn. Hist. nat., Paris 16:141–172.

Carvalho, J. C. M. 1954. Neotropical Miridae, LXXVII: Miscellaneous observations in some European museums (Hemiptera). Anais Acad. bras. Cienc. 26:423–427.

Carvalho, J. C. M. 1955*a*. Keys to the genera of Miridae of the world (Hemiptera). Bolm Mus. para. 'Emilio Goeldi' 11:1–151.

Carvalho, J. C. M. 1955*b*. Analecta Miridologica. Miscellaneous observations in some American museums and bibliography. Revta chil. Ent. 4:221–226.

Carvalho, J. C. M. 1957-1959. A catalogue of the Miridae of the world. Parts I–IV. Archos. Mus. nac., Rio de J., Brazil.

Carvalho, J. C. M., and Jurberg, J. 1974. Neotropical Miridae, CLXXX: On the *Horcias* complex (Hemiptera). Revta bras. Biol. 34:49–65.

Carvalho, J. C. M., Knight, H. H., and Usinger, R. L. 1961. *Lygus* Hahn, 1833 (Insecta, Hemiptera); proposed designation under the plenary powers of a type-species in harmony with accustomed usage. Bull. zool. Nom. 18:281–284.

Carvalho, J. C. M., and Southwood, T. R. E. 1955. Revisao do Complexo *Cyrtorhinus* Fieber-*Mecomma* Fieber (Hemiptera, Heteroptera, Miridae). Bolm Mus. para. 'Emilio Goeldi' II. 72 pp., 25 figs.

Carvalho, J. C. M., and Wagner, E. 1957. A world revision of the genus *Trigonotylus* Fieber (Hemiptera-Heteroptera, Miridae). Archos Mus. nac., Rio de J., Brazil. 43:122–125.

Cockerell, T. D. A. 1910. Some insects from Steamboat Springs, Colo. II. Can. Ent. 42:366–370.

Crevecoeur, F. F. 1905. Additions to the list of the hemipterous fauna of Kansas. Trans. Kans. Acad. Sci., 19:232–237.

De Geer, C. 1773. Mémoirs pour servir à l'histoire des insectes. 3:290–295. Stockholm.

Douglas, J. W., and Scott, J. 1865. The British Hemiptera. Vol. 1. Hemiptera-Heteroptera. London. 627 pp.

Douglas, J. W., and Scott, J. 1869. British Hemiptera: additions and corrections. Entomologist's mon. Mag. 5:259–268.

Douglas, J. W., and Scott, J. 1871. British Hemiptera: additions and corrections. Entomologist's mon. Mag. 8:60–63.

Downes, W. 1924. New records of Hemiptera from British Columbia. Proc. ent. Soc. Br. Columb. 21:27–30.

Downes, W. 1927. A preliminary list of the Heteroptera and Homoptera of British Columbia. Proc. ent. Soc. Br. Columb. 23:1–22.

Emmons, E. 1854. Natural History of New York. Agriculture, Vol. 5. Insects. Albany. 272 pp.

Fabricius, J. C. 1794. Entomologia systematica 4. Hafniae. 472 pp.

Fabricius, J. C. 1798. Supplementum entomologiae systematicae. Hafniae. 572 pp.

Fallén, C. F. 1807. Monographia Cimicum Sueciae. Apud C. G. Proft. 123 pp.

Fallén, C. F. 1828–29. Hemiptera Sueciae. London. 188 pp.

Fieber, F. X. 1861. Die europäischen Hemiptera. Halbflüger. (Rhynchota Heteroptera). Wien. 444 pp.

Fieber, F. X. 1864. Neuere Entdeckungen in europäischen Hemipteren. Wien. Ent. Mon. 8:65–86.

Flor, G. 1860–1861. Die Rhynchoten Livlands in systematischer Folge beschrieben. 826 pp.

Geoffroy, E. T. 1785. Pages 1–544 *in* A. F. Fourcroy. 1785. Entomologia Parisiensis. Paris. 2 vol.

Goeze, J. A. E. 1778. Entomologische Beiträge zu des Ritter Linné Zwölften Ausgabe des Natursystems 2. Leipzig. 352 pp.

Heidemann, O. 1892. Note on the food plants of some Capsidae from the vicinity of Washington, D.C. Proc. ent. Soc. Wash. 11:224–226.

Heidemann, O. 1910. Description of a new Capsid. Proc. ent. Soc. Wash. 12:200–201.

Herrich-Schaffer, G. A. W. 1836–53. Die Wanzenartigen Insecten. Nürnberg. pp. 35–116.

Hussey, R. F. 1954. Some new or little-known Miridae from northeastern United States (Hemiptera). Proc. ent. Soc. Wash. 56:196–202.

Kelton, L. A. 1955*a*. Genera and subgenera of the *Lygus* complex (Hemiptera: Miridae). Can. Ent. 87:277–301.

Kelton, L. A. 1955*b*. New species of *Liocoris* from North America (Hemiptera: Miridae). Can. Ent. 87:484–490.

Kelton, L. A. 1955*c*. Species of *Lygus*, *Liocoris*, and their allies in the Prairie Provinces of Canada (Hemiptera: Miridae). Can. Ent. 87:531–556.

Kelton, L. A. 1959. Male genitalia as taxonomic characters in the Miridae (Hemiptera). Can. Ent. 91. No. 11. 72 pp.

Kelton, L. A. 1961*a*. Synopsis of the Nearctic species of *Stenodema* Laporte, and description of a new species from western Canada (Hemiptera: Miridae). Can. Ent. 93:450–455.

Kelton, L. A. 1961*b*. A new Nearctic genus of Miridae, with notes on *Diaphnidia* Uhler 1895 and *Brachynotocoris* Reuter 1880 (Hemiptera). Can. Ent. 93:566–568.

Kelton, L. A. 1965*a*. *Diaphnidia* Uhler and *Diaphnocoris* Kelton in North America (Hemiptera: Miridae). Can. Ent. 97:1025–1030.

Kelton, L. A. 1965*b*. *Chlamydatus* Curtis in North America (Hemiptera: Miridae). Can. Ent. 97:1132–1144.

Kelton, L. A. 1966*a*. Review of the species of *Teratocoris* Fieber, with description of a new species from the Nearctic region (Hemiptera: Miridae). Can. Ent. 98:1265–1271.

Kelton, L. A. 1966*b*. *Pithanus maerkeli* (Herrick-Schaffer) and *Actitocoris signatus* Reuter in North America (Hemiptera: Miridae). Can. Ent. 98:1305–1307.

Kelton, L. A. 1968*a*. On the Heteroptera in the Provancher collection (Hemiptera). *Naturaliste can.* 95:1065–1080.

Kelton, L. A. 1968*b*. Revision of the North American species of *Slaterocoris* with descriptions of new species (Heteroptera: Miridae). Can. Ent. 100:1121–1137.

Kelton, L. A. 1970. Four new species of *Trigonotylus* from North America (Heteroptera: Miridae). Can. Ent. 102:334–338.

Kelton, L. A. 1971*a*. Revision of the species of *Trigonotylus* in North America (Heteroptera: Miridae). Can. Ent. 103:685–705.

Kelton, L. A. 1971*b*. Four new species of *Lygocoris* from Canada (Heteroptera: Miridae). Can. Ent. 103:1107–1110.

Kelton, L. A. 1971*c*. Review of *Lygocoris* species found in Canada and Alaska (Heteroptera: Miridae). Mem. ent. Soc. Canada. No. 83. 87 pp.

Kelton, L. A. 1972*a*. Species of *Dichrooscytus* found in Canada, with descriptions of four new species (Heteroptera: Miridae). Can. Ent. 104:1033–1049.

Kelton, L. A. 1972*b*. Descriptions of nine new species of *Dichrooscytus* from North America (Heteroptera: Miridae). Can. Ent. 104:1457–1464.

Kelton, L. A. 1973. Two new species of *Lygus* from North America, and a note on the status of *Lygus abroniae* (Heteroptera: Miridae). Can. Ent. 105:1545–1548.

Kelton, L. A. 1975. The Lygus bugs (Genus *Lygus* Hahn) in North America (Heteroptera: Miridae). Mem. ent. Soc. Canada. No. 95. 101 pp.

Kelton, L. A. 1977. Species of the genus *Pinalitus* Kelton found in North America (Heteroptera: Miridae). Can. Ent. 109:1549–1554.

Kelton, L. A. 1978. *Xenoborus* Reuter (1908): a new synonym of *Tropidosteptes* Uhler (1878) with description of a new species (Heteroptera: Miridae). Can. Ent. 110:471–473.

Kelton, L. A. 1979*a*. Two new species of *Phytocoris* from western Canada (Heteroptera: Miridae). Can. Ent. 111:689–692.

Kelton, L. A. 1979*b*. *Labopidea* Uhler in North America, with descriptions of a new species and a new genus (Heteroptera: Miridae). Can. Ent. 111:753–758.

Kelton, L. A. 1979*c*. A new genus *Brooksetta*, near *Ilnacora*, with new synonymy and new combinations for 15 species currently placed in *Melanotrichus* Reuter (Heteroptera: Miridae). Can. Ent. 111:949–954.

Kelton, L. A. 1979*d*. Replacement name for *Brooksella* Kelton (Heteroptera: Miridae). Can. Ent. 111:1423.

Kelton, L. A. 1980*a*. Descriptions of three new species of Miridae from the Prairie Provinces and a new record of European Phylini in the Nearctic region (Heteroptera). Can. Ent. 112:285–292.

Kelton, L. A. 1980*b*. Two new species of *Melanotrichus* Reuter from western Canada and a description of the male of *M. atriplicis* (Heteroptera: Miridae). Can. Ent. 112:337–339.

Kelton, L. A. 1980*c*. Description of a new species of *Parthenicus* Reuter, new records of Holarctic Orthotylini in Canada, and new synonomy for *Diaphnocoris pellucida* (Heteroptera: Miridae). Can. Ent. 112:341–344.

Kelton, L. A. 1980*d*. Lectotype designation for *Idolocoris agilis*, and descriptions of three new species of *Dicyphus* Fieber from North America (Heteroptera: Miridae). Can. Ent. 112:387–392.

Kelton, L. A., and Herring, J. L. 1978. Two new species of *Neoborella* Knight (Heteroptera: Miridae) found on dwarf mistletoe, *Arceuthobium* spp. Can. Ent. 110:779–789.

Kelton, L. A., and Knight, H. H. 1962. *Mecomma* Fieber in North America (Hemiptera: Miridae). Can. Ent. 94:1296–1302.

Kelton, L. A., and Knight, H. H. 1970. Revision of the genus *Platylygus* Van Duzee, 1915, with descriptions of twenty-six new species. Can. Ent. 102:1429–1460.

Kirkaldy, G. W. 1902. Memoir upon the Rhynchotal family Capsidae Auctt. Trans. R. ent. Soc. Lond. 1902:243–272.

Kirkaldy, G. W. 1909. Hemiptera, old and new, no. 2. Can. Ent. 41:388–390.

Knight, H. H. 1916. Remarks on *Lygus invitus* Say, with descriptions of a new species and a variety of *Lygus* (Hemiptera: Miridae). Can. Ent. 48:345–349.

Knight, H. H. 1917*a*. New and noteworthy forms of North American Miridae (Hemip.). Ent. News 28:3–8.

Knight, H. H. 1917*b*. A revision of the genus *Lygus* as it occurs in America north of Mexico, with biological data on the species from New York. Bull. Cornell Univ. agric. Exp. Stn 391:555–645.

Knight, H. H. 1917*c*. Notes on the species of Miridae inhabiting ash trees (*Fraxinus*) with description of a new species (Hemip.). Bull. Brooklyn ent. Soc. 12:80–82.

Knight, H. H. 1918*a*. Synoptic key to the subfamilies of Miridae (Hemiptera-Heteroptera). Jl N.Y. ent. Soc. 26:40–44.

Knight, H. H. 1918*b*. New species of *Lopidea* from Arizona (Hemip. Miridae). Ent. News 29:172–176.

Knight, H. H. 1918*c*. Old and new species of *Lopidea* from the United States (Hemip., Miridae). Ent. News 29:210–216.

Knight, H. H. 1920. New and little-known species of *Phytocoris* from the eastern United States. Bull. Brooklyn ent. Soc. 15:49–66.

Knight, H. H. 1921*a*. Scientific results of the Katmai Expedition of the National Geographic Society: XIV. Hemiptera of the family Miridae. Ohio J. Sci. 21:107–112.

Knight, H. H. 1921*b*. Monograph of the North American species of *Deraeocoris* (Hemiptera, Miridae). Rep. Minn. St. Ent. pp. 76–210.

Knight, H. H. 1922*a*. Nearctic records of species of Miridae known heretofore only from the Palaearctic region (Heterop.). Can. Ent. 53:280–288.

Knight, H. H. 1922*b*. The North American species of *Labops* (Heteroptera, Miridae). Can. Ent. 54:258–261.

Knight, H. H. 1923*a*. A fourth paper on the species of *Lopidea* (Heteroptera, Miridae). Ent. News 34:65–72.

Knight, H. H. 1923*b*. The Miridae (or Capsidae) of Connecticut. *In* W. Britton. The Hemiptera, or sucking insects, of Connecticut. Bull. Conn. St. geol. nat. Hist. Surv. No. 34.

Knight, H. H. 1925*a*. Descriptions of a new genus and eleven new species of North American Miridae (Hemiptera). Can. Ent. 57:89–97.

Knight, H. H. 1925*b*. A list of Miridae and Anthocoridae from Alberta, Canada (Hemiptera). Can. Ent. 57:181–182.

Knight, H. H. 1925*c*. Descriptions of twelve new species of *Polymerus* from North America (Hemiptera, Miridae). Can. Ent. 57:244–255.

Knight, H. H. 1926*a*. Descriptions of four new species of *Plagiognathus* from eastern United States (Hemiptera, Miridae). Ent. News 37:9–12.

Knight, H. H. 1926*b*. *Capsus simulans* (Stål) and *Labops burmeisteri* Stål recognized from the Nearctic region (Hemiptera, Miridae). Can. Ent. 58:59–60.

Knight, H. H. 1926*c*. Descriptions of seven new species of *Pilophorus* (Hemiptera, Miridae). Bull. Brooklyn ent. Soc. 21:18–26.

Knight, H. H. 1926*d*. Notes on species of *Polymerus* with descriptions of four new species and two new varieties (Hemiptera, Miridae). Can. Ent. 57:164–168.

Knight, H. H. 1926*e*. Descriptions of eleven new species of *Phytocoris* from eastern North America (Hemiptera, Miridae). Bull. Brooklyn ent. Soc. 21:158–168.

Knight, H. H. 1926*f*. Descriptions of six new Miridae from eastern North America (Hemiptera, Miridae). Can. Ent. 58:252–256.

Knight, H. H. 1927*a*. Notes on the distribution and host plants of some North American Miridae (Hemiptera). Can. Ent. 59:34–44.

Knight, H. H. 1927*b*. Descriptions of twelve new species of Miridae from the District of Columbia and vicinity (Hemiptera). Proc. biol. Soc. Wash. 40:9–18.

Knight, H. H. 1927*c*. Descriptions of nine new species of *Melanotrichus* Reuter from North America (Hemiptera, Miridae). Can. Ent. 59:142–147.

Knight, H. H. 1927*d*. *Megalopsallus*, a new genus of Miridae with five new species from North America (Hemiptera). Ann. ent. Soc. Am. 20:224–228.

Knight, H. H. 1927*e*. New species and a new genus of Deraeocorinae from North America (Hemiptera, Miridae). Bull. Brooklyn ent. Soc. 22:136–143.

Knight, H. H. 1927*f*. Descriptions of seven new species of the genus *Orthotylus* Fieber (Hemiptera, Miridae). Can. Ent. 59:176–181.

Knight, H. H. 1928*a*. New species of *Phytocoris* from North America (Hemiptera, Miridae). Bull. Brooklyn ent. Soc. 23:28–46.

Knight, H. H. 1928*b*. Key to the species of *Hadronema* Uhler with descriptions of five new species (Hemiptera, Miridae). Can. Ent. 60:177–182.

Knight, H. H. 1928*c*. New species of *Labopidea* and *Macrotyloides* (Hemiptera, Miridae). Can. Ent. 60:233–236.

Knight, H. H. 1928*d*. New species of *Halticotoma* and *Sixeonotus* (Hemiptera, Miridae). Bull. Brooklyn ent. Soc. 23:241–249.

Knight, H. H. 1929*a*. Descriptions of five new species of *Plagiognathus* from North America (Hemip.: Miridae). Ent. News 40:69–74.

Knight, H. H. 1929*b*. New species and varieties of *Platytytellus* from North America (Hemiptera: Miridae). Ent. News 40:189–192.

Knight, H. H. 1929*c*. The fourth paper on new species of *Plagiognathus* (Hemiptera, Miridae). Ent. News 40:263–268.

Knight, H. H. 1929*d*. *Labops verae*, new species, with *Labopella, Nicholia*, and *Pronotocrepis*, new genera of North American Miridae (Hemiptera). Can. Ent. 61:214–218.

Knight, H. H. 1930*a*. New species of *Pseudopsallus* Van D., with an allied new genus described (Hemiptera, Miridae). Bull. Brooklyn ent. Soc. 25:1–8.

Knight, H. H. 1930*b*. New species of *Ceratocapsus* (Hemiptera, Miridae). Bull. Brooklyn ent. Soc. 25:187–198.

Knight, H. H. 1930*c*. A new key to *Paracalocoris* with descriptions of eight new species (Hemiptera, Miridae). Ann. ent. Soc. Am. 23:810–827.

Knight, H. H. 1931*a*. *Dacota hesperia* Uhler referred to *Atractotomus*, also descriptions of new species (Hemiptera, Miridae). Bull. Brooklyn ent. Soc. 26:36–38.

Knight, H. H. 1931*b*. Three new species of *Cyrtorhinus* from North America (Hemiptera, Miridae). Bull. Brooklyn ent. Soc. 26:171–173.

Knight, H. H. 1938. *Strongylocoris Blanchard*: six new species from North America (Hemiptera, Miridae). Iowa St. Coll. J. Sci. 13:1–7.

Knight, H. H. 1939*a*. Three new species of Miridae from North America (Hemiptera). Bull. Brooklyn ent. Soc. 34:21–23.

Knight, H. H. 1939*b*. *Conostethus americanus* new species from Colorado, Montana and South Dakota (Hemiptera, Miridae). Ent. News 50:132–133.

Knight, H. H. 1939*c*. *Reuteria* Puton; four new species from United States (Hemiptera, Miridae). Iowa St. Coll. J. Sci. 13:129–133.

Knight, H. H. 1941*a*. New species of *Irbisia* Reuter (Hemiptera, Miridae). Bull. Brooklyn ent. Soc. 36:75–79.

Knight, H. H. 1941*b*. New species of *Lygus* from western United States (Hemiptera, Miridae). Iowa St. Coll. J. Sci. 15:269–273.

Knight, H. H. 1941*c*. The plant bugs, or Miridae, of Illinois. Bull. Ill. St. nat. Hist. Surv. No. 22.

Knight, H. H. 1943*a*. Five new species of *Dicyphus* from western North America and one new *Cyrtopeltus* (Hemiptera, Miridae). Pan-Pacif. Ent. 19:53–58.

Knight, H. H. 1943*b*. New species of *Polymerus* Westwood from the United States (Hemiptera, Miridae). Can. Ent. 75:179–182.

Knight, H. H. 1944. *Lygus* Hahn; six new species from western North America (Hemiptera, Miridae). Iowa St. Coll. J. Sci. 18:471–477.

Knight, H. H. 1953. New species of Miridae from Missouri (Hemiptera). Iowa St. Coll. J. Sci. 27:509–518.

Knight, H. H. 1959. New genera and species of North American Miridae (Hemiptera). Iowa St. Coll. J. Sci. 33:421–426.

Knight, H. H. 1963. Review of the genus *Ilnacora* Reuter with descriptions of ten new species (Hemiptera, Miridae). Iowa St. Coll. J. Sci. 38:161–178.

Knight, H. H. 1964. *Phymatopsallus* new genus, and new species of Phylinae from North America (Hemiptera, Miridae). Iowa St. Coll. J. Sci. 39:127–152.

Knight, H. H. 1965. Old and new species of *Lopidea* Uhler and *Lopidella* Knight (Hemiptera, Miridae). Iowa St. Coll. J. Sci. 40:1–126.

Knight, H. H. 1966. *Schaffneria*, a new genus of ground dwelling plant bugs (Hemiptera, Miridae). Iowa St. Coll. J. Sci. 41:1–6.

Knight, H. H. 1968*a*. Taxonomic review: Miridae of the Nevada test site and the western United States. Brigham Young Univ. Sci. Bull: 9. 264 pp.

Knight, H. H. 1968*b*. Review of the genus *Orectoderus* Uhler with a key to the species (Hemiptera, Miridae). Iowa St. Coll. J. Sci. 42:311–318.

Knight, H. H. 1970. Review of the genus *Slaterocoris* Wagner with a key and descriptions of new species (Hemiptera, Miradae). Iowa St. Coll. J. Sci. 45:233–267.

Knight, H. H. 1973. A key to the North American species of *Pilophorus* Hahn with descriptions of new species (Hemiptera, Miridae). Iowa St. J. Res. 48:129–145.

Knight, H. H. 1974. A key to species of *Phytocoris* Fallén belonging to the *Phytocoris junceus* Kngt. group of species (Hemiptera, Miridae). Iowa St. J. Res. 49:123–135.

Linnaeus, C. 1758. Systema naturae. Laurentii Salvii, Holmiae. 824 pp.

Linnaeus, C. 1761. Fauna Suecica. Stockholmiae. 578 pp.

Linnaeus, C. 1767. Systema naturae. Holmiae. 1:533–1327.

Martin, J. E. H. 1977. The insects and arachnids of Canada. Part 1. Collecting, preparing, and preserving insects, mites, and spiders. Agric. Can. Publ. 1643. 182 pp.

McAtee, W. L. 1916. Key to the Nearctic species of *Paracalocoris* (Heteroptera; Miridae). Ann. ent. Soc. Am. 9:366–390.

Moore, G. A. 1907. List of Hemiptera taken at Como, Que. 1906. Can. Ent. 39:161–163.

Moore, G. A. 1950. Catalogue des Hémiptères de la Province de Québec. Naturaliste can. 77:233–271.

Moore, T. E. 1955. A new species of *Agnocoris* from Illinois, and a synopsis of the genus in North America (Hemiptera, Miridae). Proc. ent. Soc. Wash. 57:175–180.

Moore, T. E. 1956. *Agnocoris rubicunda* in North America (Hemiptera, Miridae). J. Kans. ent. Soc. 29:37–39.

Osborn, H. 1898. Additions to the list of Hemiptera of Iowa, with descriptions of new species. Proc. Iowa Acad. Sci. 5:232–247.

Oshanin, B. 1906–1910. Verzeichnis palaearctischen Hemipteren mit besonderer Berücksichtigung ihrer Verteilung im russichen Reiche. Ezheg. Zool. Muz. 1:587–1087.

Palisot de Beauvois, A. M. F. J. 1805–1821. Insectes recueillis en Afrique et en Amérique, dans les royaumes d'Oware et de Benin, à Saint-Dominique et dans les États-Unis pendant les années 1781–1797. Paris. 267 pp.

Parshley, H. M. 1914. List of the Hemiptera-Heteroptera of Maine. Psyche Camb. 22:139–149.

Parshley, H. M. 1919. On some Hemiptera from western Canada. Occ. Pap. Mus. Zool. Univ. Mich., No. 71. 36 pp.

Poppius, B. R. 1911. Zwei neue nearktische Miriden Gattungen. Annls. Soc. ent. Belg. 55:84–87.

Poppius, B. R. 1914. Einige neue Miriden- Gattungen und Arten aus Nord-Amerika und Cuba. Ann. ent. Soc. Belg. 58:255–261.

Provancher, L. 1872. Descriptions de plusieurs Hémiptères nouveaux. Naturaliste can. IV:73–79; 103:108.

Provancher, L. 1886–89. Petite faune entomologique du Canada. Les Hémiptères. 3. Quebec. 354 pp.

Reuter, O. M. 1870. Pargas sockens Heteroptera, förtecknade. Notis. Sällsk. Faun. Flor. fenn. 11:309–326.

Reuter, O. M. 1875*a*. Revisio critica Capsinarum, praecipue Scandinaviae et Fenniae. Akad. Afh. Helsingfors 2:1–190.

Reuter, O. M. 1875*b*. Genera Cimicidarum Europae. Bih. Svenska Vet. Ak. Handl. 3:1–66.

Reuter, O. M. 1875*c*. Hémiptères nouveaux. Pet. Nouv. Ent. 1:544–545.

Reuter, O. M. 1875*d*. Capsidae ex America boreali in Museo Holmiensi asservatae, descriptae. Öfvers. K. VetenskAkad. Förh. 32:59–92.

Reuter, O. M. 1878. *Actitocoris*, novum Hemipterorum genus e Fennia Australi. Meddn Soc. Fauna Flora fenn. 2:194–197.

Reuter, O. M. 1878–1896. Hemiptera Gymnocerata Europae. Acta Soc. Sci. fenn. 13–33. (1883), 3:313–496. (1896), 5:1–392.

Reuter, O. M. 1888. Revisio synonymica Heteropterorum palaearcticorum. Acta Soc. Sci. fenn. 15:443–812.

Reuter, O. M. 1903. Capsidae chinenses et Thibetanae hactenus cognitae enumeratae novaeque species descriptae. Öfvers. finska Vetensk, Soc. Förh. 45:1–23.

Reuter, O. M. 1904. Uebersicht der paläarktischen Stenodema-Arten. Öfvers. finska Vetensk. Soc. Förh. 46:1–21.

Reuter, O. M. 1908. *Neoborus* (*Xenoborus*, n. subg.) *commissuralis*, n. sp. Pages 109–116 *in* E. P. Van Duzee, List of Hemiptera taken by W. J. Palmer, about Quinze Lake, P. Que., in 1907. Can. Ent. Vol. 40.

Reuter, O. M. 1909. Bermerkungen über nearktische Capsiden nebst Beschreibung neuer Arten. Acta Soc. Sci. fenn. 36:1–86.

Reuter, O. M. 1910. Neue Beiträge zur Phylogenie und Systematik der Miriden nebst einleitenden Bemerkungen über die Phylogenie du Heteropteren-Familien. Acta Soc. Sci. fenn. 37:1–171.

Reuter, O. M. 1912. Hemipterologische miscellen. Öfvers. finska Vetensk. Soc. Förh. 54:1–76.

Sahlberg, J. R. 1870. Hemiptera Heteroptera samlade under en resa i ryska Karelen sammaren 1869. Notis. Sällsk. Faun. Flor. fenn. 11:277–307.

Sahlberg, J. R. 1878. Bidrag till Nordvestra Sibiriens Insektfauna, Hemiptera-Heteroptera. K. svenska VetenskAkad. Handl. 16:1–39.

Sahlberg, R. F. 1842. Nova species generis *Phytocoris* (Fallén), ex ordine Hemipterorum descripta. Acta Soc. Sci. fenn. 1:411–412.

Say, T. 1832. Descriptions of new species of heteropterous Hemiptera of North America. New Harmony. 39 pp.

Schuh, T. 1976. Pretarsal structure in the Miridae (Hemiptera) with a cladistic analysis of relationships within the family. Am. Mus. Novit. No. 2601. pp. 1–39.

Slater, J. A. 1954. Notes on the genus *Labops* Burmeister in North America, with the descriptions of three new species (Hemiptera: Miridae). Bull. Brooklyn ent. Soc. 49:57–65.

Slater, J. A. 1956. *Megaloceraea recticornis* (Geoffr.), a mirid new to the eastern United States, with the description of a new genus of Stenodemini (Hemiptera, Miridae). Proc. ent. Soc. Wash. 58:116–120.

Slater, J. A., and Davis, N. T. 1952. The scientific name of the tarnished plant bug (Hemiptera, Miridae). Proc. ent. Soc. Wash. 54:194–197.

Slosson, A. T. 1898. Additional list of insects taken in alpine region of Mt. Washington. Ent. News 9:251–253.

Smith, J. B. 1909. A report on the insects of New Jersey. Order Hemiptera. Rep. New Jers. St. Mus. pp. 131–170.

Stål, C. 1858. Beitrag zur Hemipteren-Fauna Sibirens und des russischen Nord-Amerika. Stettin ent. Ztg. 19:175–198.

Stål, C. 1862. Hemiptera Mexicana enumeravit speciesque novas descripsit. Stettin. ent. Ztg. 23:289–325.

Stanger, N. W. 1942. New species of *Lygus* from California (Hemiptera, Miridae). Univ. Calif. Publs Ent. 7:161–178.

Stevenson, C. 1903. A new Capsid. Can. Ent. 35:214.

Strickland, E. H. 1953. An annotated list of the Hemiptera (S. L.) of Alberta. Can. Ent. 85:193–214.

Thomson, C. G. 1871. Ofversigt af de i Sverige funna arter af gruppen Capsina. Opusc. ent. 4:410–452.

Tucker, E. S. 1907. Some results of desultory collecting of insects in Kansas and Colorado. Kans. Univ. Sci. Bull. 4:51–112.

Uhler, P. R. 1872. Notices of the Hemiptera of the western territories of the United States, chiefly from the surveys of Dr. F. V. Hayden. Pages 392–423 *in* F. V. Hayden. Prelim. Rep. U.S. Geol. Surv. Montana.

Uhler, P. R. 1875. Report upon the collections of Hemiptera made in portions of Nevada, Utah, Colorado, New Mexico, and Arizona, during the years 1871, 1873 and 1874. Rep. U.S. geol. Surv. 5:829–842.

Uhler, P. R. 1876. List of Hemiptera of the region west of the Mississippi River, including those collected during the Hayden explorations of 1873. Bull. U.S. geol. Surv. 1:269–361.

Uhler, P. R. 1877. Report upon the insects collected by P. R. Uhler during the explorations of 1875, including monographs of the families Cydnidae and Saldae, and the Hemiptera collected by A. S. Packard, Jr., M.D. Bull. U.S. geol. Surv. 3:355–475.

Uhler, P. R. 1878*a*. Notices of the Hemiptera Heteroptera in the collection of the late T. W. Harris, M.D. Proc. Boston Soc. nat. Hist. 19:365–446.

Uhler, P. R. 1878*b*. On the Hemiptera collected by Dr. Elliott Coues, U.S.A., in Dakota and Montana, during 1873–74. Bull. U.S. geol. Surv. 4:503–512.

Uhler, P. R. 1886. Check list of the Hemiptera Heteroptera of North America. Bull. Brooklyn ent. Soc. 32 pp.

Uhler, P. R. 1887*a*. Observations on some Capsidae with descriptions of a few new species. Entomologica am. 3:29–35.

Uhler, P. R. 1887*b*. Observations on North American Capsidae, with descriptions of new species. Entomologica am. 3:29–35; 67–72.

Uhler, P. R. 1890. Observations on North American Capsidae, with descriptions of new species. Trans. Md Acad. Sci. 1:73–88.

Uhler, P. R. 1892. Observations on some remarkable Heteroptera of North America. Trans. Md Acad. Sci. 1:179–184.

Uhler, P. R. 1893. Summary of the collection of Hemiptera secured by Mr. E. A. Schwarz in Utah. Proc. ent. Soc. Wash. 2:366–385.

Uhler, P. R. 1894. Observations upon the heteropterous Hemiptera of Lower California, with descriptions of new species. Proc. Calif. Acad. Sci. 4:223–295.

Uhler, P. R. 1895. A preliminary list of the Hemiptera of Colorado. Pages 1–137 *in* C. P. Gillette, and C. F. Baker. Bull. Colo. St. Univ. agric. Exp. Stn No. 31.

Uhler, P. R. 1904. List of Hemiptera-Heteroptera of Las Vegas Hot Springs, New Mexico, collected by Messrs. E. S. Schwarz and Herbert S. Barber. Proc. U.S. natn. Mus. 27:349–364.

Van Duzee, E. P. 1887. Partial list of Capsidae taken at Buffalo, N.Y. Can. Ent. 19:69–73.

Van Duzee, E. P. 1889. List of the Hemiptera from the Muskoka Lake district, Canada. Can. Ent. 21:1–11.

Van Duzee, E. P. 1905. List of Hemiptera taken in the Adirondack Mountains. Pages 546–556 *in* Felt, Fifth Rept. New York St. Entomol. for 1905.

Van Duzee 1908. List of Hemiptera taken by W. J. Palmer, about Quinze Lake, P. Que., in 1907. Can. Ent. 40:109–116.

Van Duzee, E. P. 1912*a*. Hemipterological gleanings. Bull. Buffalo Soc. nat. Sci. 10:477–512.

Van Duzee, 1912*b*. Synonymy of the Provancher Collection of Hemiptera. Can. Ent. 44:317–329.

Van Duzee, E. P. 1914. A preliminary list of the Hemiptera of San Diego County, California. Trans. S. Diego Soc. nat. Hist. 2:1–57.

Van Duzee, E. P. 1915. New genera and new species of North American Hemiptera. Pomona Coll. J. Ent. 7:109–121.

Van Duzee, E. P. 1916*a*. New or little known genera and species of Orthotylini (Hemiptera). Univ. Calif. Publs Ent. 1:217–227.

Van Duzee, E. P. 1916*b*. Monograph of the North American species of *Orthotylus* (Hemiptera). Proc. Calif. Acad. Sci. 6:87–128.

Van Duzee, E. P. 1917. Catalogue of the Hemiptera of America north of Mexico (excepting the Aphididae, Coccidae, and Aleurodidae). Univ. Calif. Publs Ent. Vol.2.

Van Duzee, E. P. 1918. New species of Hemiptera chiefly from California. Proc. Calif. Acad. Sci. 8:271–308.

Van Duzee, E. P. 1919. *Orthotylus pacificus* Van Duzee, sp. nov. Pages 1–35 *in* H. M. Parshley. On some Hemiptera of western Canada. Occ. Pap. Mus. Zool. Univ. Mich. No. 71.

Van Duzee, E. P. 1920. New hemipterous insects of the genera *Aradus, Phytocoris* and *Camptobroshis*. Proc. Calif. Acad. Sci. 9:331–356.

Van Duzee, E. P. 1921. A study of North American grass-bugs of the genus *Irbisia*. Proc. Calif. Acad. Sci. 11:145–152.

Van Duzee, E. P. 1925. New Hemiptera from western North America. Proc. Calif. Acad. Sci. 14:391–425.

Wagner, E. 1939. Die Wanzen der Sammlung Kirschbaum. Jb. Nassau. Ver. Naturk. 86:34–75.

Wagner, E. 1955. Was ist *Capsus opacus* Zett. 1839? (Hem. Het. Miridae). Opusc. ent. 20:2–3.

Wagner, E. 1956. On the genus *Strongylocoris* Blanchard, 1840 (Hemiptera Heteroptera, Miridae). Proc. ent. Soc. Wash. 58:277–281.

Wagner, E., and Slater, J. A. 1952. Concerning some Holarctic Miridae (Hemiptera, Heteroptera). Proc. ent. Soc. Wash. 54:273–281.

Walker, F. 1873. Catalogue of the specimens of Hemiptera Heteroptera in the collection of the British Museum. Part VI. London.

Walley, G. S. 1928. Key to species of *Dicyphus* occurring in eastern North America (Hemipt., Miridae) Can. Ent. 60:119.

Walley, G. S. 1932. A second report on Hemiptera from the north shore of the Gulf of St. Lawrence. Can. Ent. 64:152–155.

Westwood, J. O. 1840. An introduction to the modern classificastion of insects. 11. Synopsis of the genera of British insects. London.

Wheeler, A. G., Jr., and Henry, T. J. 1975. Recognition of seven Uhler manuscript names, with notes on thirteen other species used by Heidemann (1892) (Hemiptera: Miridae). Trans. Am. ent. Soc. 101:355–369.

Zetterstedt, J. W. 1840. Insecta Lapponica descripta. Leipzig. 1140 pp.

Index

(Page numbers of principal entries are in boldface; synonyms of species are in italic type.)